T0135046

Rechargeable Lithium-ion Batteries: Trends and Progress in Electric Vehicles

Edited by

Thandavarayan Maiyalagan

SRM Institute of Science and Technology
Kattankulathur, India

Perumal Elumalai

Pondicherry University, Puducherry, India

CRC Press
Taylor & Francis Group
Boca Raton   London   New York

CRC Press Is an imprint of the
Taylor & Francis Group, an **informa** business

A SCIENCE PUBLISHERS BOOK

CRC Press
Taylor & Francis Group
6000 Broken Sound Parkway NW, Suite 300
Boca Raton, FL 33487-2742

© 2021 by Taylor & Francis Group, LLC
CRC Press is an imprint of Taylor & Francis Group, an Informa business

No claim to original U.S. Government works

Printed on acid-free paper
Version Date: 20200728

ISBN-13: 978-0-367-51013-8 (pbk)
ISBN-13: 978-1-138-48409-2 (hbk)
ISBN-13: 978-1-351-05270-2 (ebk)

DOI: 10.1201/9781351052702

Visit the Taylor & Francis Web site at
http://www.taylorandfrancis.com

and the CRC Press Web site at
http://www.routledge.com

Preface

Modern lifestyles demand increased utilization of the fuels such as oil, coal and gases for domestic, industrial and consumer electronics applications to meet the fast-growing energy needs. Additionally, combustion of these resources releases toxic gases, which leads to adverse effects on the environmental ecosystem consisting of both aquatic and non-aquatic systems. Thus, researchers put efforts to go for utilization of clean and renewable energy resources for various applications. Unfortunately, for vehicular applications it is not practically viable to use any renewable energy source directly. However, energy from the renewable sources such as wind and the sun can be stored in efficient energy storage devices such as batteries and supercapacitors for later usage. Among them, lithium-ion based batteries are the best power sources for electric vehicle applications, because of their high energy density even in small size and light weight compared to other rechargeable batteries. Thus, with the help of battery-powered automobiles, the environment can become more sustainable and an eco-friendly niche for a healthy life in future.

Battery operated vehicles are classified into three types: namely, hybrid vehicles, plug-in hybrid vehicles and complete electric vehicles. A typical battery-operated electric vehicle requires at least 20 kWh energy to cover a distance of about 100 km. Amongst available technologies, lithium-ion batteries are considered as the most apt technology for vehicular application. The conventional lithium-ion battery (LIB) operates on the basis of intercalation chemistry, which contains graphite as anode, lithium transition metal-oxide as cathode and carbonate-based electrolyte solution. Moreover, several parameters like voltage, capacity, energy density, power density and affordability, etc. have to be considered and optimized before utilisation for vehicular applications. Therefore, electric vehicle policies have been formulated by several nations, in order to electrify the automobiles for the next generation transportation. Currently, most of the automobile manufacturers are devoting efforts to develop battery-operated vehicles. Hence, this is the right time to introduce this edited book entitled "Rechargeable Lithium-ion Batteries: Trends and Progress in Electric Vehicles". This book contains ten chapters, each chapter focusing on a specific topic pertaining to lithium-ion batteries for EV applications. To name a few, Chapter 1 focuses on the basis of LIBs, Chapter 2 focuses on overview of the various issues and challenges of rechargeable lithium batteries. Moreover, on the basis of recent reports, various strategies have been presented and described with specific attention paid to suppress the dendrite growth and stable solid electrolyte interphase (SEI) formation during Li plating/stripping processes. Chapter 3 and Chapter 4 present a brief overview on developing advanced Li-ion battery systems with different electrode materials and electrolyte solutions. Chapter 5 gives an insight into the high voltage cathode materials for Li-ion battery, their structure and working principle. Chapter 6 brings out strategies over using nickel-based cathodes in Li-ion batteries to be used in next generation electric powered vehicles. In Chapter 7, one can clearly count on the benefits of using graphene-based composite electrodes in Li-ion batteries for boosting the performance of energy storage devices. Such an approach not only leads to a mass production of potential electrodes but also can reduce the cost of electric vehicles. Similarly, Chapter 8 outlines the role of polymer-based electrodes in Li-ion batteries. Most interestingly, in this book, Chapter 9 focuses on the recycling and recovery of spent Li-ion batteries to prevent accumulation of e-waste in the environment in future. In the end Chapter 10 confirms that there is still much scope for improvement in each technological aspect especially in battery charge controllers, and plug in electric hybrid vehicles.

We hope this book stands unique among the countable books focusing on the lithium-ion battery technologies for vehicular applications. We assure that the contents presented in the book will provide strong basics and working knowledge on lithium-ion battery for vehicular application. Students, scholars, academicians and battery industries are the major beneficiaries from this volume. We earnestly thank all the authors and co-authors who have immensely contributed for this edited volume. We also would like to thank CRC Press for their editorial support and bringing out the book in time. We would like to appreciate and thank all authorities of our institution/University for providing a conducive and pleasing environment to create this book.

Dr. T. Maiyalagan
Dr. P. Elumalai

Contents

About the Editors

Dr Thandavarayan Maiyalagan received his PhD in Physical Chemistry from the Indian Institute of Technology Madras, and completed postdoctoral programs at the Newcastle University (UK), Nanyang Technological University (Singapore) and at the University of Texas, Austin (USA). Presently he is an Associate Professor of Chemistry at SRM Institute of Science and Technology, India. His main research interests focus on design and development of electrode nanomaterials for energy conversion and storage applications,

particularly fuel cells, supercapacitors and batteries. His research and collaborations have resulted in 180 papers in peer reviewed publications (with 7000 citations and a H-index of 35) and has edited a fuel cell book for Wiley. He received Young Scientist Award for the year 2017 from Academy of Sciences, Chennai. He has been invited to present his work at various international and national meetings.

Dr. Perumal Elumalai is Professor in Department of Green Energy Technology, Pondicherry University, Puducherry India. He received his PhD from Indian Institute of Science Bangalore. He had his postdoctoral research at Kyushu University, Japan for more than six years. He was a recipient of a prestigious JSPS fellowship (Japan Society for the Promotion of Science fellowship). He has won

special recognition award for Young Ceramist by Ceramic Society of Japan for his contribution to ceramics applications. He has visited USA, Italy,

Singapore, China and Japan for his professional developments. He has published more than 80 research papers in reputed international journals including book chapters, and holds six patents. Currently, his H-index is 22. His research work concentrates on batteries, supercapacitors, fuel cells and sensors.

List of Corresponding Authors

C. Bharatiraja
Department of Electrical and Electronics Engineering,
SRM Institute of Science and Technology, Kattankulathur, India

F. Caballero-Briones
Instituto Politécnico Nacional, Materials and Technologies for
Energy, Health and Environment (GESMAT),
CICATA Altamira, 89600 Altamira, México

Jun Lu
Chemical Sciences and Engineering Division,
Argonne National Laboratory,
9700 Cass Ave, Lemont, IL 60439, USA

N. Sivakumar
Department of Physics,
Chikkaiah Naicker College, Erode,
Tamilnadu, India

T. Maiyalagan
Department of Chemistry,
SRM Institute of Science and Technology, Kattankulathur, India

O. Padmaraj
Department of Nuclear Physics,
University of Madras, Guindy Campus,
Chennai – 600 025, Tamil Nadu, India

Perumal Elumalai
Department of Green Energy Technology,
Pondicherry University, Puducherry, India

S.K. Kamaraj
Instituto Tecnológico de El Llano,
Km. 18 Carretera Aguascalientes-San Luis Potosi, 2
0330 El Llano, Aguascalientes, México

Vinodkumar Etacheri
Faculty of Science, Universidad Autónoma de Madrid,
C/ Francisco Tomás y Valiente,
7, Madrid 28049, Spain

Weixiang Shen
Faculty of Science, Engineering and Technology,
Swinburne University of Technology,
Hawthorn, Australia

Y.S. Lee
Department of Chemical Engineering,
Chonnam National University,
Gwang-ji 500-757, Republic of Korea

Zhi Sun
Division of Environment Technology and Engineering,
Institute of Process Engineering,
Chinese Academy of Sciences,
Beijing 100190, China

Zhiqun Lin
School of Materials Science and Engineering,
Georgia Institute of Technology,
Atlanta, GA 30332

Fundamental Principles of Lithium Ion Batteries

A. Selva Sharma[1], A. Prasath[1], E. Duraisamy[1], T. Maiyalagan[2] and P. Elumalai[1]*

[1] Madhanjeet School of Green Energy Technologies, Department of Green Energy Technology, Pondicherry University, Pondicherry, India
[2] SRM Institute of Science and Technology, Chennai, India

1. Introduction

In recent years, the increased reliance on the advanced technology has led to the sudden spike in the demand for energy. In fact, at present almost the entire world economy is dependent on technologically driven automation processes leading to severe depletion of non-renewable fossil fuel reserves at a much faster rate than expected. Consequently, this has resulted in undesirable emissions of toxic greenhouse gases into the environment causing unpredictable adverse climatic conditions resulting in significant health hazards to the general public. Thus far, to mitigate the imminent threat arising due to global warming, several countries have started deploying renewable energy sources such as solar, wind and water. Although renewable energy is available in abundance, its effective utilisation is hindered by several factors such as fluctuations in energy outputs and their intermittent availability. For instance, with an increased reliance on solar energy the need to store it for periods when the sunlight is not available is of utmost importance. In such situations, an electrochemical energy storage device such as a battery is considered as a viable option to store energy over a specific period of time for utilisation when needed. Efforts are currently underway to directly integrate energy storage devices with the grid systems of renewable energy sources [1-3]. The advent of electrochemical devices has led to the concept of portable

*Corresponding author: drperumalelumalai@gmail.com; elumalai.get@pondiuni.
edu.in

electricity and this has paved the way for the miniaturisation of many electronic devices. The discovery of high energy and power density energy storage devices has opened up new options in the automobile industry and this has resulted in the development of electric bikes, cars and buses. The use of electric powered automobiles is expected to reduce the emission of toxic gases thus contributing to a sustainable green economy [3-6].

By definition, batteries are devices that directly convert the chemical energy stored within the constituent electrode materials into electrical energy by electrochemical redox reactions. In the case of rechargeable or secondary batteries, the electrochemical reactions responsible for the generation of electricity are reversible. While, in the case of primary batteries the reactions are unidirectional or irreversible. Ever since the breakthrough works of Galvani and Volta, the discoveries on batteries have evolved a long way [7, 8]. In fact, the first commercial primary battery was first invented by Leclanche way back in 1865 comprising a zinc anode, a manganese dioxide cathode, and an acidic aqueous electrolyte of ammonium chloride/zinc chloride, with an operating voltage of 1.5 V [9]. Subsequently, the acidic electrolyte was replaced with alkaline potassium hydroxide to dry alkaline cells with enhanced discharge capacity within the same voltage range. Later, a series of efficient primary batteries such as zinc air batteries (1.4 V), silver oxide batteries (1.5 V) and 3 V lithium primary batteries with lithium as an anode were developed and commercialised in the late 1980s [1-10].

The advent of rechargeable secondary batteries began with the discovery of the lead-acid battery by Plante in early 1859, in which the cathode was lead peroxide, lead acted as anode and sulphuric acid (6 M) was used as the electrolyte with a reasonable voltage window of 2 V. Of late, in 1899 W. Jungner introduced nickel cadmium batteries, nevertheless, owing to their harmful environmental effects their practical use was soon curtailed. The roll out of nickel metal hydride batteries was seen as an effective alternative to nickel cadmium batteries. However, the arrival of lithium ion battery technologies in the early 1980s and the consequent commercialisation of lithium ion batteries in late 1991 had given the lithium based secondary batteries an unassailable position in the field of portable energy storage devices [10-21]. In lithium-ion secondary batteries, it is possible to achieve an average discharge voltage of 3.7 V with higher energy density and longer cycle life compared to that of other available batteries. All these unique capabilities have made lithium-ion batteries (LIB) ideal energy storage device for portable electronic goods. At present, the battery market is vastly dominated by LIB with a worldwide market value of around 30 million USD. Many research initiatives are currently underway to improve the electrochemical properties of LIB by various means to improve performance for use in potential applications. In this chapter, the basic configuration of the lithium-ion battery, its thermodynamic aspects, critical battery parameters such as voltage, capacity, energy demand, power density, C-rate, columbic efficiency, cycle life etc are extensively discussed. The ideal requirements of anode, cathode and electrolyte characteristics are also presented in detail. The general merits and demerits of conventional anodes and cathodes are also discussed.

2. Lithium Ion Batteries

An ideal electrode material for a high performance secondary battery should possess low electrode potential and high electrochemical equivalence. It is apparent from Table 1 that lithium possess the lowest redox potential (–3.05 V) and highest electrochemical equivalence (3.86 Ahg^{-1}) among metals. Lithium, being an alkali metal is the lightest among all the metallic elements with superior theoretical gravimetric charge density with a fairly large potential window. LIBs basically function on the basis of electrochemical intercalation and de-intercalation (Fig. 1) in which both the anode and cathode essentially undergo insertion and de-insertion of lithium ions reversibly over several cycles. A typical LIB consists of a graphite anode (negative electrode, having a theoretical capacity of 372 mAh/g), a layered $LiCoO_2$ as the cathode (positive electrode, having a theoretical capacity of 274 mAh/g) and the electrolyte is usually $LiPF_6$ dissolved in non-aqueous organic solvent. During the charging processes, the Li ions from the host anode material $LiCoO_2$ de-intercalate and move forth through the electrolyte and intercalate within the graphite layers (Fig. 1). On the other hand during the discharge step, the reverse will happen and will result in the continuous movement of Li ions during the charging and discharging processes. LIBs are also referred to as rocking chair batteries. In a LIB, it is usually observed that during the cycling process, the electrolyte undergoes a limited reaction with lithium to form a multi-layered film at the interface of both the electrodes which is referred to as the solid electrolyte interphase (SEI). This SEI play a vital role in preventing the electrolyte from undergoing side reactions with the electrode surface. Thus, in the subsequent cycling process the movement of Li ions will occur through the surface films, in a reversible electrochemical reaction as given in the following electrode reactions [10-15]:

At the anode:

$$xLi^+ + Li_{1-x}CoO_2 + xe^- \rightleftharpoons LiCoO_2 \qquad\qquad E^0 = 0.6$$

At the cathode:

$$LiC_6 \rightleftharpoons Li_{1-x}C_6 + x\,Li^+ + xe^- \qquad E^0 = -3.0\ V$$

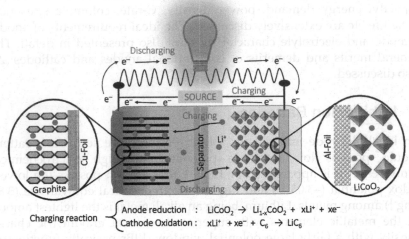

$$
\text{Charging reaction} \begin{cases} \text{Anode reduction} & : \quad LiCoO_2 \rightarrow Li_{1-x}CoO_2 + xLi^+ + xe^- \\ \text{Cathode Oxidation} & : \quad xLi^+ + xe^- + C_6 \rightarrow LiC_6 \end{cases}
$$

Fig. 1. Schematic representation of the intercalation/de-intercalation mechanism in LIB.

Net cell reaction:

$$
LiC_6 + Li_{1-x}CoO_2 \rightleftharpoons Li_{1-x}C_6 + LiCoO_2 \qquad E_{cell} = 3.6 \text{ V}
$$

2.1 Mechanisms of LIBs

The electrochemical reactions involved in LIBs depend on the nature of the electrode materials. Based on the nature of the electrochemical redox reactions of the active material in lithium storage, the underlying operational mechanisms of LIBs can be classified.

2.2 Intercalation/De-intercalation Mechanism

In this type of mechanism, the electrode materials act as a host to accommodate Li ions in a highly ordered one/two- or three-dimensional structure without any drastic loss in the crystal structure. In general, host materials are characterised by a number of vacant host sites in the crystal lattice and the Li ions undergo insertion within these sites in a systematic manner. For instance, intercalation/de-intercalation reactions occur by the insertion of lithium ions within one-dimensional (graphites or oxides $LiCoO_2$), two-dimensional ($LiMn_2O_4$ or $Li_4Ti_5O_{12}$) and three-dimensional ($LiFePO_4$) structures.

2.3 Alloying Mechanism

The electrode materials such as pure metals Si, Sn, Bi and Cd that are capable of undergoing reversible alloying/dealloying reactions with Li have been extensively utilised as potential anodes in LIBs. It has been well established that the Li alloying-dealloying reactions proceed at lower potentials (≤ 1.0 V) vs Li) and this in turn has immensely contributed to the reversible capacity during the Li cycling process. Alloying-based anode materials have gained traction in recent years owing to their exceptional property to dissolve large amount of lithium, as this has been evident from the higher theoretical capacity of 4200 mAh/g for $Li_{22}Si_5$ and 991 mAh/g for $Li_{22}Sn_5$ which is much higher than the conventional anode materials (C : 372 mA h g^{-1}) [12-19]. Despite possessing these favourable characteristics, alloying-based anode materials are prone to huge volume expansion leading to mechanical stress and subsequent cracking of the electrode material. The higher initial capacity delivered by alloy-based electrode materials tend to fade out during the remaining course of the cycling process. It is pertinent to note that in the case of intercalation-based electrode materials, the observed volume change is much lesser and as a result they showed robust performance in longer cycles. Apart from pure metals, metal oxide-based electrode materials such as TiO_2, SnO and Bi_2O_3 etc are known to undergo alloying reactions in a series of steps. In the initial step, the metal oxide gets converted to metallic nanoparticles irreversibly within the matrix of Li_2O and in the final step the dispersed metallic species participate in the reversible alloying/de-alloying reactions.

2.4 Conversion Mechanism

The conversion/displacement mechanism of LIBs involves formation and decomposition of Li_2O from transition metal oxides with concomitant reduction/oxidation of nano metal particles. In general, transition metal compounds such as transition metal oxides, sulfides, fluorides, phosphides and nitrides are known to undergo conversion reactions. For instances, the reaction of transition metal oxides with Li metal involves the amorphization of the crystal lattice to form metallic nanoparticles and Li_2O. During the course of the charging process, the metal oxides are reformed with subsequent decomposition of Li_2O. The different LIB operating mechanisms are summarised in Fig. 2.

3. Battery Components and Its Fabrication

3.1 Electrodes

In recent decades, LIBs have been increasingly studied and utilised in automobile industries. At present electric vehicles are poised to replace gasoline-based internal combustion (IC) engines and the electric powered vehicles are classified into three types namely electric vehicle (EV), plug-in hybrid vehicle (PHEV) and hybrid electric vehicles. EVs and PHEVs utilise electric motors for propulsion, in which the latter are equipped with both

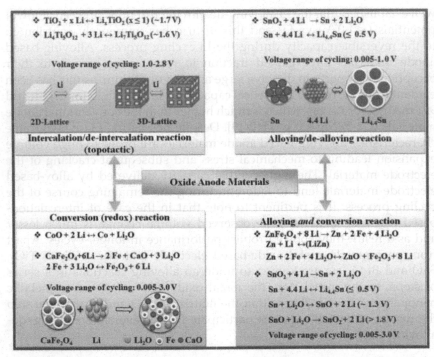

Fig. 2. Classification of operational mechanism of LIBs on the basis of the reversible Li insertion and extraction process [18].

electric and IC engines. In PHEVs, the IC engines swing into action when the battery power is completely used up and this serves the purpose of a generator to run the electric motor. In the case of a HEV, the electric motor is used for low speed momentum, whereas the IC engine gets activated for high speed and long distance propulsion. To meet these end results, LIBs with high power and high energy density are required. A typical EV requires a battery that has the ability to deliver 25-45 KW hrs to match the performance of the IC engine powered vehicle. In addition to this, LIBs are considered as an ideal energy storage device for power tools, back-up power supply units, and off-peak energy storage from the electric grid. So far, commercial HEVs make use of Ni-MH battery owing to their superior safety features and other factors. The major roadblock that prevents the use of the LIB technologies in high power applications is the cost, safety concerns and lower capacities associated with the electrode materials. For instance, the use of cobalt-based cathodes and microcrystalline purified graphite anode along with the Li-based electrolyte contribute to higher operational costs of LIBs. However, to realise potential applications the energy density of LIBs needs drastic improvement to the tune of ~120 to ~250 W·h kg^{-1}. The inherent capacities and operating potential of the electrode materials are keys to achieving desired energy density for

LIBs. So far, less than half of the theoretical capacity has been realised in practical applications. Consequently, the need to explore cheaper alternative electrode materials is of utmost importance to achieve the full potential of LIBs in a vast array of applications [17-24].

3.2 Battery Components

The size of the LIB varies depending upon the type of its application and power specifications. LIBs come in different forms such as coin, cylindrical, prismatic and pouch types. However, the basic components used in all of the LIBs are the same. The basic components of a battery are an external casing, the current collector, binder and separator. In the case of typical coin and cylindrical cells, stainless steel casing is usually used as external packaging material with a resealable vent to release pressure under excessive charge. While expensive metallic casings are used in the cylindrical and prismatic cell, a relatively cheaper multilayer foil is used in the case of pouch cell. Another most important component of LIB is the current collector. An ideal current collector should not react with lithium even at low electrode potential and it must be resistant to the standard oxidising potential of LIBs. Thus, in general, aluminium and copper foils are used as the current collector for cathode and anode, respectively. Binders play a crucial part in the process of adhesion of active materials on to the current collector. For example, during the intercalation/de-intercalation of Li ions the electrode materials experience volume expansion and contraction over prolonged cycling processes. In such a physical process, the contact at the interface of the active material and the current collector weakens and this leads to contact resistance thereby ultimately resulting in poor performance of the battery. Such adverse complications arising due to the mechanical changes of the active materials can be mitigated by the use of ideal binder materials. A perfect binder should have superior binding properties, excellent thermal stability and resistance to oxidation without dissolution in non-aqueous organic electrolytes. So far, fluorinated polymer polyvinylidene fluoride (PVDF) is widely used for binding in LIBs apart from other binders such as styrene butadiene rubber (SBR) and sodium carboxymethyl cellulose (CMC). The separators used in LIBs are placed between the two electrodes to avoid direct contact while allowing ionic conductivity and isolating electron flow. The choice of the separator is important as it is known to influence the battery performance and safety. Materials with good chemical stability, porosity, pore size, and permeability are generally used as separators. For example, polyethylene (PE)-polypropylene (PP) multi-layer microporous structures have been widely used in LIBs. In the case of gel-polymer LIBs, the conventional multilayer microporous separators are replaced by a thin layer of PVDF-HFP, or other polymer. It is pertinent to note that the components used in the LIB fabrication does not participate in the electrochemical reaction.

Therefore, care must be taken to minimize the "dead weight" of the components to as low as possible with higher active material loading to achieve better energy density and power density.

3.3 Electrode Fabrication

The fabrication of electrodes in LIBs involves a series of processes. In the initial step, the active material is mixed with the conductive additive (namely acetylene black) along with the binders (PVDF) in N-methyl-2-pyrrolidinone (NMP) solvent to make a slurry. The mixing and mechanical grinding of the active material have to be carried out separately for a specified period of time to ensure good homogeneity. The resulting slurry is then coated on to the current collectors and the coated substrates are then subjected to vacuum drying followed by roll pressing to ensure a firm coating. In the ensuing step, the substrates are cut into desired dimensions in an electrode cutting machine and the resulting electrode materials are then assembled in an Ar-filled glove box having O_2 and H_2O concentration less than 0.1 ppm. After the stacking of electrodes, electrolytes are injected on to the separator and then the fabricated cell is sealed within the metal casing by means of crimping machine. The various steps involved in the fabrication of typical Li-ion coin cell is demonstrated here in Fig. 3.

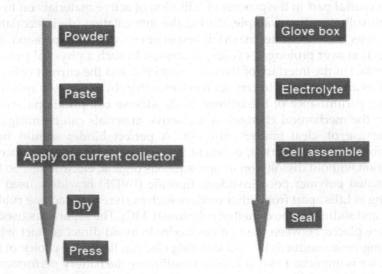

Fig. 3. Flow chart depicting the various steps involved in the fabrication of electrode and assembling of the LIBs.

3.4 Merits and Demerits of LIBs

LIBs are fast growing in consumer electronics and find application in many other fields. This humongous rise of LIBs is attributed to their extremely low weight, large working voltage window of up to 3.6 V, high theoretical energy and power density of 150 Wh/kg and 400 Wh/kg, respectively. In addition, LIBs are characterised by very low self-discharge rate of 2%-8% per month, absence of memory effect, high coulombic efficiency over longer cycle life of more than 1000-3000 cycles, capacities from less than 500 mAh to 1000 Ah can be achieved. Apart from this, LIBs can be customized into varying sizes and shapes to suit the wide array applications across different fields. Despite possessing many merits LIBs suffer from array of demerits such as short life of 2 to 3 years from the date of manufacture whether they are used or not. LIBs are prone to degradation much faster if they are exposed to heat as compared to the normal temperatures because they are extremely sensitive to high temperatures. For instance, LIBs are susceptible to losing capacity permanently at higher temperatures compared to that of NiCd or NiMH batteries. The cost of LIBs is high and they get ruined if they are completely discharged. In addition, there is small risk of LIBs getting busted in to flames if they are not properly manufactured, though the likelihood is as low as 2 to 3 packs per million batteries produced.

4. Battery Characteristics

The various characteristics that are used to validate the battery performances are listed below:

Voltage: Voltage refers to the electric potential difference between the two terminals (electrodes) in an electric circuit. It is essentially the electrical driving force and measured in volts (V). Theoretical or thermodynamic voltage can never be achieved practically, due to electrode polarization, resistance of electrode/electrolyte and sluggish kinetics.

$$E^0 - E_a^0 - E_c^0 ; \quad E = E^0 - 0.054/n \log Q$$

Open circuit voltage: It is defined as the difference in the electric potential between the two electrodes at zero current. Open circuit voltage V_{oc} can be estimated from the respective chemical potential values of the anode/cathode and it is given by the following relation:

$$V_{oc} = \mu^A - \mu^C /F$$

where μ^A and μ^C represent the chemical potential of the anode and cathode, respectively, and F is the Faraday constant (96485 JK^{-1}).

Current: It is the quantity of electrode charge stored in a given material. It represents the rate of flow of electric charge and it is dependent on the rate of the electrochemical reactions taking place at the electrodes.

Capacity: It is given by the product of the total amount of charge, under a completely discharged state in specified conditions and time. Capacity is directly associated with the atomic weight of the active material and number of electrons involved in the electrochemical reaction. The theoretical capacity (C_T) per gram of the electrode materials is given by the equation:

$$C_T = n \, F/M; \; C/g = n \times 26.8/m \, \text{Ah g}^{-1}$$

where n denotes the number of moles of electrons exchanged during the electrochemical reaction, F is the Faraday constant (96485 C/mol) and M is the molecular mass of the active material (g/mol). The theoretical capacity or discharge capacity is $I \times t/m$, where I is current, t is discharge time and m is mass of active material.

Current rate (C-rate): It refers to the rate of charging/discharging process and it can be better explained on the basis of the relation between the battery capacity and current involved during the charging/discharging process as given below:

$$t = C_T/i$$
$$nC = (1/n) \, h$$

where t is the time taken to complete charging or discharging in hours, current rate (C_T) corresponds to the battery capacity in ampere hour (Ah) and i is the current drawn in amperes (A). Thus, the C-rate is given by the reciprocal of t and this explains the fact that as the C-rate increases the battery requires less time to charge or discharge or as the C-rate increases the battery capacity decreases. C-rate gives a measure how fast the battery can be charged and discharged in a given period.

Energy density: It is defined as the energy stored per unit mass or volume of the active material. Thus, the gravimetric energy density is given by the ratio of the energy deliverable with a given current to the mass of the battery and is expressed in Wh/kg with the following equation:

Gravimetric energy density = Specific capacity (kg) × cell voltage (V)

In a similar way, the volumetric energy density can be defined as the ratio of the energy deliverable under a specified current to the volume of the battery and is expressed in Wh/m^3 or Wh/dm^3 with the following equation:

Volumetric energy density = Specific capacity (dm^3) × cell voltage (V)

Power density: It is defined as the power per unit mass that can be delivered by the battery at specified conditions. The gravimetric power density is expressed in W/kg. While, the volumetric power density is expressed in W/m^3 or W/L and it is given by the expression:

$$\text{Power density} = \text{Current (A/kg or A/L)} \times \text{Voltage (V)}$$

$$= \text{Energy density/Time}$$

Depth of discharge: It corresponds to the amount of electricity discharged from a battery in relation to its capacity and is given by the expression:

$$\text{Depth of discharge} = \int \text{Depth of discharge} \frac{\int_0^t i(t)dt}{\text{Capacity}}$$

The depth of discharge is usually represented in percentage and it is a dimensionless value.

State of charge: It is the measure of the total electricity still available in the battery with respect to its capacity under specified conditions and it is given by the equation:

$$\text{State of charge} = \frac{\text{Amount of charge available}}{\text{Capacity}}$$

Coulombic efficiency: It is defined as the ratio between the number of charges delivered by and the number of charges injected into the battery. During the course of the electrochemical reactions several factors such as resistance, side reactions and instability of electrode could lead to loss of charge and lessen the coulombic efficiency. An ideal battery must possess a coulombic efficiency of more than 100%.

Table 1. Desirable battery characteristics for LIBs

Parameter	Desirable characteristics
Energy density (Volumetric - Wh/L and Gravimetric - Wh/kg	High
Power density (Volumetric - W/L and Gravimetric - W/kg	High
Mechanical, Chemical and Structural stability	Good
Discharge curve	Flat
Rate capability	High, fast recharge
Cycle-life	Long
Self-discharge	Low
Temperature range	Wide
Cost	Low
Safety	High

Cycle life: The cycle life or capacity retention refers to the number of cycles that a battery can undergo before its capacity falls below 80%. Apart from the battery chemistry, the cycle of LIBs is influenced by depth of discharge (DOD) and state of charge (SOC) and as well as the operating temperature.

5. Existing Challenges in LIBs

The existing LIBs though successfully used in many portable electronics, suffer from certain pitfalls. For instance, the extremely low electrochemical potential and higher reactivity of Li-metal lead to side reactions with the electrolytes in LIBs (Fig. 4). As a result, Li deposition will start to occur at the surface of the electrodes. Such deposit of Li on the electrode surface is referred to as dendrite formation. The word dendrite refers to the non-uniform Li accumulation with branched needle-like-feature. During the discharge process, dendrites present at the interface of the electrode liberates the Li metal which further moves to the top of the dendrite i.e., away from the electrode to form highly reactive and electrochemically inert Li. Consequently, the highly reactive Li present at the tip of the dendrite further undergoes parasitic reactions with electrolytes and contributes to lower Coulombic efficiency. The dendrites are capable of penetrating through the separator and short-circuit the working LIBs by

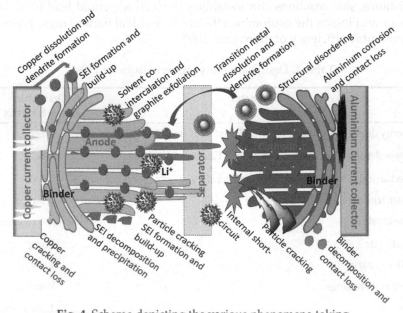

Fig. 4. Scheme depicting the various phenomena taking place during the cycling of LIBs.

establishing direct electrical contact between the two electrodes. Moreover, the dendrites formed at the surface of anode/solid electrolyte interface would continuously consume the incoming Li ions and such reactions are highly exothermic and may cause undesired thermal runaways in LIBs.

Graphite is an ideal anode material for LIBs owing to its excellent properties such as low voltage hysteresis, good rate capability, low irreversible capacity and good thermal stability. However, owing to the overlap between the electrochemical potential of the graphite and the lowest unoccupied molecular orbital (LUMO) of the electrolyte (LiPF$_6$ in 1:1 ethylene carbonate (EC)/diethyl carbonate (DEC)), transfer of electrons from the graphite anode reduces the electrolytes at the anode surface leading to the formation of a thin layered solid-electrolyte interface film. The SEI layer serves as a barrier to prevent further promotion of electron from the LUMO of the electrolyte, but the difficulty in Li ion diffusion through this layer leads to plating of metallic lithium on graphite. The plating of Li over the graphite would further contribute to the formation of dendrites and this could ultimately result in safety issues. Thus, the shortfalls associated with the existing electrode materials have paved the way for the development of new promising cathode and anode materials. In recent years, researchers have unveiled several high energy, power, robust and safer alternate electrode materials. However, none of these materials have been commercialised for practical applications. The ensuing section will discuss the new and existing electrode materials and the wide array of challenges in realising potential applications of future LIBs.

6. Cathode Materials for LIBs

In early 1972, Exxon started working on primary batteries by utilising Li metal as the anode, lithium perchlorate in dioxolane as the electrolyte and TiS$_2$ as the cathode that exhibited superior intercalation properties. Despite the excellent performance of the cathode, the system encountered several inadequacies such as the parasitic reaction of Li-metal with electrolytes resulting in dendrite formation, which further created undesirable safety concerns such as volume expansion, thermal breakaways and explosion. Meanwhile, to mitigate the safety concerns pertaining to the use of Li metal, researchers have used different strategies by adopting alternative electrolytes or cathode materials. The concept of present day intercalation/de-intercalation mechanisms based on cathode materials dates back to the early 1970s, when Matsuchita first demonstrated the use of carbon monofluoride lattice to intercalate lithium in Li/(CF)$_n$ primary batteries. It was reported that lithium intercalates within the lattice of carbon monofluoride to form lithium fluoride. However, the challenges associated with the use of fluorine were not addressed despite elaborate research on the carbon fluorides. Finally, the concept of rocking chair LIB technology was realised in laboratory conditions by Murphy et al. and then by Scrosati et al., which involves the utilisation of Li ions instead of the Li metal. The primary role played by the insertion/de-insertion of Li ions alleviated the growth of dendrites and this has outperformed Li metal cells on the safety front. Nevertheless, to account for the increase in the potential of the cathode, a relatively higher potential anode material is required. Consequently, the use of layered or three-dimensional-type transition-metal oxides gained more importance compared to the layered-type transition-metal disulphides. The quest for robust LIB electrodes led to extensive research, commercialising the Li-ion battery based on the intercalation-de-intercalation concept by incorporating a carbonaceous material as the anode and $LiCoO_2$ as the cathode in an organic electrolyte [25-34]. This Li-ion battery possessed a very high potential of 3.6 V and energy densities thrice that of an alkaline battery and Ni-Cd batteries. Typical cathode materials in LIBs are primarily lithiated transition metal oxides with an ability to undergo reversible Li ion extraction. Based on morphology and intercalation phenomena the commonly used cathode materials in LIBs are discussed below.

6.1 Layered Oxides

6.1.1 Vanadium Pentoxide

Among the layered oxides, vanadium pentoxide, V_2O_5, and molybdenum trioxide, MoO_3, are some of the earliest studied oxides for LIB application. Molybdenum oxides are known to react with 1.5 lithium ions per molybdenum ions but owing to their sluggish reaction kinetics it did not garner enough attention in LIBs. Vanadium pentoxide is characterised by layered structure consisting of V_2O_5 layers along the c-axis of the orthorhombic cell and the each layer is built up of VO_5 square pyramids sharing edges and corners with four V–O distances being close to 2 A (Fig. 5). V_2O_5 is shown to undergo intercalation mechanism with Li as given below [35-38]:

$$x\,Li + V_2O_5 \rightarrow Li_xV_2O_5$$

However, the intercalation of Li into the crystal lattice of V_2O_5 induce a series of phase changes. For instance at the onset of the Li insertion, the α-phase (x < 0.01) and then the ε-phase (0.35 < x < 0.7) of V_2O_5 is formed with puckered layers. Upon the insertion of one more Li, considerable structural change from ε-phase to γ-phase of V_2O_5 take place. As compared to that of the α-, ε-, and δ-phases where the VO_5 square pyramids that make up the structure of V_2O_5 are arranged in rows the γ-phase is highly puckered. When more than two Li ions are intercalated, a rock salt tetragonal structure is formed namely, ω-$Li_3V_2O_5$ phase, beyond which

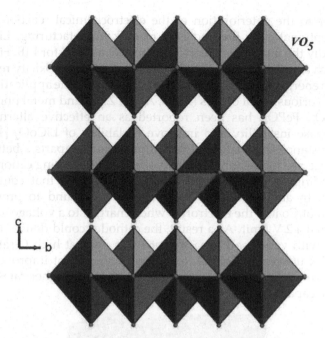

Fig. 5. Crystal structure of layered V_2O_5 showing the square pyramids of VO_5 sharing edges and corners [35].

reversible extraction of Li ion is no longer possible after a continued cycling process.

6.1.2 Lithium Cobalt Oxide

In the year 1980, Goodenough reported the existence of reversible Li insertion in the host structure of $LiCoO_2$ and recognized its remarkable structure similarity with the layered dichalcogenides. Subsequent to the commercialization of Li-ion battery, $LiCoO_2$ continues to dominate the LIB market for a long period of time. $LiCoO_2$ are oxides with a general formula $LiMO_2$ that adapt the rhombohedral α-$NaFeO_2$ structure with Rm3 space group in which the oxygens are arranged in a cubic closely-packed manner (Fig. 6). During the charging process, reversible extraction of Li ions from the crystal lattice of the $LiCoO_2$ takes place along with the simultaneous oxidation of Co^{3+} to Co^{4+}. This process is further accompanied by the rearrangement of oxygen to give hexagonal close packing of the oxygen in CoO_2. In the reverse direction, i.e. during discharge process, Co^{4+} of $Li_{1-x}CoO_2$ is reduced to Co^{3+} and the Li ions are again inserted into the lattice site. The practical capacity of $LiCoO_2$ (172 mAh g^{-1}) is limited to almost half of the theoretical capacity 274 mAh g^{-1} owing to the phase transformation from monoclinic to hexagonal at 4.2 V and other factors such as dissolution of Co^{4+}, oxygen evolution at higher voltages further contributes to the deterioration of the electrochemical reaction. Owing to its exceptional cycle life and ease of bulk manufacturing, $LiCoO_2$ is undoubtedly the most extensively used anode material for LIB. However, factors such as scarce availability, toxicity and material toxicity restrict its use in high energy and high power density hybrid vehicle applications. The coating of various metal oxides (MgO, Al_2O_3, ZnO) and metal phosphates (e.g., $AlPO_4$, $FePO_4$) has been reported as an effective alternative to prevent phase instability and improve cyclability of $LiCoO_2$ [10], [23], [31]. For instance, the coating of vanadium oxides imparts a better cycle performance at a high-charge cut-off voltage by preventing cation mixing during cycling and reducing the active surface area that contacts the electrolyte. In another study, CuO coating was found to prevent the dissolution of Co into the electrolyte when charged to a voltage above the conventional 4.2 V limit. As a result, the cathodes could deliver excellent capacities with very good capacity retention even at high C-rates. At a rate of 5C, CuO coated $LiCoO_2$ displayed a significant improvement in performance as compared to $LiCoO_2$. Even at much higher rates of 50C, an average capacity of 100 mAh g^{-1} was achieved.

$$\bullet Li \qquad \bullet M$$

Fig. 6. Crystal structure of layered $LiMO_2$ cathode material (Reprinted with permission from [39]. Copyright (2015) American Chemical Society.

$$Li_{(1-x)}CoO_2 + xLi^+ + xe^- \underset{charge}{\overset{\longrightarrow}{\longleftarrow}} LiCoO_2$$

6.1.3 Lithium Nickel Oxide

Of late, the higher input costs associated with the use of $LiCoO_2$ was compensated by replacing Co with other metals namely Ni, Mn or Al within the lattice structure of $LiMO_2$. $LiNiO_2$ shares identical crystal lattice pattern with $LiCoO_2$. The advent of $LiNiO_2$ was touted to be an ideal replacement for $LiCoO_2$ anodes owing to their excellent electrical conductivity, low cost and better discharge capacity of 200 mAh g^{-1}, which is almost 40% higher than that of $LiCoO_2$. However, it was later found that the preparation of cation-free $LiNiO_2$ was extremely difficult and it was reported that even a trace amount of cation within the crystal lattice decreased the electrode performance drastically. The insertion of excess Li ions results in phase change from hexagonal to monoclinic phase as it was observed in the case of $LiCoO_2$. Further factors such as thermodynamic instability and higher cation mixing of Li^+ and Ni^{3+} without any dimensional mismatch results in the presence of trace amounts of nickel over the lithium layer which makes the diffusion of Li ion difficult ultimately leading to poor power capability. The materials with lower Li content are highly unstable and are bound to react exothermically with organic electrolyte [22], [40].

The superior theoretical capacity of $LiMnO_2$ (285 mAh g^{-1}) and environment friendly character makes them an interesting candidate among the layered $LiMO_2$ cathode materials. However, the stability of manganese within the lattice of $LiMnO_2$ is not much. Further, owing to the identical ionic radii of Li^+ and Mn^{3+} the straight forward solid state synthesis that affords layered structure is not possible in the case of $LiMnO_2$. Consequently, $LiMnO_2$ showed a dismal discharge capacity of about 130 mAh g^{-1} in the subsequent cycling processes with severe irreversible phase changes. As a result $LiMnO_2$ was not considered a suitable cathode material.

6.1.4 Spinel Oxides

The general formula for spinel structure is given as AB_2O_4 and it is closely related to the structure of α_r-$NaFeO_2$ phase with slight difference in the distribution of the cations within the octahedral and tetrahedral sites. Among the various spinels, $LiMn_2O_4$ is thermodynamically stable with a space group of Fd3m comprising Li ions at the tetrahedral site and at the octahedral site manganese cations are accommodated. Both the tetrahedral and octahedral sites together act as three dimensional pathways for reversible Li insertion process (Fig. 7). $LiMn_2O_4$ spinels are capable of intercalating one Li ion per formula unit at ambient temperature and it is estimated to deliver a theoretical capacity of 148 mAh.g^{-1}. However, $LiMn_2O_4$ spinels are known to undergo fading over prolonged cycling at 4 V owing to the factors such as Jahn-Teller distortion and decomposition of electrolyte at higher potential. Moreover, at elevated temperatures dissolution of Mn^{3+} occurs as a result of disproportional reaction, and the formed Mn^{3+} readily dissolve in electrolytes and this in turn further aggravates capacity retention. To overcome the shortfalls associated with the $LiMn_2O_4$ spinels many potential alternatives have been put forward by researchers. For example, substitution of Mn with other metal ions has shown dramatic improvement in the cycling performance of spinel materials. Coating with Al_2O_3 and doping with multiple inactive ions such as Mg, Al, and Zn, was employed to increase the structural stability of $LiMn_2O_4$ spinels. Modification of spinels with the first row transition metal ions such as Ti, Cr, Fe, Co, Ni, and Cu was found to be effective in curbing the John-Teller distortion by Mn^{3+} ions and showed better phase integrity [41-45]. Amongst them, $LiNi_{0.5}Mn_{1.5}O_4$ showed the best overall electrochemical performance. Spinel oxides modified by the above substitution methods can serve as ideal high-power lithium secondary batteries in the automobile industry.

$$Li_{(1-x)}Mn_2O_4 + xLi^+ + xe^- \underset{charge}{\rightleftharpoons} LiMn_2O_4$$

(x = lithium exchanged; $0 < x \leq 1$)

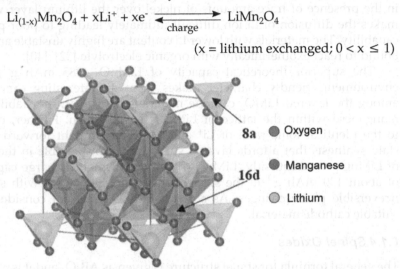

Fig. 7. Crystal lattice points of the spinel $LiMn_2O_4$ Reprinted with permission from [46]. Copyright (2019) American Chemical Society.

6.1.5 *Olivine Materials*

In recent years, the olivine type compound $LiFePO_4$ commonly referred to as triphylite has gained significant interest as a cathode material in LIB owing to its extremely low cost, environment friendliness and structural stability at harsh conditions. The crystal lattice of $LiFePO_4$ is comprised of corners shared FeO_6 octahedra and edge shared LiO_6 octahedra linked together by PO_4 tetrahedra (Fig. 8). During the cycling process, the intercalation of Li ions within the crystal lattice of $LiFePO_4$ take places in two phase steps namely between the lithium-rich and -poor phase. In the extraction process of Li ions, the oxidation of Fe^{2+} ions into Fe^{3+} ions forms $FePO_4$, which has a crystal structure similar to that of $LiFePO_4$. Such formations of similar phase structures help in excellent reversibility of Li insertions and contribute to a longer life cycle of the cathode. The diffusion of Li ions in olivine structures takes place through a one dimensional

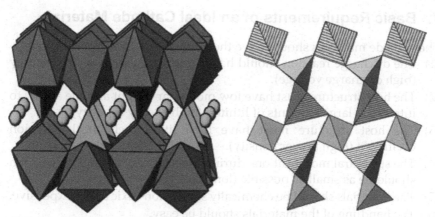

Fig. 8. Structures of orthorhombic $LiFePO_4$ and trigonal quartz-like $FePO_4$ [40].

tunnel because of the one dimensional crystal morphology. Thus, the presence of other cations along the Li ion pathway hinders the movement of Li ions and as a result the Li channel no longer remains active in the electrochemical reactions. Therefore, to prevent such an undesirable event, care must be taken to achieve a high purity crystal structure of $LiFePO_4$. Despite offering several advantages such as stability at elevated temperature and longer cycle life, $LiFePO_4$ suffers from poor inherent electronic and ionic conductivity and delivers a much lower capacity. It has been reported that a carbon coating on $LiFePO_4$ was found to improve the electrochemical performance of the material and in a similar way many other strategies have been used to improve the electronic conductivity of the $LiFePO_4$ particles [40-52].

$$Li_{(1-x)}FePO_4 + xLi^+ + xe^- \underset{charge}{\overset{\longrightarrow}{\longleftarrow}} LiFePO_4$$

(x = lithium exchanged; $0 < x \leq 1$)

Table 2. Gravimetric capacities of different cathode materials

Materials	Theoretical gravimetric capacity (mAh/g)	Theoretical gravimetric capacity (mAh/g) with account taken of the degree of insertion	Average practical gravimetric capacity (mAh/g)	Cost
$LiCoO_2$	274	137	120	High
$LiNiO_2$	275	275	220	Medium
$LiMn_2O_4$	148	148	120	Low
$LiCo_{0.2}Ni_{0.8}O_2$	274	247	180	Medium
$LiFePO_4$	170	170	150	Low

7. Basic Requirements of an Ideal Cathode Material

The cathode materials should have the following requisite properties:
(1) The discharge reaction should have large negative Gibbs free energy (high discharge voltage).
(2) The host structure must have low molecular weight and the ability to intercalate large amounts of lithium (high energy capacity).
(3) The host structure must have high lithium chemical diffusion coefficient (high power density).
(4) The structural modifications during intercalation and deintercalation should be as small as possible (long life cycle).
(5) The materials should be chemically stable, non-toxic and inexpensive.
(6) The handling of the materials should be easy.

Table 3. Comparison of various physical parameters of selected cathode materials

Battery	$LiFePO_4$	$LiCoO_2$	$LiMn_2O_4$	$Li(NiCo)O_2$
Stability	Stable	Not stable	Acceptable	Not stable
Environmental concern	Most environ-friendly	Very dangerous	Acceptable	Very dangerous
Cycle life	Best/Excellent	Acceptable	Acceptable	Acceptable
Power/Weight density	Acceptable	Good	Acceptable	Best
Long term cost	Most economic/ Excellent	High	Acceptable	High
Temperature range	Excellent (–20 to 70° C)	Decay beyond (–20 to 55° C)	Decay extremely fast over 50° C	–20 to 55° C)

8. Anode Materials

In recent years, the quest for alternative anode materials are extensively investigated and in comparison to cathode materials, the choice for anode materials is more. With the advent of new age technologies, there is an urgent need to improve the battery performances by fine tuning the chemical and morphological properties of potential anode materials. The following section will discuss the challenges and pitfalls of existing and prospective anode materials.

8.1 Carbon Based Anodes

At the nascent stage of LIB technologies, pristine Li metal was used as the anode material in the secondary batteries. Although Li metal was able to deliver high capacity, the formation of dendrites at the surface of the metal anode resulted in safety issues such as short circuits. Over a period of time the concept of intercalation/de-intercalation of Li ions paved the way for carbon based anode materials. Even now in almost all the portable electronic goods graphitic carbon is used as the anode material. Graphitic carbon possesses a layered structure which can accommodate the Li ion movement to and fro within its lattice space with almost no restriction. Despite having many favourable properties such as high electrical conductivity, low voltage hysteresis, good rate capability, low irreversible capacity, good cycle life, high Coulombic efficiency, inexpensive and sufficient raw material reserves, the graphitic carbon anodes with their theoretical maximum capacity of 372 mAh g^{-1} are not sufficient enough to meet high power and energy density requirements of future LIBs in automobile applications. The mechanism of Li insertion and extraction within the carbon matrix is highly dependent on the active surface area, structure and morphology of the carbonaceous material. For example, in the case of graphitic carbon the inter-layer separation distance is about 3.35 A. During the Li ion intercalation process this interlayer spacing increases to about 3.5 A. This small expansion (<5%) enables the graphite to keep its structural integrity intact. In fact, when Li is inserted, graphite's planar conductivity is actually increased which further enhances its ability to act as an ideal anode. In the case of disordered carbon, high initial capacity is observed with issues of huge first-cycle irreversible capacity loss and capacity fading. Moreover, the decomposition of organic electrolytes on the surface of carbon results in the formation of solid electrolyte interphase (SEI). The incomplete formation of SEIs would continue with increased cycling processes thereby contributing to the internal resistance and irreversibility of Li ion insertion. The side reactions of organic electrolytes and the insertion of Li ions along with the solvent molecules is known to contribute to the expansion of graphene layers, thereby resulting in the reduction of charge storing capability. Apart from graphite and disordered carbon, other carbon based anode materials such as buckminsterfullerene and carbon nanotubes (CNTs) are being extensively studied. In particular, CNTs owing to their exceptional conductivity and linear dimensionality can act as an ideal material for reversible Li insertion. It was found out that the formation of defects is responsible for lithium diffusion into CNTs. Lithium particularly diffuses into CNTs with 9 membered ring defects rather than those with none or 8 or 10 membered defects (Fig. 9). As the carbon atoms begin to be removed, a hole begins to occur in the wall of the carbon nanotube as each of the carbon atoms attempts to remain bonded with its neighbours. The lithium is able to move through the interior of the CNT and can be absorbed successfully, indicating that lithium ions can be accumulated in the interior of CNTs in addition to the exterior. Although the introduction of defects into CNTs generally improves reversible capacity, it also contributes to irreversible capacity. This means that while more lithium ions are stored and retrieved later, the increased number of defects also causes a larger amount of lithium ions stored in the initial charge cycle to be wasted. High irreversible capacity means that lithium ions are essentially consumed by the first cycle and are never returned to the cathode, and so no network can be done with the lithium ions lost to irreversible capacity. As a result, most of the present day research is currently focussed on CNT, carbon and graphene based composites instead of pristine carbon, CNTs or graphene to achieve much higher capacity than that of pristine carbon [12, 13, 16].

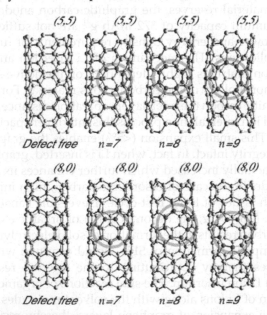

Fig. 9. Defect-free and the defective (5, 5) (upper panel) and (8, 0) (lower panel) Carbon nanotubes. Reprinted with permission from [53]. Copyright (2005) American Physical Society.

8.2 Li Alloying Anodes

Researchers have managed to attain a reversible capacity of as high as 450 mAh g^{-1} for carbon materials obtained using improved chemical and physical treatments. However, the possibility of achieving a higher capacity for carbon anodes is low and as a result intense researches are currently underway to find suitable anode materials with superior electrochemical properties. For example, materials based on Li metal alloys including Cu-Sn-Li, Cu-Sb-Li, In-Sn-Li, Si-Li, and Si-C-Li and metal oxides such as Co_3O_4, CoO, CuO, and FeO have demonstrated excellent electrochemical performances. In particular, Si and Sn are known to react with lithium to form alloys with higher capacity than the graphite anodes. These metals undergo reactions with Li at a specific voltage during the charging process to form alloys and while discharging return back to their actual state resulting in reversible charge/discharge. Other metals such as Al, Ge, and Pb showed less reversible reaction efficiency making them inadequate as anode materials. Silicon is considered as one of the most attractive Li alloying based anode material [15, 18, 20-22].

Table 4. Physical characteristics of graphite and alloying anodes

Materials	Li	C	$Li_4Ti_6O_{12}$	Si	Sn	Sb	Al	Mg	Bi
Density (g cm^{-3})	0.53	2.25	3.5	2.33	7.29	6.7	2.7	1.3	9.78
Lithiated phase	Li	LiC_6	$Li_7Ti_6O_2$	$Li_{4.4}Si$	$Li_{4.4}Sn$	Li_3Sb	LiAl	Li_3Mg	Li_3Bi
Theoretical specific capacity (mAh g^{-1})	3862	372	175	4200	994	660	993	3350	385
Theoretical charge density (mAh cm^{-3})	2047	837	613	9786	7246	4422	2681	4355	3765
Volume change (%)	100	12	1	320	260	200	96	100	215
Potential vs. Li (V)	0	0.05	1.6	0.4	0.6	0.9	0.3	0.1	0.8

8.2.1 Silicon Anodes

Depending on the two lithiated phases namely $Li_{4.4}Si$ and $Li_{3.75}Si$, silicon (Si) exhibits theoretical capacity of 3572 and 4000 mAh g^{-1}, respectively, which is eventually higher than the capacities reported for existing graphite and metal oxide anodes. While charging, the alloying process proceeds by receiving electrons and in the reverse de-alloying reaction the liberation of Li ions to the anode material takes place and the reaction involving Si is expressed as follows:

$$5Si + 22Li + 22e^- \longrightarrow Li_{22}Si_5$$

$$Si + xLi^+ + xe^- \longrightarrow Li_xSi$$

In the initial charging, the vacant sites within the Si lattice are filled with Li ions and this contributes to huge volume expansion. As single Si atom is capable of reacting with up to 4.4 Li ions the volume expansion can go up to 400% in Si. It is believed that the increase in the lattice volume easily breaks the weak ionic bonds between the Li and Si leading to cracking, pulverisation and electrode disintegration accompanied by a dismal performance. Compared to its bulk counterparts nanosized Si demonstrated better electrochemical performances. Cui and colleagues prepared Si nanowires and demonstrated that Si nanowires showed much better Li storage ability as compared to that of Si thin films and Si particles. It was inferred that the use of Si nanowires resulted in much better structure stability delivering a discharge capacity 10 times that of graphite anode. However, it is difficult to prepare nanosized Si on a large scale by simple cost-effective approaches. Several other approaches have been adopted to suppress Si volume expansion such as by creating a composite structure consisting of an inactive host matrix in which Si is finely dispersed. The inactive material matrix serves the purpose of a cushion to offset the volume change and prevent pulverisation of the Si anode. Highly conductive metals with good structural integrity such as Ni, Fe, Co, Ca, B, Ti, etc are used as inactive support, but such composites suffer from quick fading of capacity. Despite the shortfall, Si based nano composites are widely expected to replace the commercial carbon anodes soon [15, 18, 20-22].

8.2.2 Tin Anodes

Tin (Sn) is considered as another prospective alloyed anode owing to its favourable properties such as high operating voltage and theoretical capacity of 994 mAh g^{-1}. In the initial lithiation step, SnO_2 first undergoes conversion to Sn, then the Sn phase undergoes alloying reaction with Li as given below:

$$SnO_2 + 4\,Li \longrightarrow Sn + 2Li_2O$$

$$Sn + 4.4\,Li \longleftrightarrow Li_{4.4}Sn \; (0.5\,V)$$

The diffusivity of Li ions in Sn is significantly high (5.9107 cm^2 s^{-1}) which helps in rapid reversible alloying/de-alloying reaction cycles. However, Sn also suffers from poor cycle life owing to tremendous volume expansion of as much as 300% during the electrochemical reactions. To counter this, recently a novel three dimensional porous Sn-Cu alloy anode was proposed and it demonstrated a stable capacities and high rate capabilities. It was found out that the Cu acts as an inactive material in the anode and does not participate directly in the Li alloying reaction. However, reactions involving inter metallic Cu_6Sn_5 form a ductile matrix that allows to offset the large volume expansion associated with Sn-Li alloying reactions, thereby improving the cycle life of the anodes. Apart from metallic Sn- or SnO_2-based anodes, other Sn-based materials namely, tin phosphate, SnO_2 filled mesoporous tin phosphate, SnP, $AlPO_4$-coated SnO_2, and Zn_2SnO_4 have been studied and found to possess reasonable electrochemical performances [54-58].

Table 5. Theoretical capacity and lithiated phases of various metal oxide anodes

Metal Oxide	Lithiated phase	Theoretical specific capacity (mAh g^{-1})
SnO	$Li_{4.4}Sn/Li_2O$	875
SnO_2	$Li_{4.4}Sn/Li_2O$	782
TiO_2	$Li_4Ti_5O_{12}$	175
In_2O_3	Li_3In/Li_2O	289
Sb_2O_3	Li_3Sb/Li_2O	275
PbO	$Li_{4.4}Pb/Li_2O$	528
ZnO	$Li_{1.0}Zn/Li_2O$	329
CdO	$Li_{0.6}Cd/Li_2O$	125
FeO	Fe/Li_2O	746
CoO	Co/Li_2O	715
NiO	Ni/Li_2O	718
CuO	Cu/Li_2O	674
Cr_2O_3	Cr/Li_2O	529

8.3 Transition Metal Oxide Anodes

In general, transition metal oxides with the rock salt structure (CoO, CuO, NiO, FeO) without any free voids to host lithium and metallic elements do not undergo alloying reaction with Li. In contrast, metal oxide based anode materials are known to undergo Li storage reactions by two different pathways namely by intercalation/deintercalation and conversion reactions. It is well known that TiO_2 oxide exists in different phases namely anatase, rutile, ramsdellite TiO_2, and TiO_2-B [59-62]. Among the various phases of TiO_2, anatase TiO_2 has gained more attention due to its ability to undergo reversible Li insertion based on the following electrochemical Li insertion-extraction reactions:

$$TiO_2 + x\ Li \longleftrightarrow Li_xTiO_2\ (x \leq 1)\ (\sim1.7\ V)$$

It was revealed that the anatase TiO_2 is capable of participating in the electrochemical reaction without undergoing electrolyte decomposition and SEI formation, delivering a considerable specific capacity of 168 mAh/g. However, anatase TiO_2 finds limited use in LIB due to the large irreversible capacity as a result of intercalation reactions between TiO_2 and Li at the particle surface. To its advantage, the use of anatase TiO_2 as anode material could mitigate potential risk arising out of Li deposition namely dendrite formation, battery short circuit and thermal runaway as observed in carbon-based batteries. Thus, TiO_2-based anodes can be utilised in aviation and aerospace applications which require utmost safety.

Another important class of TiO_2 based anode materials is the $Li_4Ti_5O_{12}$ which exists in spinel phase in which the Li occupies the 8a position and while titanium is located at the 16d position. $Li_4Ti_5O_{12}$ is known to undergo intercalation with 3 Li to afford Li_7Ti_5O12 thereby delivering a capacity of 175 mAh g^{-1}. During the cycling process, the native crystal lattice of $Li_4Ti_5O_{12}$ remains intact without any strain owing to the similarity of the crystal structure with the Li intercalated $Li_7Ti_5O_{12}$ phase. As a result, the $Li_4Ti_5O_{12}$ shows excellent rate capability over 1000 cycles without any capacity fading. However, to counter the inherent low electronic conductivity of $Li_4Ti_5O_{12}$ strategies such as coating with conductive carbon and other doping methods were adopted [63].

$$Li_4Ti_5O_{12} + 3Li \longleftrightarrow Li_7Ti_5O_{12} \ (\sim 1.6 \text{ V})$$

In recent years, the conversion reaction based transition metal oxide anode materials attracted great deal of interest and showed great promise for the realisation of next generation high energy LIBs. The use of transition metal oxides in battery applications dates back to 1980, during which CuO was used as a cathode in Li-CuO primary batteries. Of late, nanostructured transition metal oxides have been increasingly exploited for their conversion type reaction as anode materials in LIBs. Unlike the alloying and intercalation type materials, conversion type metal oxide anodes (such as CoO, NiO, Fe_2O_3 etc.) undergo electrochemical reduction of metal oxides to ultrafine metal particles and Li_2O [64-71]. The metal nanoparticles embedded within the Li_2O matrix are highly reactive and hence readily decompose back to Li and respective metal oxide. The general reaction can be expressed as follows:

$$MO + 2Li^+ + 2e^- \longleftrightarrow M + Li_2O$$

$$(M = Mn, Fe, Co, Ni, Cu)$$

Tarascon first reported the systematic studies on metal oxides with M = Fe, Co, Ni, and Cu as anode materials in LIBs. It was reported that when cycled at 0.2 C, CoO delivered a reversible capacity of 700 mA·h g^{-1} and exhibited good capacity retention over several cycles within the voltage window of 0.01-3.0 V. The use of ultrafine CoO in different morphologies such as micrometer-size particles, nanosize particles, nanoplatelets, porous nanowires, and composites with carbon/graphene were found to exhibit good electrochemical properties. Poizot et al first reported the Li storage properties of NiO through conversion reaction with an initial reversible capacity of 600 mA·h g^{-1} after which it slowly degrades to 200 mA·h g^{-1} at the end of 50 cycles. Since then several groups reported studies on nanoparticles and thin films of NiO, and composites with carbon in the voltage range 0.01-3.0 V [72-74].

Owing to their abundant availability and reasonable theoretical reversible capacities, both copper oxides, Cu_2O (375 mA·h g^{-1}) and CuO (674 mA·h g^{-1}) have been extensively studied for their Li storage porperties. For instance, Tarascon studied these two oxides and found that the particle size plays a significant role in giving stable capacities of ~400 mA·h g^{-1} up to 70 cycles at 0.2C rate are obtained for both Cu_2O and CuO having a grain size of 1 m. Several studies on CuO in different sizes and morphologies namely thin films, sisal-like CuO/C films, polycrystalline powders, nanowires, nanoribbon arrays, leaf-like nanoplates and hierarchical nanostructures have also been investigated [75-82].

The initial report on the MnO based anode was not encouraging. However, in the subsequent studies by Tirado [83] it was found that the sub-micrometer MnO showed a reversible capacity of 650 mA·h g^{-1}, which then slowly degrades to 390 mA·h g^{-1} at the end of 25 cycles. MnO_x nanoparticles confined to porous carbon nanofibers were reported to deliver initial reversible capacity of 785 mA·h g^{-1}, which further stabilized to 600 mA·h g^{-1} at the end of 10 cycles. Several reports on the MnO based anode materials have been studied such as MnO/C core shell nanorods, porous MnO microspheres, nanoflakes, coaxial MnO/C nanotubes, MnO/C composites, MnO/graphene composites, and MnO, MnO/CoO, and MnO/Co composites [84-98].

Co_3O_4 is an inverse spinel structured anode material having a theoretical capacity of 890 mA·h g^{-1}. Tarascon et al. first studied the Li cycling properties of Co_3O_4 and since then, this mixed-valent oxide has been extensively investigated as anode material. The ease of preparation of Co_3O_4 in a much simpler way than the CoO or Co_2O_3 makes this material an ideal choice for Li storage application. Poizot et al. showed that Co_3O_4 could deliver a stable reversible capacity of 800-900 mA·h g^{-1} at current rates ranging from 50 to 200 mA g^{-1} [72]. It is well established that Co_3O_4 can readily undergo conversion reactions and it is capable of consuming 8 mol of Li per mole of Co_3O_4.

$$Co_3O_4 + 8Li^+ + 8e^- \longleftrightarrow 3Co + 4Li_2O$$

In the initial discharge reaction Li intercalation takes place in a single-phase process to afford $Li_x[Co_3O_4]$, $x \approx$ 0.2-0.5, which is then followed by crystal structure destruction and formation of Co metal nanoparticles dispersed within the matrix of Li_2O. The ultrafine and unusually distinct morphology of the Co_3O_4 particles was found to give stable and near theoretical capacities on long-term cycling. Composites with varying amounts of carbon or Ni or Cu metal also perform well in comparison to the bare-Co_3O_4. Iron oxides have several advantages such as abundance and a much lower cost than other 3d metal oxides, such as Co and Ni.

Table 6. General overview of graphite and prospective future anodes

Materials	Graphite	Sn	Si	Transition metal (Cu, Co, Ni, Fe) oxides	$Li_4Ti_5O_{12}$
Theoretical capacity (mAh/g)	372	990	4200	400-700	175
Reaction mechanism	Intercalation	Alloying	Alloying	Conversion (replacement)	Intercalation
Advantages	Low cost, abundance	Large capacity	Large capacity, abundance	Large capacity	High power density, safety
Disadvantages	Low capacity, limited safely	Poor cycleability	Poor cycleability	Large over-potential	Low capacity, potential too high
Comments	Most commonly used and commercialised		Future high performance anode		Ideal for safe battery

However, the use of iron oxides in practical applications is limited due to their poor conductivity and large volume change cycling processes. The preparation of nanosized iron oxides and their composites was found to alleviate the problems associated with volume expansion. Thus, many iron oxides with tailored morphologies have been prepared, including nanoparticles [99], nanowires [100-103], nanobelts [100, 104], nanorods [105-108], nanotubes [109, 110], nanoflakes [111], nanodisks [112] and nanorings [114], nanocubes [115-117], nanospheres [119-122], nanospindles [119, 123], nanourchins [124], and nanoflowers [98-125]. Hollow structured iron oxides are of great interest as the space would, ideally, accommodate the volume expansion during Li insertion [99-102]. Despite having many favourable properties, the use of metal oxide based conversion anodes is still confined to laboratory level experiments and is yet to be commercialised. Thus, the quest for the alternative anode material that matches all the exceptional properties of carbon anode remains elusive and daunting.

9. Basic Requirements of Ideal Anode Materials

The basic requirements of a material to qualify as an ideal anode material are given below:

1. A typical anode material must possess inherent ability to accommodate large fraction of Li per formula unit (high Gibbs energy with Li) with excellent stability.
2. It must possess an electrode potential as close as possible to that of Li metal.
3. The material must have good electronic and Li conductivity.
4. The material should not undergo any side reactions with the electrolyte and should remain stable in the solvents of the electrolyte.
5. It must be cheap and environmentally friendly.

References

1. Deng, D., Kim, M.G., Lee, J.Y. and Cho, J. 2009. Green energy storage materials: Nanostructured TiO_2 and Sn-based anodes for lithium-ion batteries. Energy & Environmental Science 2(8), 818-837.
2. LeVine, S. 2010. The Great Battery Race. Foreign Policy 182, 88.
3. Yoshino, A. 2012. The Birth of the Lithium-Ion Battery. Angewandte Chemie International Edition 51(24), 5798-5800.
4. New York Times 2013, 162, B5 B5.
5. Deng, D. 2015. Li-ion batteries: basics, progress, and challenges. Energy Science & Engineering 3(5), 385-418.

6. Tarascon, J.M. and Armand, M. 2011. Issues and challenges facing rechargeable lithium batteries. Nature 171-179.

7. Whittingham, M.S. 1976. Electrical Energy Storage and Intercalation Chemistry. Science 192(4244), 1126-1127.

8. Besenhard, J.O. and Eichinger, G. 1976. High energy density lithium cells: Part I. Electrolytes and anodes. Journal of Electroanalytical Chemistry and Interfacial Electrochemistry 68(1), 1-18.

9. Eichinger, G. and Besenhard, J.O. 1976. High energy density lithium cells: Part II. Cathodes and complete cells. Journal of Electroanalytical Chemistry and Interfacial Electrochemistry 72(1), 1-31.

10. Mizushima, K., Jones, P.C., Wiseman, P.J. and Goodenough, J.B. 1981. Li_xCoO_2 (0< x ≤ 1): A new cathode material for batteries of high energy density. Solid State Ionics 3, 171-174.

11. Thackeray, M.M., David, W.I.F., Bruce, P.G. and Goodenough, J.B. 1983. Lithium insertion into manganese spinels. Materials Research Bulletin 18(4), 461-472.

12. Yazami, R. and Touzain, P. 1983. A reversible graphite-lithium negative electrode for electrochemical generators. Journal of Power Sources 9(3), 365-371.

13. Basu, S., Zeller, C., Flanders, P.J., Fuerst, C.D., Johnson, W.D. and Fischer, J.E. 1979. Synthesis and properties of lithium-graphite intercalation compounds. Materials Science and Engineering 38(3), 275-283.

14. Yoshino, A., Sanechika, K. and Nakajima, T. 1987. Art. Name. USP4, 668, 595.

15. Chan, C.K., Peng, H., Liu, G., McIlwrath, K., Zhang, X.F., Huggins, R.A. and Cui, Y. 2008. High-performance lithium battery anodes using silicon nanowires. Nature Nanotechnology 3(1), 31.

16. Bates, J.B., Dudney, N.J., Neudecker, B., Ueda, A. and Evans, C.D. 2000. Thin-flm lithium and lithium-ion batteries. Solid State Ionics 135, 33-45.

17. Li, H., Huang, X., Chen, L., Wu, Z. and Liang, Y. 1999. A high capacity nano-Si composite anode material for lithium rechargeable batteries. Electrochemical and Solid-State Letters 2(11), 547-549.

18. Reddy, M.V., Subba Rao, G.V. and Chowdari, B.V.R. 2013. Metal oxides and oxysalts as anode materials for Li ion batteries. Chem. Rev. 113, 5364-5457.

19. Xing, W., Wilson, A.M., Eguchi, K., Zank, G. and Dahn, J.R. 1997. Pyrolyzed polysiloxanes for use as anode materials in lithium-ion batteries. Journal of the Electrochemical Society 144(7), 2410-2416.

20. Wilson, A.M., Zank, G., Eguchi, K., Xing, W. and Dahn, J.R. 1997. Pyrolysed silicon-containing polymers as high capacity anodes for lithium-ion batteries. Journal of Power Sources 68(2), 195-200.

21. Xue, J.S., Myrtle, K. and Dahn, J.R. 1995. An epoxy-silane approach to prepare anode materials for rechargeable lithium ion batteries. Journal of the Electrochemical Society 142(9), 2927-2935.

22. Goodenough, J.B. and Kim, Y. 2009. Challenges of Rechargeable Lithium Batteries. Chemistry of Materials 22(3), 587-603.

23. Chebiam, R.V., Kannan, A.M., Prado, F. and Manthiram, A. 2001. A. Comparison of the Chemical Stability of High Energy Density Cathodes of Lithium-Ion Batteries. Electrochemistry Communications 3(11), 624-627.

24. Wakihara, M. and Yamamoto, O. 2008. Lithium Ion Batteries: Fundamentals and Performance. John Wiley & Sons.
25. Van Schalkwijk, W. and Scrosati, B. 2002. Advances in Lithium-Ion Batteries. Springer, Boston, MA.
26. Shi, Y., Wang, J.Z., Chou, S.L., Wexler, D., Li, H.J., Ozawa, K., Liu, H.K. and Wu, Y.P. 2013. Hollow Structured Li_3VO_4 Wrapped with Graphene Nanosheets in Situ Prepared by a One-Pot Template-Free Method as an Anode for Lithium-Ion Batteries. Nano Letters 13(10), 4715-4720.
27. Ragupathy, P., Shivakumara, S., Vasan, H.N. and Munichandraiah, N.J. 2008. Preparation of Nanostrip V2O5 by the Polyol Method and Its Electrochemical Characterization as Cathode Material for Rechargeable Lithium Batteries. Phys. Chem. C 112(42), 16700-16707.
28. Arico, A.S., Bruce, P., Scrosati, B., Tarascon, J.M. and Van Schalkwijk, W. 2011. Nanostructured materials for advanced energy conversion and storage devices. In: Materials for Sustainable Energy: A collection of Peer-Reviewed Research and Review articles from Nature Publishing Group, World Scientific, Singapore. pp. 148-159.
29. Armand, M. and Tarascon, J.M. 2008. Building better batteries. Nature 451(7179), 652.
30. Ki Park, J. 2012. Principles and Applications of Lithium Secondary Batteries. Wiley-VCH Verlag & Co. KGaA, Germany.
31. Jacobson, A.J., Chianelli, R.R. and Whittingham, M.S. 1979. Amorphous Molybdenum Disulfde Cathodes. Journal of the Electrochemical Society 126(12), 2277-2278.
32. Watanabe, N. and Fukuda, M. 1970. Panasonic Corp. Primary cell for electric batteries. U.S. Patent 3,536,532.
33. Whittingham, M.S. 1975. Mechanism of Reduction of the Fluorographite Cathode. Journal of the Electrochemical Society 122(4), 526-527.
34. Armand, M.B. 1980. Intercalation electrodes. pp. 145-161. In: Materials for Advanced Batteries. Springer, Boston, MA.
35. Ragupathy, P., Shivakumara, S., Vasan, H.N. and Munichandraiah, N. 2008. Preparation of Nanostrip V_2O_5 by the Polyol Method and Its Electrochemical Characterization as Cathode Material for Rechargeable Lithium Batteries. Journal of Physical Chemistry C 112(42), 16700-16707.
36. Whittingham, M.S. 1976. Role of Ternary Phases in Cathode Reactions. Journal of the Electrochemical Society 123(3), 315-320.
37. Walk, C.R. and Margalit, N. 1997. δ-LiV_2O_5 As a Positive Electrode Material for LithiumIon Cells. Journal of Power Sources 68(2), 723-725.
38. Delmas, C., Cognac-Auradou, H., Cocciantelli, J.M., Menetrier, M. and Doumerc. 1994. The $Li_xV_2O_5$ System: An Overview of the Structure Modifications Induced by The Lithium Intercalation. Solid State Ionics 69(3-4), 257-264.
39. Croy, J.R., Balasubramanian, M., Gallagher, K.G., and Burrell, A.K. 2015. Review of the U.S. Department of Energy's "Deep Dive" Effort to Understand Voltage Fade in Li- and Mn-Rich Cathodes. Accounts of Chemical Research 48(11), 2813-2821.
40. Whittingham, M.S. 2004. Lithium batteries and cathode materials. Chemical Reviews 104(10), 4271-4302.

41. Garau, C., Frontera, A., Quinonero, D., Costa, A., Ballester, P. and Deya, P.M. 2003. Lithium diffusion in single-walled carbon nanotubes: a theoretical study. Chem. Phys. Lett. 374, 548-555.

42. Park, O.K., Cho, Y., Lee, S., Yoo, H.C., Song, H.K. and Cho, J. 2011. Who will drive electric vehicles, olivine or spinel? Energy & Environmental Science 4(5), 1621-1633.

43. Ellis, B.L., Lee, K.T. and Nazar, L.F. 2010. Positive electrode materials for Li-ion and Li-batteries. Chemistry of Materials 22(3), 691-714.

44. Yoon, Y.K., Park, C.W., Ahn, H.Y., Kim, D.H., Lee, Y.S. and Kim, J. 2007. Synthesis and characterization of spinel type high-power cathode materials Li $M_x Mn_2-x O_4$ (M= Ni, Co, Cr). Journal of Physics and Chemistry of Solids 68(5-6), 780-784.

45. Hassoun, J., Lee, K.S., Sun, Y.K. and Scrosati, B. 2011. An advanced lithium ion battery based on high performance electrode materials. Journal of the American Chemical Society 133(9), 3139-3143.

46. Nieminen, H. E., Miikkulainen, V., Settipani, D., Simonelli, D., Hönicke, P., Zech, C., Kayser, Y., Beckhoff, B., Honkanen, A., Heikkilä, M.J., Mizohata, K., Meinander, K., Ylivaara, M.E., Huotari, S., and Ritala, M. 2019. Intercalation of Lithium Ions from Gaseous Precursors into 2-MnO_2 Thin Films Deposited by Atomic Layer Deposition. Journal of Physical Chemistry C 123(25), 15802-15814.

47. Assaaoudi, H., Fang, Z., Ryan, D.H., Butler, I.S. and Kozinski, J.A. 2006. Hydrothermal synthesis, crystal structure, and vibrational and Mössbauer spectra of a new tricationic orthophosphate $KCo_3Fe(PO_4)_3$. Canadian Journal of Chemistry 84(2), 124-133.

48. Yang, S., Song, Y., Zavalij, P.Y. and Whittingham, M.S. 2002. Reactivity, stability and electrochemical behavior of lithium iron phosphates. Electrochemistry Communications 4(3), 239-244.

49. Morgan, D., Van der Ven, A. and Ceder, G. 2004. Li conductivity in Li_xMPO_4 (M= Mn, Fe, Co, Ni) olivine materials. Electrochemical and Solid-State Letters 7(2), A30-A32.

50. Gibot, P., Cabanas, M.C., Laffont, L., Levasseur, S., Carlach, P., Hamelet, S., Tarascon, J.M. and Masquelier, C. 2008. Room-temperature single-phase Li insertion/extraction in nanoscale Li_xFePO_4. Nat. Mater. 7, 741-747.

51. Li, H. and Zhou, H. 2012. Enhancing the performances of Li-ion batteries by carbon-coating: present and future. Chemical Communications 48(9), 1201-1217.

52. Chung, S.Y., Bloking, J.T. and Chiang, Y.T. 2002. Electronically conductive phospho-olivines as lithium storage electrodes. Nat. Mater. 1, 123-128.

53. Nishidate, K. and Hasegawa, M. 2005. Energetics of lithium ion adsorption on defective carbon nanotubes. Physics Review B 71, 245418.

54. Kim, J. and Cho, J. 2006. SnO_2 filled mesoporous tin phosphate - High capacity negative electrode for lithium secondary battery. Electrochemical and Solid-State Letters 9(8), A373-A375.

55. Kim, E., Son, D., Kim, T.G., Cho, J., Park, B., Ryu, K.S. and Chang, S.H. 2004. A mesoporous/crystalline composite material containing tin phosphate for use as the anode in lithium-ion batteries. Angewandte Chemie International Edition 43(44), 5987-5990.

56. Kim, Y., Hwang, H., Yoon, C.S., Kim, M.G. and Cho, J. 2007. Reversible lithium intercalation in teardrop-shaped ultrafne $SnP_{0.94}$ particles. Advanced Materials 19(1), 92-96.

57. Kim, T.J., Son, D., Cho, J., Park, B. and Yang, H. 2004. Enhanced electrochemical properties of SnO_2 anode by $AlPO_4$ coating. Electrochimica Acta 49, 4405-4410.

58. Addu, S.K., Zhu, J., Ng, K.Y.S. and Deng, D. 2014. A Family of Mesocubes. Chemistry of Materials 26, 4472-4485.

59. Julien, C.M., Massot, M. and Zaghib, K. 2004. Structural studies of $Li_4/3Me_5/3O_4$ (Me= Ti, Mn) electrode materials: local structure and electrochemical aspects. Journal of Power Sources 136(1), 72-79.

60. Yakubovich, O.V., Simonov, M.A. and Belov, N.V. 1977. The crystal structure of a synthetic triphylite LiFe [PO_4]. Soviet Physics Doklady 2, 347.

61. Ariyoshi, K., Yamato, R. and Ohzuku, T. 2005. Zero-strain insertion mechanism of Li [$Li_{1/3}Ti_{5/3}$] O_4 for advanced lithium-ion (shuttlecock) batteries. Electrochimica Acta 51(6), 1125-1129.

62. Ohzuku, T., Takeda, S. and Iwanaga, M.J. 1999. Solid-state redox potentials for Li[$Me_{1/2}Mn_{3/2}$]O_4 (Me: 3d-transition metal) having spinel-framework structures: a series of 5 volt materials for advanced lithium-ion batteries. Power Sources 81-82, 90.

63. Kavan, L. and Grätzel, M. 2002. Facile synthesis of nanocrystalline $Li_4Ti_5O_{12}$(spinel) exhibiting fast Li insertion. Electrochemical and Solid-State Letters 5(2), A39-A42.

64. Wang, G.X., Chen, Y., Konstantinov, K., Lindsay, M., Liu, H.K., Dou, S.X. 2002. Investigation of cobalt oxides as anode materials for Li-ion batteries. Journal of Power Sources 109, 142.

65. Yao, W., Yang, J., Wang, J. and Nuli, Y. 2008. Multilayered cobalt oxide platelets for negative electrode material of a lithium-ion battery. Journal of the Electrochemical Society 155(12), A903-A908.

66. Jiang, J., Liu, J.P., Ding, R. M., Ji, X.X., Hu, Y.Y., Li, X., Hu, A.Z., Wu, F., Zhu, Z.H. and Huang, X.T. 2010. Direct synthesis of CoO porous nanowire arrays on Ti substrate and their application as lithium-ion battery electrodes. Journal of Physical Chemistry C 114, 929.

67. Zhang, H.J., Tao, H.H., Jiang, Y., Jiao, Z., Wu, M.H. and Zhao, B.J. 2010. Ordered CoO/CMK-3 nanocomposites as the anode materials for lithium-ion batteries. Journal of Power Sources 195, 2950.

68. Qiao, H., Xiao, L., Zheng, Z., Liu, H., Jia, F. and Zhang, L. 2008. One-pot synthesis of CoO/C hybrid microspheres as anode materials for lithium-ion batteries. Journal of Power Sources 185(1), 486-491.

69. Sun, B., Liu, H., Munroe, P., Ahn, H. and Wang, G.X. 2012. Nanocomposites of CoO and a mesoporous carbon (CMK-3) as a high performance cathode catalyst for lithium-oxygen batteries. Nano Research 5, 460.

70. Xiong, S., Chen, J.S., Lou, X.W. and Zeng, H.C. 2012. Mesoporous Co_3O_4 and CoO@C Topotactically Transformed from Chrysanthemum-like $Co(CO_3)0.5(OH)\cdot0.11H_2O$ and Their Lithium-Storage Properties Advanced Functional Materials 22(4), 861-871.

71. Peng, C.X., Chen, B.D., Qin, Y., Yang, S.H., Li, C.Z., Zuo, Y.H., Liu, S.Y. and

Yang, J. H. 2012. Facile ultrasonic synthesis of CoO quantum dot/graphene nanosheet composites with high lithium storage capacity. ACS Nano 6, 1074.

72. Poizot, P.L.S.G., Laruelle, S., Grugeon, S., Dupont, L. and Tarascon, J.M. 2000. Nano-sized transition-metal oxides as negative-electrode materials for lithium-ion batteries. Nature 407, 496.

73. Grugeon, S., Laruelle, S., Herrera-Urbina, R., Dupont, L., Poizot, P. and Tarascon, J.M. 2001. Particle size effects on the electrochemical performance of copper oxides toward lithium. Journal of the Electrochemical Society 148(4), A285-A292.

74. Debart, A., Dupont, L., Poizot, P., Leriche, J.B. and Tarascon, J.M. 2001. A transmission electron microscopy study of the reactivity mechanism of tailor-made CuO particles toward lithium. Journal of the Electrochemical Society, 148(11), A1266-A1274.

75. Wang, H., Pan, Q., Zhao, J., Yin, G. and Zuo, P. 2007. Fabrication of CuO film with network-like architectures through solution-immersion and their application in lithium ion batteries. Journal of Power Sources 167(1), 206-211.

76. Pan, Q., Jin, H., Wang, H. and Yin, G. 2007. Flower-like CuO film-electrode for lithium ion batteries and the effect of surface morphology on electrochemical performance. Electrochimica Acta 53(2), 951-956.

77. Wang, H., Pan, Q., Zhao, J. and Chen, W. 2009. Fabrication of CuO/C films with sisal-like hierarchical microstructures and its application in lithium ion batteries. Journal of Alloys and Compounds 476(1-2), 408-413.

78. Chen, L.B., Lu, N., Xu, C.M., Yu, H.C. and Wang, T.H. 2009. Electrochemical performance of polycrystalline CuO nanowires as anode material for Li ion batteries. Electrochimica Acta 54(17), 4198-4201.

79. Ke, F.S., Huang, L., Wei, G.Z., Xue, L.J., Li, J.T., Zhang, B., Chen, S.R., Fan, X.Y. and Sun, S.G. 2009. One-step fabrication of CuO nanoribbons array electrode and its excellent lithium storage performance. Electrochimica Acta 54(24), 5825-5829.

80. Xiang, J.Y., Tu, J.P., Zhang, J., Zhong, J., Zhang, D. and Cheng, J.P. 2010. Incorporation of MWCNTs into leaf-like CuO nanoplates for superior reversible Li-ion storage. Electrochemistry Communications 12(8), 1103-1107.

81. Dar, M.A., Nam, S.H., Kim, Y.S. and Kim, W.B. 2010. Synthesis, characterization, and electrochemical properties of self-assembled leaf-like CuO nanostructures. Journal of Solid State Electrochemistry 14(9), 1719-1726.

82. Xiang, J.Y., Tu, J.P., Zhang, L., Zhou, Y., Wang, X.L. and Shi, S.J. 2010. Self-assembled synthesis of hierarchical nanostructured CuO with various morphologies and their application as anodes for lithium ion batteries. Journal of Power Sources 195(1), 313-319.

83. Aragón, M.J., Pérez-Vicente, C. and Tirado, J.L. 2007. Submicronic particles of manganese carbonate prepared in reverse micelles: A new electrode material for lithium-ion batteries. Electrochemistry Communications 9(7), 1744-1748.

84. Sun, B., Chen, Z., Kim, H.S., Ahn, H. and Wang, G. 2011. MnO/C core–shell nanorods as high capacity anode materials for lithium-ion batteries. Journal of Power Sources 196(6), 3346-3349.

85. Zhong, K., Xia, X., Zhang, B., Li, H., Wang, Z. and Chen, L. 2010. MnO powder as anode active materials for lithium ion batteries. Journal of Power Sources 195(10), 3300-3308.

86. Li, X., Li, D., Qiao, L., Wang, X., Sun, X., Wang, P. and He, D. 2012. Interconnected porous MnO nanoflakes for high-performance lithium ion battery anodes. Journal of Materials Chemistry 22(18), 9189-9194.

87. Ding, Y.L., Wu, C.Y., Yu, H.M., Xie, J., Cao, G.S., Zhu, T.J., Zhao, X.B. and Zeng, Y.W. 2011. Coaxial MnO/C nanotubes as anodes for lithium-ion batteries. Electrochimica Acta 56(16), 5844-5848.

88. Liu, Y., Zhao, X., Li, F. and Xia, D. 2011. Facile synthesis of MnO/C anode materials for lithium-ion batteries. Electrochimica Acta 56(18), 6448-6452.

89. Li, S.R., Sun, Y., Ge, S.Y., Qiao, Y., Chen, Y.M., Lieberwirth, I., Yu, Y. and Chen, C.H. 2012. A facile route to synthesize nano-MnO/C composites and their application in lithium ion batteries. Chemical Engineering Journal 192, 226-231.

90. Xu, G.L., Xu, Y.F., Sun, H., Fu, F., Zheng, X.M., Huang, L., Li, J.T., Yang, S.H. and Sun, S.G. 2012. Facile synthesis of porous MnO/C nanotubes as a high capacity anode material for lithium ion batteries. Chemical Communications 48(68), 8502-8504.

91. Zhang, X., Xing, Z., Wang, L., Zhu, Y., Li, Q., Liang, J., Yu, Y., Huang, T., Tang, K., Qian, Y. and Shen, X. 2012. Synthesis of MnO@ C core–shell nanoplates with controllable shell thickness and their electrochemical performance for lithium-ion batteries. Journal of Materials Chemistry 22(34), 17864-17869.

92. Yang, G., Li, Y., Ji, H., Wang, H., Gao, P., Wang, L., Liu, H., Pinto, J. and Jiang, X. 2012. Influence of Mn content on the morphology and improved electrochemical properties of Mn_3O_4 | MnO@ carbon nanofiber as anode material for lithium batteries. Journal of Power Sources 216, 353-362.

93. Hsieh, C.T., Lin, C.Y. and Lin, J.Y. 2011. High reversibility of Li intercalation and de-intercalation in MnO-attached graphene anodes for Li-ion batteries. Electrochimica Acta 56(24), 8861-8867.

94. Zhang, K., Han, P., Gu, L., Zhang, L., Liu, Z., Kong, Q., Zhang, C., Dong, S., Zhang, Z., Yao, J. and Xu, H. 2012. Synthesis of nitrogen-doped MnO/graphene nanosheets hybrid material for lithium ion batteries. ACS Applied Materials & Interfaces 4(2), 658-664.

95. Mai, Y.J., Zhang, D., Qiao, Y.Q., Gu, C.D., Wang, X.L. and Tu, J.P. 2012. MnO/reduced graphene oxide sheet hybrid as an anode for Li-ion batteries with enhanced lithium storage performance. Journal of Power Sources 216, 201-207.

96. Qiu, D., Ma, L., Zheng, M., Lin, Z., Zhao, B., Wen, Z., Hu, Z., Pu, L. and Shi, Y. 2012. MnO nanoparticles anchored on graphene nanosheets via in situ carbothermal reduction as high-performance anode materials for lithium-ion batteries. Materials Letters 84, 9-12.

97. Tang, Q., Shan, Z., Wang, L. and Qin, X. 2012. MoO_2–graphene nanocomposite as anode material for lithium-ion batteries. Electrochimica Acta 79, 148-153.

98. Bronstein, L.M., Huang, X., Retrum, J., Schmucker, A., Pink, M., Stein, B.D. and Dragnea, B. 2007. Influence of Iron Oleate Complex Structure on Iron Oxide Nanoparticle Formation. Chemistry of Materials 19(15), 3624-3632.

99. Xiong, Y., Li, Z., Li, X., Hu, B. and Xie, Y. 2004. Thermally stable hematite hollow nanowires. Inorganic Chemistry 43(21), 6540-6542.

100. Wen, X., Wang, S., Ding, Y., Wang, Z.L. and Yang, S. 2005. Controlled growth of large-area, uniform, vertically aligned arrays of α-Fe_2O_3 nanobelts and nanowires Journal of Physical Chemistry B 109(1), 215-220.

101. Ling, Y., Wang, G., Wheeler, D.A., Zhang, J.Z. and Li, Y. 2011. Sn-Doped Hematite Nanostructures for Photoelectrochemical Water Splitting. Nano Letters 11(5), 2119-2125.

102. Yuan, L., Jiang, Q., Wang, J. and Zhou, G. 2012. The growth of hematite nanobelts and nanowires—tune the shape via oxygen gas pressure. Journal of Materials Research 27(7), 1014-1021.

103. Zhou, W., Tang, K., Zeng, S. and Qi, Y. 2008. Room temperature synthesis of rod-like $FeC_2O_4 \cdot 2H_2O$ and its transition to maghemite, magnetite and hematite nanorods through controlled thermal decomposition. Nanotechnology 19(6), 065602.

104. Ganguli, A.K. and Ahmad, T. 2007. Nanorods of Iron Oxalate Synthesized Using Reverse Micelles: Facile Route for-Fe_2O_3 and Fe_3O_4 Nanoparticles. Journal of Nanoscience and Nanotechnology 7(6), 2029-2035.

105. Cho, W., Park, S. and Oh, M. 2011. Coordination polymer nanorods of Fe-MIL-88B and their utilization for selective preparation of hematite and magnetite nanorods. Chemical Communications 47(14), 4138-4140.

106. Chen, J., Xu, L., Li, W. and Gou, X. 2005. α-Fe_2O_3 Nanotubes in Gas Sensor and Lithium-Ion Battery Applications. Advanced Materials 17(5), 582-586.

107. Kang, N., Park, J.H., Choi, J., Jin, J., Chun, J., Jung, I.G., Jeong, J., Park, J.G., Lee, S.M., Kim, H.J. and Son, S.U. 2012. Nanoparticulate Iron Oxide Tubes from Microporous Organic Nanotubes as Stable Anode Materials for Lithium Ion Batteries. Angewandte Chemie International Edition 51(27), 6626-6630.

108. Reddy, M.V., Yu, T., Sow, C.H., Shen, Z.X., Lim, C.T., Subba Rao, G.V. and Chowdari, B.V.R. 2007. α-Fe_2O_3 Nanoflakes as an Anode Material for Li-Ion Batteries. Advanced Functional Materials 17(15), 2792-2799.

109. Chen, J.S., Zhu, T., Yang, X.H., Yang, H.G. and Lou, X.W. 2010. Top-Down Fabrication of α-Fe_2O_3 Single-Crystal Nanodiscs and Microparticles with Tunable Porosity for Largely Improved Lithium Storage Properties. Journal of the American Chemical Society 132(38), 13162-13164.

110. Hu, X., Yu, J.C., Gong, J., Li, Q. and Li, G. 2007. α-Fe_2O_3 Nanorings Prepared by a Microwave-Assisted Hydrothermal Process and Their Sensing Properties Advanced Materials 19(17), 2324-2329.

111. Xiong, S., Xu, J., Chen, D., Wang, R., Hu, X., Shen, G. and Wang, Z.L. 2011. Controlled synthesis of monodispersed hematite microcubes and their properties. Cryst Eng Comm, 13(23), 7114-7120.

112. Liang, X., Wang, X., Zhuang, J., Chen, Y., Wang, D. and Li, Y. 2006. Synthesis of nearly monodisperse iron oxide and oxyhydroxide nanocrystals. Advanced Functional Materials 16(14), 1805-1813.

113. Cao, H., Wang, G., Warner, J.H. and Watt, A.A. 2008. Amino-acid-assisted synthesis and size-dependent magnetic behaviors of hematite nanocubes. Applied Physics Letters 92: 013110-013113.

114. Zeng, S., Tang, K., Li, T., Liang, Z., Wang, D., Wang, Y. and Zhou, W. 2007. Hematite hollow spindles and microspheres: selective synthesis, growth mechanisms, and application in lithium ion battery and water treatment. Journal of Physical Chemistry C 111(28), 10217-10225.

115. Jia, X.H. and Song, H.J. 2012. Facile synthesis of monodispersed α-Fe$_2$O$_3$ microspheres through template-free hydrothermal route. Journal of Nanoparticle Research 14(1), 663.

116. Cao, S.W. and Zhu, Y.J. 2008. Hierarchically Nanostructured α-Fe$_2$O$_3$ Hollow Spheres: Preparation,Growth Mechanism, Photocatalytic Property, and Application in Water Treatment. Journal of Physical Chemistry C 112(16), 6253-6257.

117. Wang, B., Chen, J.S., Wu, H.B., Wang, Z. and Lou, X.W. 2011. Quasiemulsion-Templated Formation of α-Fe$_2$O$_3$ Hollow Spheres with Enhanced Lithium Storage Properties. Journal of the American Chemical Society 133(43), 17146-17148.

118. Xu, X., Cao, R., Jeong, S. and Cho, J. 2012. Spindle-like Mesoporous α-Fe$_2$O$_3$ Anode Material Prepared from MOF Template for High-Rate Lithium Batteries. Nano Letters 12(9), 4988-4991.

119. Du, D. and Cao, M. 2008. Ligand-assisted hydrothermal synthesis of hollow Fe$_2$O$_3$ urchin-like microstructures and their magnetic properties. Journal of Physical Chemistry C 112(29), 10754-10758.

120. Zeng, S., Tang, K., Li, T., Liang, Z., Wang, D., Wang, Y., Qi, Y. and Zhou, W. 2008. Facile route for the fabrication of porous hematite nanoflowers: its synthesis, growth mechanism, application in the lithium ion battery, and magnetic and photocatalytic properties. Journal of Physical Chemistry C 112(13), 4836-4843.

121. Wu, Z., Yu, K., Zhang, S. and Xie, Y. 2008. Hematite hollow spheres with a mesoporous shell: controlled synthesis and applications in gas sensor and lithium ion batteries. Journal of Physical Chemistry C 112: 11307-11313.

122. Kim, H.J., Choi, K.I., Pan, A., Kim, I.D., Kim, H.R., Kim, K.M., Na, C.W., Cao, G. and Lee, J.H. 2011. Template-free solvothermal synthesis of hollow hematite spheres and their applications in gas sensors and Li-ion batteries. Journal of Materials Chemistry 21(18), 6549-6555.

123. Song, H.J., Li, N. and Shen, X.Q. 2011. Template free synthesis of hollow α-Fe$_2$O$_3$ microspheres. Applied Physics A 102(3), 559-563.

124. Hu, C.Y., Xu, Y.J., Duo, S.W., Li, W.K., Xiang, J.H., Li, M.S. and Zhang, R.F. 2010. Preparation of Inorganic Hollow Spheres Based on Different Methods. Journal of the Chinese Chemical Society 57(5A), 1091-1098.

125. Liu, J., Li, Y., Fan, H., Zhu, Z., Jiang, J., Ding, R., Hu, Y. and Huang, X. 2009. Iron oxide-based nanotube arrays derived from sacrifcial template-accelerated hydrolysis: large-area design and reversible lithium storage. Chemistry of Materials 22(1), 212-217.

Issues and Challenges of Rechargeable Lithium Batteries

O. Padmaraj[1*], N. Satyanarayana[2], S. Austin Suthanthiraraj[3] and
C. Venkateswaran[1]

[1] Department of Nuclear Physics, University of Madras, Guindy Campus,
    Chennai – 600 025, Tamil Nadu, India
[2] Department of Physics, Pondicherry University, Kalapet,
    Pondicherry – 605 014, India
[3] Department of Energy, University of Madras, Guindy Campus,
    Chennai – 600 025, Tamil Nadu, India

1. Introduction

In recent years, there have been ever-increasing energy demand and
environmental pollution associated with the usage of fossil fuels, due
to high energy consumption in different fields such as industries,
transportation and household purposes for the development of economic
growth and advances in lifestyle. Hence, there is a need to develop
alternative eco-friendly renewable energy technologies including
solar, wind, bio, hydropower energies and also electrochemical energy
conversion and storage systems, etc., for effective utilization of energy
consumption to fulfill our basic needs. Among them, electrochemical
energy conversion and storage systems are one of the best viable options
for a wide variety of applications with attractive characteristics over
solar and wind energies. Of the numerous rechargeable battery systems,
lithium-ion batteries (LIBs) are having high energy and power densities,
even in small size and light weight compared to other rechargeable battery
systems. Thus, it has been used as the best power sources for all portable
and stationary electronic devices, automobiles, etc., applications in the
past decades [1-3]. Nevertheless, the current Li-ion battery technology

*Corresponding author: padmarajphysics@gmail.com

based on the lithium-insertion anode material limit their charge storage capacity and energy density for long-range electric vehicles and smart-grid applications.

Thereby, lithium-metal has received considerable attention as anode material in the lithium metal batteries (LMBs), including lithium-air (Li-air) and lithium-sulfur (Li-S) batteries for next-generation electric vehicle (EV) applications, because of its high theoretical specific energy density (3505 Wh kg^{-1} for Li-air and 2600 Wh kg^{-1} for Li-S) relative to the current LIBs and lowest negative electrochemical potential (–3.040 V *vs.* standard hydrogen electrode). Nevertheless, Lithium-metal has been considered as "unsafe" for practical applications, owing to the major problems such as dendrite formation, high chemical reactivity with an organic liquid electrolyte and an infinite volume expansion/contraction during lithium plating and stripping processes, which lead to pose safety hazards and poor cycling performances in practical applications [4-10]. Consequently, various approaches are being employed to modify the electrode (anode), electrolyte and separator through different techniques such as interface protections, a new class of organic liquid electrolyte with a high concentration of lithium salts and/or electrolyte additives, ionic liquids and ceramic particles coated or filled separators, for improving the safety and stability of rechargeable lithium batteries [11, 12]. Hence, in this chapter, authors have briefly described the various factors related to safety issues and an effective approach through different strategies for improving the safety and cyclic stability of rechargeable lithium-metal batteries for next-generation hybrid electric vehicle applications.

2. Working Principles

In general, the lithium battery contains four functional components, namely anode, cathode, electrolyte and separator. When the battery is being discharged, the lithium ion (Li$^+$) moves from Li metal anode to the Li$_x$MO$_2$ (M = Co, Ni or Mn) cathode through an organic liquid electrolyte. While the battery is charging by applying an external electrical power source, Li$^+$ ions move back from cathode to anode through an electrolyte medium and the current to pass in reverse direction. The schematic illustration of a typical lithium battery operation is shown in Fig. 1 [4]. The electrolyte is a medium to transfer the ions from one electrode to another electrode. The separator is one of the most important components to guarantee the cell safety by preventing the electrical contact between the two electrodes, which is typically a thin microporous polyolefin (PP, PE or laminates of PP and PE) membrane and also acts as an electrolyte reservoir to facilitate the Li$^+$ ionic transport in the cell. Importantly, it should be a good electronic insulator and have the capability of conducting ions by either intrinsic ionic conductor or by soaking in an aprotic liquid electrolyte. Additionally, it should have good chemical stability for enhancing the battery performances, including safety, cyclic stability, cell resistance, rate capability, energy density, etc. Hence, worldwide, researchers have been focusing to develop on advanced battery components with versatile features to restrain the safety issues associated with the usage of lithium metal anode and improve the cyclic performances of rechargeable Li metal batteries.

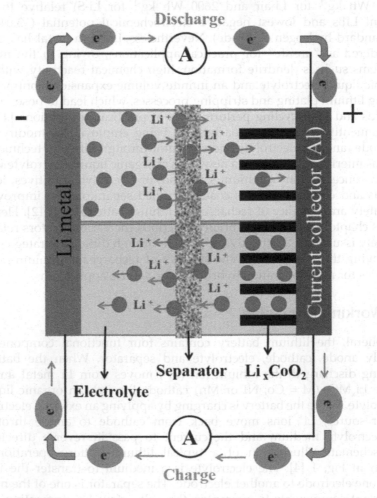

Fig. 1. Schematic illustration of Li metal battery.

3. Issues

Recently, rechargeable lithium-metal batteries have been focussed as one of the best power sources for the next generation electric vehicle applications, because of its attractive high energy density than that of commercial rechargeable LIBs. However, the formation of Li dendrite is one of the major shortcoming to the practical applications, due to the instability or of solid electrolyte interface (SEI) layer produced between lithium-metal electrode and liquid electrolyte, an infinite relative volume expansion/contraction changes during Li plating/striping processes, which leads to serious explosion hazards and poor cycling performances caused by internal shorting and capacity fading of the cells [4, 13]. All these issues, associated with the use of lithium-metal are presented in Fig. 2, and also discussed as well in the upcoming sections.

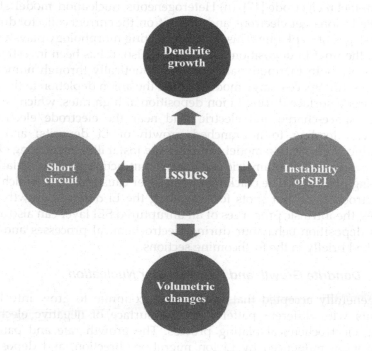

Fig. 2. Major issues of rechargeable lithium-metal batteries.

3.1 Li Dendrite Growth

3.1.1 Nucleation Location and Models of Li Dendrites

As we know, the nucleation sites occur in each cycle repeatedly during Li plating/stripping processes, which play a vital role for subsequent Li deposition behaviour. In order to understand the lithium deposition behaviour, various nucleation sites have been proposed, including tip-induced, bottom-induced and multidirectional induced nucleation sites in a rechargeable Li metal battery [11]. In the tip-induced nucleation sites, Li ions are more likely to deposit on the spherical tips due to the enhanced electronic and ionic fields [14-16]. Whereas, in the bottom-induced nucleation sites, Li dendrites are mostly caused by defects of SEI films for Li deposits. Furthermore, it has been demonstrated by several models, including surface nucleating and diffusion model, heterogeneous nucleation model, space-charge model and SEI-induced nucleation model through theoretical derivations and experimental observation in the past decades [11]. The summary of all these four models were presented as follows: (i) Surface nucleating and diffusion model describes the two important features, such as surface energy and diffusion barriers are responsible for the tendency of uniform Li deposition over the surface of lithium-metal electrode [17]; (ii) Heterogeneous nucleation model states that the Li ions get electrons and deposit on the current collector during initial stages of Li plating. This initial nucleating morphology plays a vital role in the final Li deposition patterns and also, it has been investigated extensively both thermodynamically and kinetically through numerical solutions; (iii) Space-charge model implies the anion depletion in the area of electrode surface during Li ion deposition at high rates, which creates a large space-charge and electric field near the electrode/electrolyte interfaces, leading to a branched growth of Li deposits; and (iv) SEI-induced nucleation model expresses the instability or cracking of SEI layer, owing to the volumetric expansion/contraction during Li plating/stripping processes. The cracking on SEI layer causes defects, which acts as electrochemical hot spots for triggering the Li dendrite growth [11]. Besides, the intrinsic properties of an unruptured SEI layer can also affect the Li deposition behaviour during electrochemical processes and it is described briefly in the forthcoming sections.

3.1.2 Dendrite Growth and Patterns after Nucleation

It is generally accepted that Li dendrites continue to grow into large deposits with different patterns on the surface of negative electrode during electrochemical plating process. The growth rate and patterns are strongly influenced by Li ion migration direction, and deposition mechanisms, respectively, which are driven by several impacting factors, including external applied electric field conditions (current density, charging time, etc.), the properties of SEI layer (ionic conductivity, stability, etc.), electrolytes (ionic mobility, concentration, viscosity, etc.), operating temperature and inner pressure of the LMBs. Besides, several models have been proposed to describe the effect of an applied electric field and the Li ion concentration gradients in liquid electrolytes on the growth of dendrites [11]. Mostly, the electric field plays a vital role for dendrite growth during Li plating process, which acts as the driving force for lithium ion migration. For instance, Chazalviel et al. [18] found that the space-charge influences the growth of a ramified Li dendrites in dilute salt solutions, owing to the depletion of anion near the anode surface based on the electric field. Likewise, Akolkar et al. [15] developed a mathematical diffusion-reaction model to describe the dendritic growth with respect to current density during Li electrodeposition process near the Li surface at both the flat surface and the dendritic tip. These models also described the transport or migration of Li ions in a diffusion boundary layer, which tends to estimate a dendrite growth rate of about 0.02 mm s^{-1} at an operating current density of 10 mA cm^{-2}. This result is relatively comparable with an experimentally observed dendrite growth rates, which is reported in Nishikawa et al. [19].

Based on the electrodeposition mechanisms and impacting factors, the Li dendrites exhibit various morphologies, such as needle, moss and tree like patterns. Needle like dendrites grow simultaneously in both the length and diameter directions without branches from all the nucleation sites, such as tip, base and kinks, which may be induced by crystalline defects owing to an instability of thin SEI layer, grain boundaries, and chemical inhomogeneities [20, 21]. Such needle like Li dendrites can maintain the features of 1D structure, it could easily penetrate the separator, causing cell short circuit and safety hazards in LMBs [22]. At certain circumstances, the 1D needle like dendrites grow into the 3D moss (bush) like patterns with small diameters, due to the high interface area and electrolyte decompositions in cells. The growth of mossy like patterns are a very dynamic process involving nonlinear, apparent random growth of the motion through Li filaments, which are not dominated by the direction of the electric field in the electrolyte solution. During the Li stripping process, large parts of the mossy can get isolated from the current collectors to form "dead Li" causing an instability of SEI layer [16, 23]. The growth of mossy patterns was demonstrated by the raisin bread expansion model [20]. In this model, there is no preferred direction and distance between each raisin in the bread as the bread expands. The growth of the raisin bread has no growth centre, but the movement of the parts can be restricted due to its support.

Tree- like dendrites grow in all directions including length, diameter, and branches with a neat hierarchical structure, but are not common like 1D needle and 3D moss like dendrites in experimental LMBs. It was generally accepted that, under most experimental conditions, metallic Li dendrites cannot grow in all directions at the proper rates with stable branch structures, since the Li plating process is strongly influenced by the migration of Li ions in an electrolyte. Thus, the dendrites may grow into 1D needle or 3D moss like patterns during electrodeposition process. However, tree like dendrites are considered as a textbook dendrite pattern, it has been widely investigated through modelling and simulations in recent years [24]. Naturally, such tree-like dendrites will also bring all kinds of safety hazards, including short circuits, cycle life, and capacity loss during electrochemical processes.

3.2 Instability of Solid Electrolyte Interphase (SEI)

The solid electrolyte interface (SEI) layer forms instantaneously by the parasitic reactions between lithium-metal anode and an organic liquid electrolyte, due to the most negative electrochemical potential of Li metal. It was first observed by Dey in 1970 [25] and named as SEI layer by Peled in 1979 [26]. Extensive studies on SEI film have been evolving continuously by the theoretical calculations, simulations and experimental results, in order to understand the formation of SEI films over Li metal surfaces, Li ion diffusion through SEI film, structure and components, etc. For instance, Goodenough et al. [5, 27] demonstrated the relationship between the electrochemical potential of anode (μ_A)/cathode (μ_C) electrodes and the lowest unoccupied molecular orbital (LUMO)/highest occupied molecular orbital (HOMO) of electrolytes. If, $\mu_A > E_{LUMO}$, electrons on the anode are disposed to transfer to the unoccupied molecular orbital of the electrolyte, inducing the intrinsic reduction reactions of electrolyte. Likewise, in the case of $\mu_C < E_{HOMO}$, redox or parasitic reactions contribute to the formation of SEI layer between electrode and electrolyte. Importantly, it serves as a potential barrier to prevent further decomposition of an organic liquid electrolyte for improving the safety and cycle stability, ought to have the following attractive features such as, i) high ionic conductivity with an appropriate thickness to reduce the Li ion diffusion resistance, ii) strong mechanical strength to adapt an non-uniform volumetric change, and iii) high stability in structure, components and morphology during Li depositing/stripping processes [28-31]. Nevertheless, a practical SEI film is not perfect affirmatively, owed to poor ionic conductivity (huge diffusion resistance), and low shear modulus of heterogeneous components with dual-layer structured SEI film, which will be discussed briefly in the upcoming sections. Therefore, extensive research activities have been devoted to seeking the proper Li salts, solvents, concentrations, and electrolyte additives for stabilizing the electrode/electrolyte interface film to achieve the dendrite free electrode for the commercialization of an efficient secondary LMBs.

3.2.1 SEI Structure and Components

It is generally accepted that the SEI layer comprises heterogeneous insoluble products with dual-layer (organic/inorganic) structure like mosaic morphology, due to the numerous reductive decompositions proceed and deposit on the negatively charged anode surface shown in Fig. 3. The layer close to the Li metal surface covers inorganic species with lower oxidation states such as, Li_2O, Li_3N, LiF, $LiOH$, and Li_2CO_3, labelled as the inorganic layer.

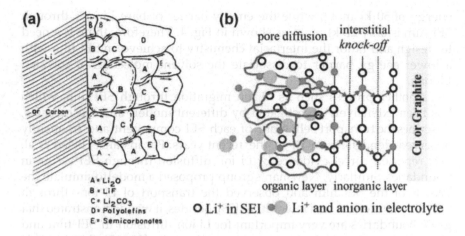

Fig. 3. Schematic diagrams of (a) Polyhetero microphase SEI on Li or carbon [32], and (b) Dual (organic and inorganic) layered SEI structure [33].

On the other hand, the outer part of the SEI film contains organic species with higher oxidation states such as, $ROCO_2Li$, $ROLi$, and $RCOO_2Li$ (R is an organic group related to the solvent), labelled as the organic layer. The formation of various organic/inorganic components with dual-layer structured in SEI layer strongly depend on the electrolyte compositions, including Li salts and solvents, which decides the primary features of SEI film [12, 29, 31-33]. In most cases, the organic species are formed by solvent reactions, and inorganic species are originated from the reactions between salt and solvents, respectively. Therefore, the forming mechanism, distribution and clear identification of each SEI species with respect to liquid electrolyte compositions can be investigated for achieving a better chemical/electrochemical stability with lithium metal anode, low ion-diffusion resistance and high shear modulus of SEI film.

3.2.2 Ionic Conductivity of SEI Film

SEI film is an essential route for two distinct processes, such as desolvation and migration of Li ions to improve the ionic conductivity, which can render a dendrite-free Li depositing morphology. The desolvation process takes place gradually at the pore diffusion stage in the outer (organic) layer of SEI film. It is strongly depending on the structure of solvent species and Li ion concentrations [34-36]. After that, the solvated Li ions must shed solvent molecules and migrate through an interstitial knock-off diffusion in the inner (inorganic) layer to obtain electrons and subsequently deposit on the current collector [37]. The stripping of the solvation sheath of Li ions is the rate determining step with an activation energy of 50 kJ mol^{-1}, while the energy barrier of bare Li ions through SEI film is only 20 kJ mol^{-1} as shown in Fig. 4. Therefore, there is a need to design and tailor the interfacial chemistry to achieve an SEI film with a lower energy barrier to dissociate the solvent molecules of solvated Li ions.

Another process such as Li ion migration through SEI film, which has been extensively investigated by different models and mechanisms, describes the transport behaviour of each SEI component and frequently observed dendrite growth in the recent years. For instance, Peled et al. [32] reported direct evidence of Li ion diffusion through SEI *via* grain boundaries. Similarly, Newman's group proposed a model to simulate the growth of the SEI film and observed the transport of Li ions through SEI film *via* defects and interstitials [38]. Besides, it was demonstrated that grain boundaries are very important for Li ion diffusion in SEI film and some of them are verified by experimental results [39, 40]. Considerable experiments are conducted to probe the ionic diffusion in some specific SEI components as mentioned above. Among them, Li_2CO_3, Li_2O, LiF and Li alkylcarbonate are four important inorganic components that built up the

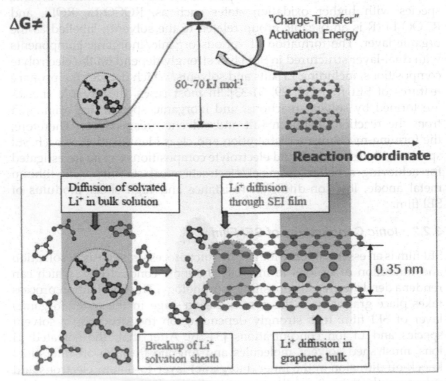

Fig. 4. Schematic diagram of energetic coordinates for two distinct desolvation and Li$^+$ ion migration processes [37].

dense layer of SEI film. The electronic structures of these materials are all insulated with a large forbidden band gap. Li ion diffusion can be very rapid when Li ion vacancies are easily available in Li_2CO_3 and Li_2O. The energy barriers of Li migration in Li_2CO_3 (ranging from 0.227 to 0.491 eV) and Li_2O (0.152 eV) are comparable, while Li migration in LiF (energy barrier of 0.729 eV) is much slower even when there are Li vacancies in the lattice [41, 42]. Hence, the ionic conductivity is determined by interfacial defects and topological distributions of these composite phases, due to the complex mosaic and heterogeneous structural nature of formed SEI film. A large ionic conductivity of SEI film, with typically in the range of 10 to 1000 Ω cm^2, can render a dendrite-free Li depositing morphology. This knowledge can present an instructional approach when designing the electrolyte additives and artificial layers to construct a stable and an efficient SEI film for improving the cycle life, faradic efficiency and the morphology of the lithium deposit.

3.2.3 Mechanical Strength of SEI Film

In practice, the SEI layer formed on lithium metal can be easily cracked as a result of an infinite relative volume change during lithium deposition/ stripping processes, which lead to rapid dendrite growth and a significant loss of lithium and solution species through repeated breakdown and reparation, owing to low and heterogeneous shear modulus. In order to suppress the dendrite growth, the shear modulus of SEI film is high enough relative to that of the Li anode ($\sim 10^9$ Pa) to achieve a safe and efficient LMBs [43, 44]. Interestingly, *in situ* scanning electron and atomic force microscopy (AFM) techniques have been extensively employed to examine the mechanical performance of SEI film by monitoring the interface topography on the anode surface [45, 46]. For example, Liu et al. [47] and Zhang et al. [48] reported modulus mapping of SEI by in situ AFM investigation and demonstrated that SEI film showed a Young's modulus within the range of 50-400 MPa, due to the heterogeneous composition nature of the SEI film. Likewise, Li et al. [49] adopted AFM technique to characterize the artificial Li_3PO_4 film on Li metal anode. It was demonstrated that the artificial Li_3PO_4 film on Li anode displayed a smooth surface with Young's modulus of 10-11 GPa, which can sufficiently suppress the Li dendrite growth. Other than the AFM characterization, density functional theory (DFT) calculations provide mechanical properties as well, which cannot be achieved by state-of-the art experimental investigations in the recent years [50].

3.3 Volumetric Changes and Short-circuit

In most of the commercial rechargeable Li-ion batteries, electrode materials like graphite and alloy type anodes exhibit volume change of

~10% and ~400% for Si, respectively, during charge/discharge processes, which leads to hamper the cycling performances and capacity fading. In the case of rechargeable Li-metal batteries, the relative volume change (expansion/contraction) of Li-metal anode is virtually infinite during Li plating/stripping processes as shown in Fig. 5. An uneven expansion of the Li-metal anode leads to straining and cracking of the solid electrolyte interphase (SEI) layer on the Li-metal surface, resulting to rapid growth of high surface area Li filaments and dendrites from the cracks, owing to low impedance for Li deposition.

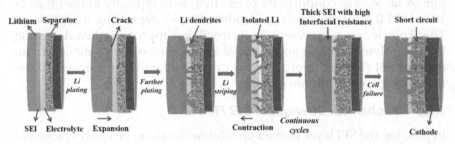

Fig. 5. Schematic showing the relative volume change and its impact during Li plating/stripping processes.

The Li dendrites can penetrate through the commercial polyolefin (PE, PP, PE/PP, PP/PE/PP) separator to form a short-circuit between the electrodes and cause serious safety issues, owing to the poor wettability with an organic electrolyte, lower mechanical strength and thermal stability. Besides further fractures in the SEI layer, the volume contraction during stripping process could break the electrical contact and produce 'dead' Li dendrites. These repeated processes can produce a thick accumulated SEI layer and excessive dead Li, which may lead to blocked ion transport and capacity fading [13]. Hence, there is a need to impose formidable challenges on the SEI stability to inhibit the chemical side-reactions between Li-metal anode and an organic liquid electrolyte and also modifications of separator for suppressing the Li dendrite growth and improve the reliability of rechargeable Li-metal batteries.

4. Challenges

In the recent years, various strategies such as Lithium-metal interface protections, a new class of organic liquid electrolyte with a high concentration of lithium salts, solvated ionic liquids, electrolyte additives and conventional polyolefin separators by inorganic ceramic particles coating or filled, have been employed for improving the safety and cyclic stability of rechargeable lithium-metal batteries as shown in Fig. 6.

Additionally, their merits and demerits are summarized in Table 1. Each approach will be described briefly in the forthcoming sections.

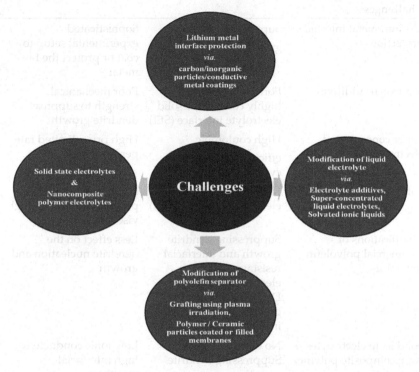

Fig. 6. Various strategies to suppress Li dendrite growth for improving the safety and cyclic stability of rechargeable lithium-metal batteries.

4.1 Lithium-metal Interface Protection

It was well-known that Li metal anode has a virtually infinite relative volume change during Li plating/stripping processes, resulting in the fracture of SEI layer and dendrite growth owed to poor mechanical strength and inhomogeneous distribution of Li ions and current density, respectively [51]–[54]. Thus, LMBs suffers capacity fading, low coulombic efficiency, poor lifespan and safety issues, etc. Therefore, a number of strategies have been employing including lithiophilic matrix, lamination of an ultrathin coatings on Li metal anode to reinforce the stable SEI layer that mitigate the volume change, parasitic reactions with an electrolyte, and also regulate Li ion plating/stripping behaviour to suppress Li dendrite growth for high energy density rechargeable Lithium metal batteries.

Table 1. Merits and demerits of proposed strategies

Proposed strategies/ Challenges	Merits	Demerits
Lithium-metal interface protection	Suppressing dendrite growth	Sophisticated experimental setup to coat or protect the Li-metal
Electrolyte additives	Forming a thin and highly conductive solid electrolyte interface (SEI)	Poor mechanical strength to suppress dendrite growth
Super concentrated electrolyte	High coulombic efficiency (CE) and cycling stability	High price, limited rate performances
Solvated ionic liquids	Highly conductive solid electrolyte interface (SEI)	High price, limited rate performance due to high viscosity
Modifications of commercial polyolefin separators	Suppressing dendrite growth and interfacial resistance, High electrolyte uptake, Detecting the dendrite growth, Improve interfacial stability	Less effect on the dendrite nucleation and growth
Solid-state electrolytes & Nanocomposite polymer electrolytes	No electrolyte leak, Suppressing dendrite growth	Low ionic conductivity, high interfacial resistance

4.1.1 Lithiophilic Matrix

Recently, lithiophilic oxidized polyacrylonitrile nanofiber matrix has been adopted to prevent volume change and regulate Li ion deposition on Li metal surface, which lead to improve the stable cycling performances with an average coulombic efficiency even upto 120 cycles at higher current density. Likewise, graphene, N-doped PAN, N-doped graphene, and metal-organic framework (MOF) matrixes can also enable to suppress Li dendrite growth by uniform Li plating/striping behaviour during operations [55-58]. Another attempt has been explored to suppress Li dendrite growth by placing glass fibers (GFs) with many polar functional groups on the anode surface, which can facilely and evenly distribute Li ions [59]. Besides, it can not only suppress Li dendrite growth, but also shed emerging strategies (melting and spark) to fabricate Li metal composite anode for high energy density Li-S, Li-O_2 batteries [60-63]. However, it is really complicated to fabricate Li metal composite anode, because the most available matrixes are poor wettability with molten Li (i.e., lithiophobic) or not able to withstand at high temperature. Thus, in this situation, it is possible to develop a universal surface functionalization method to afford lithiophilicity by means of a thin and conformal coating of Si or ZnO *via* spontaneous reaction between Si or ZnO and molten Li to form Li_xSi or Li_xZn/Li_2O, respectively. For instance, Yi Cui research group demonstrated the fabrication of lithiophilic Li metal composite anode *via* "melting strategy" to minimize volume-change and dendrite-free lithium metal anode for LMBs [54, 60-62]. Interestingly, Liu et al. (2016) [54] reported a thermally stable polyimide (PI) nanofiber with conformal atomic layer deposition (ALD) ZnO coating matrix for the development of Li metal composite anode. They found that the ALD ZnO coated PI matrix enables the wettability by molten Li or 'lithiophilic', which can serve as a lithiophilic Li-coated PI-ZnO scaffold for minimizing volume-change and effective dendrite suppression at a high current density of 5 mA cm^{-2} in both carbonate and ether electrolytes. Furthermore, Liang et al. (2016) [60] employed a conductive polyacrylonitrile (PAN) based carbon nanofiber scaffold with an ultrathin Si coated layer by chemical vapour deposition (CVD) technique to assist Li melt-infusion process into the PAN-Si scaffold. Also, investigated the effectiveness of Si coating on Li wettability of a wide variety of porous materials, including copper foam and carbon fiber matrixes. However, the fabrication of Li-scaffold composite anode is a little complex, and requires a tedious materials processing.

Alternatively, Lin et al. (2016) [63] proposed an effective "spark" strategy for the fabrication of Li-rGO composite anode with uniform nanogaps and infusing Li into the interlayer gaps between the graphene sheets. Because, graphene oxide (GO) has been used as a porous lithiophilic matrix for molten Li infusing process. When the GO film was partially put into contact with molten Li, a spark reaction was triggered across the whole film and also expanding the film into a much more porous structure due to the removal of functional groups within the GO layers. Hence, it generates the desired nanogaps for molten Li infusion and partially reduces the GO film into a porous conductive Li-rGO scaffold for dendrite-free morphology during electrochemical operations. Though, selection of proper lithiophilic matrix species and ratio in the composite anode is further optimized for practical applications.

4.1.2 *Laminations of an Ultrathin Coating*

In recent years, ultrathin coatings have been engaging as a new approach to prevent direct contact between lithium metal and an organic liquid electrolyte for the suppression of Li dendrite growth during plating/striping processes, but it was usually introduced unacceptably high impedance to the batteries. For instance, Zhang et al. [64] proposed a three dimensional carbon nanotube sponges (CNTS) as a Li deposition host for uniform distribution to suppress Li dendrite growth. Accordingly, ALD Al_2O_3 films can be prepared using alternating exposures to trimethyl aluminium (TMA) and H_2O, which deposits approximately 0.13 nm per TMA/H_2O cycle [65]. Typically, ALD is performed by injecting individual pulses of the two precursors into an evacuated chamber to be coated [66-69]. Furthermore, ALD has been studied extensively to deposit protective coatings of Al_2O_3 and other metal oxides onto lithium ion battery cathode materials in the range of nanometer scale [70, 71].

4.2 Modification of Separator

It was generally accepted that separators are broadly classified into four major types based on the structure and composition such as, i) microporous polyolefin, ii) modified microporous polyolefin, iii) non-woven polyolefin, and iv) composite polyolefin membranes, as shown in Fig. 7. These are characterized by the basic features like porosity, pore sizes, wettability, mechanical strength and thermal stability. Nevertheless, the microporous polyolefin (PE or PP) membranes have been widely used as separators in most of the commercial lithium-ion batteries over the past decades, due to their attractive features, safety and lower cost [72-75]. Besides, it has been classified further into monolayer (PP or PE) and multilayer (PE/PP or PP/PE/PP) microporous polyolefin separators based on the number of layers to achieve thermal shutdown behaviour for safety hazards [76]. Though, it has poor wettability with an organic electrolyte as a result loss of electrochemical compatibility and cycling efficiency. Thus, the commercially available microporous polyolefin (PP, PE, PE/PP, PP/PE/PP) separators have been modified *via* surface modification through grafting of hydrophilic monomers using plasma, electron beam (EB) and gamma irradiations [77-83]. Alternatively, the non-woven polyolefin membranes have been prepared by different methods namely, melt-blown [84], wet-laid [85] and electrospinning [86, 87] to develop web-like structure with fully interconnected randomly oriented multi-fibres, for improving the safety and stability of rechargeable lithium-based batteries. Likewise, composite polyolefin membrane separators are prepared by either coating of inorganic ceramic particles onto the surface of microporous/non-woven polyolefin membranes [88-90] or filling of inorganic particles into the variety of polymer host matrixes [91-95] to further increase the electrolyte wettability, thermal and electrochemical stabilities without sacrificing their basic features. All the aforementioned four major types of separators were briefly focused in this section.

4.2.1 Microporous Polyolefin Membranes

Most of the commercial microporous polyolefin separators are made up of polyethylene (PE), polypropylene (PP) and laminated PE/PP or PP/PE/

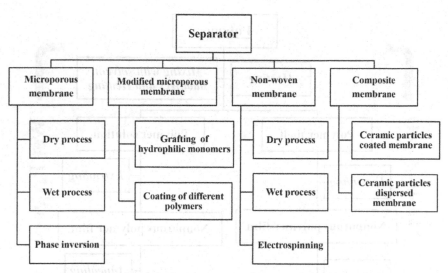

Fig. 7. Classification of separators and manufacturing processes.

PP membranes. In addition to this, other polymers like isotacticpoly(4-methyl-1-pentene), polyoxymethylene and polymer blends including polystyrene (PS)/poly(propylene) (PP), poly(ethylene terephthalate) (PET)/PP and P(VdF-co-HFP)/PAN for the preparation of microporous membranes through two different dry and wet processes as shown in Fig. 8 and also phase-inversion method [96-102].

The dry process is technologically attractive for the fabrication of microporous separator without solvents by following four major steps such as heating, extruding, annealing and stretching. In fact, the porosity, thermal stability and mechanical strength of the as-prepared microporous membrane separator depends on the surface morphology of the extruded polymer film, annealing temperature & time, and stretching ratio, etc [103]. In a typical fabrication process, first, the required amount of polymer was heated up to the melting temperature. Subsequently, the resultant polymer melt was extruded to form a nonporous polymer film with lamellae structure in the row direction as shown in Fig. 9a. Then, the as-prepared nonporous polymer film is annealed at glass transition temperature to further improve the crystallinity of the polymer film, which may lead to facilitate the formation of uniform slit-like microporous structure, while stretched along the machine direction as shown in Fig. 9b. Generally, there are two different uniaxial and biaxial stretches that have been adopted to make a microporous membrane for productions [73, 76]. Among them, the uniaxial stretching has been more effective to develop a uniform slit-like microporous structure oriented in the machine direction with high tensile strength for battery applications.

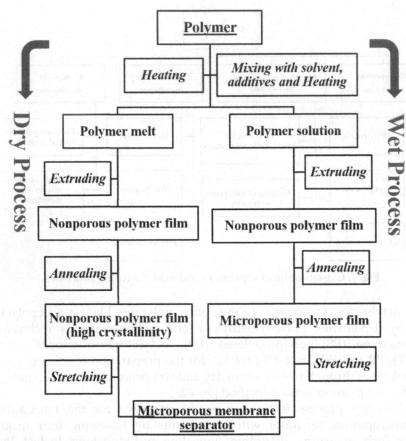

Fig. 8. Manufacturing processes of microporous separators [73, 103].

Likewise, the wet process also includes four steps but, it employs the principle of solvent extraction. First, an appropriate polymer was dissolved in suitable solvent and mixed with additives under heating to form a homogeneous solution. Then, the resultant polymer solution is extruded to form a nonporous polymer film and further, allowed to extract the solvent and additives by annealing to prepare the microporous polymer membrane. Finally, the stretching of as-prepared microporous polymer membrane can be conducted, before and after the extrusion of solvent and additives to achieve desirable porous structure and size. In the wet process, the microporous polyolefin membrane separators exhibit distinct differences in the orientation of pore structure, size and shape depending on the concentration of polymer solutions, extraction of solvents, annealing temperature and stretching conditions [76, 103, 104]. Therefore, both these processes involve common extrusion and stretching steps to make a thin polymer film and form different porous

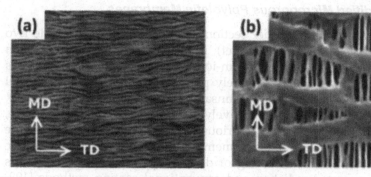

Fig. 9. SEM images of (a) before and (b) after stretched dry process polymer film. Reproduced with permission [73].

structures, respectively [105]. Furthermore, the as-prepared microporous polyolefin membranes by both dry and wet processes showed slit-like and interconnected elliptical pores structures, which are more beneficial to enhance the ion diffusion and also preventing the dendrite growth during charging/discharging processes as shown in Fig. 10.

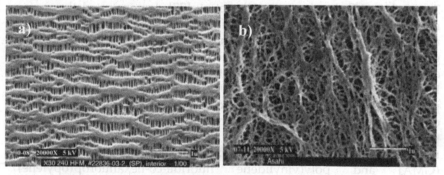

Fig. 10. SEM images of (a) dry and (b) wet processes microporous polyolefin separators [105].

Alternatively, phase inversion method has been extensively studied to prepare the microporous membranes by solvent exchange (solvent/non-solvent) for battery applications over the past decades [98, 99, 106, 107]. Nevertheless, the membranes showed the asymmetric porous structure related to the polymer concentration, additive concentration, solvent/non-solvent type, temperature and time. Thus, the asymmetric porous structure limits the overall battery performances by reducing the electrolyte absorption, which may lead to increase the internal resistance, and hinders the ionic flows.

4.2.2 Modified Microporous Polyolefin Membranes

As we discussed in an earlier section, the commercially available two different processed (dry and wet) polyolefin separators have been widely used in most of the lithium-ion batteries over the past decades [74, 103]. However, it has relatively poor wettability and low thermal stability, which lead to provide unsatisfactory electrolyte absorption/ retention and safety issue, respectively, during cell operation. Therefore, in order to inhibit these issues, various methods have been reported the modified microporous polyolefin membranes *via* grafting of hydrophilic monomers [83, 108] and coating of different polymer/polymer blends using high energy irradiation and conventional coating methods [109-112], respectively, onto the surface of commercial microporous polyolefin membranes, which are discussed briefly in the forthcoming sections.

4.2.2.1 Grafting Functionalization of Polyolefin Membrane Separators

Grafting of hydrophilic monomer is an effective way to modify the surface structure and properties of commercial microporous polyolefin membranes through plasma, ultraviolet (UV), electron beam (EB) and gamma irradiations and it has been reported over the past decades [78, 79, 81, 113-115]. For instance, Kim et al. [77] demonstrated the surface modification of PE membrane by grafting of acrylonitrile (AN) monomers *via* plasma treatment. The plasma induced acrylonitrile (AN) monomer grafted PE membrane showed high electrolyte absorption/retention during electrochemical processes, owing to the presence of hydrophilic surface. Furthermore, Gineste et al. (1995) [108] investigated the physicochemical properties of the hydrophilic acrylic acid (AA) and diethylene glycol-dimethacrylate (DEGDM) monomers grafted electron beam irradiated polypropylene (PP) separator by an influence of grafting solution temperature and monomers content. In addition, glycidyl methacrylate (GMA) and poly(vinylidene fluoride-co-hexafluoropropylene)/ poly(ethylene glycol) dimethacrylate (PVDF-HFP/ PEGDMA) mixtures also used to modify the surface properties of PE separator followed by electron beam irradiation leading to improve the electrolyte wettability and electrochemical performances of lithium-ion batteries [82]. Accordingly, Gao et al. (2006) [116] and Lee et al. (2009) [80] demonstrated the surface modification of methyl methacrylate (MMA), poly(ethylene glycol) borateacrylate (PEGBA) and 2,4,6,8-tetramethyl-2,4,6,8-tetravinylcyclote-trasiloxane (siloxane) monomers grafted PE separators *via* electron beam irradiation and also investigated the degrees of grafting (DG) with different irradiation doses 5, 10, 20 and 30 kGy. It was found that the grafted separators such as PE-g-MMA, PE-g- PEGBA and PE-g-siloxane at 10 kGy of irradiation dose showed high ionic conductivity (10^{-3} Scm^{-1}), improved electrolyte wettability (210% and 320%) and better discharge capacity (100-134 mAh g^{-1}) at different current rates (0.1C-3C) with stable electrochemical performances up to 200 cycles.

4.2.2.2 Polymer Coated Polyolefin Separators

Another simple approach has been employed to modify the conventional microporous polyolefin (PP and PE) membranes by surface coating of hydrophilic polymers in order to improve the electrolyte wettability/retention, interfacial and thermal stability during cell operations. For example, Lee et al. (2013) [110] reported hydrophilic poly(dopamine) (PDA) coating onto the surface of commercial microporous PE separator *via* dip coating method without sacrifying their porosity and permeability. As a result, the PDA coated PE separator showed high electrolyte uptake/retention, ionic conductivity (8.74×10^{-4} Scm^{-1}) and also, the electrochemical performances of fabricated Li/LiMn$_2$O$_4$ cells with PDA coated separator delivered high initial discharge capacity (118.4 mAh g^{-1} at 0.1 C) and better capacity retention of 92.8% after 500 cycles than that of uncoated PE separator. Likewise, a number of studies have reported the surface modification of polyolefin (PP and PE) separators using hydrophilic polydopamine (PDA) coating process for enhancing the electrolyte wettability, interfacial/thermal stability and high ionic transference number [112, 117, 118]. Consequently, Kang et al. (2012) [119] demonstrated a novel approach to improve the safety and electrochemical stability of lithium-ion batteries using PDA/DMAET- Silica coated PE separator. It has been found that the PDA/DMAET-Silica modified PE separator improved thermal shrinkage resistance (10% at 140 °C for 1 h), electrolyte [1 M LiPF6, EC:DEC (1:1, v/v)] wettability up to 121%, high ionic conductivity (3.5 × 10^{-4} Scm^{-1}) at room temperature compared to unmodified PE separator. Furthermore, the fabricated cells [Graphite/ LiMn$_2$O$_4$] with modified separator delivered high initial discharge capacity (115 mAh g^{-1} at 1 C rate) with stable performances up to 200 cycles and better capacity retention of 80.5% to the initial capacity than PDA/ DMAET (76.9%) and unmodified PE separator (74.4%). Nevertheless, the usage of polydopamine (PDA) polymer for surface coating onto the polyolefin separators may be limited, because of its high cost. Besides, the other polymeric materials, namely PEO [111, 120], PMMA [121], ANF [122, 123], PEG [124], PVdF [125-128] and P(VdF-co-HFP) [129, 130], etc., have been reported to modify the surface properties of microporous polyolefin separators *via* different coating techniques for improving the safety and stability of rechargeable lithium based batteries.

4.2.3 Non-woven Membranes

Non-woven membranes have web-like structures with fully interconnected multifibre layers that generate a micro/nano porous structures for absorbing a large amount of liquid electrolyte, which may lead to increase the number (n) and mobility (μ) of charge carriers without sacrificing their mechanical strength. It has been manufactured by using both natural and synthetic polymers through two different dry and wet processes, such as melt-blown method and wet-laid method, respectively, and also electrospinning method as shown in Fig. 11. The melt-blown method is described by two steps, first, the molten polymer is extruded by high-velocity hot air through the spinneret to form a polymeric fibers. In the second step, the resultant formed fibers were collected on the rotating drum collector and bonded by calendaring at an appropriate temperature and pressure to form a non-woven microporous fibrous membrane as shown in Fig. 12 [84, 131, 132]. In the case of a wet process, the resin is used as an adhesive agent, which is sprayed onto the fibrous web to bond the base fibers by applying an appropriate heat and pressure to form a non-woven fibrous membrane [85, 133, 134]. However, the structure and properties of the non-woven fibrous membranes were greatly influenced by various parameters such as polymers, their concentrations, temperature, and pressure. For example, Kritzer et al. (2006) [135] developed PE non-woven separator by the wet-laid method and investigated their mechanical strength, pore size distribution, thermal shrinkage at 180 °C and electrochemical stability of PE separator against lithium electrode. It was found that the average pore sizes and porosity of the PE separator was in the range of 20-30 μm and 55-65%, respectively. Similarly, Yi et al. (2009) [85] prepared PPTA fibrillated fiber by the wet-laid method and investigated the average pore size distribution with different pressure

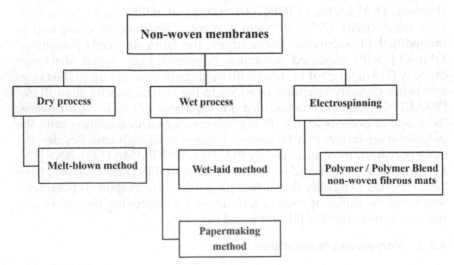

Fig. 11. Flow chart for the manufacturing processes of non-woven membrane separators.

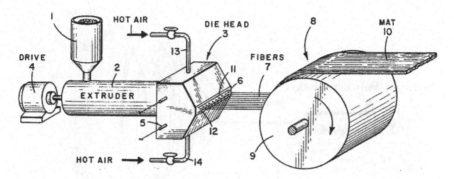

Fig. 12. Schematic representation of melt-blowing process to fabricate the non-woven membrane separators; (1) pellet hopper, (2) extruder, (3) die head, (4) drive, (5) Heater, (6) die openings, (7) polymer fibers web, (8) collecting substrate, (9) drum, (10) non-woven mat, (11 and 12) gas jets and (13 and 14) gas inlets [136].

for thermal bonding. However, the non-woven membranes prepared by dry and wet processes showed relatively large fiber diameter and pore sizes, which may lead to allow the dendrite growth owing to poor compatibility with lithium electrode, and thus, it was an inappropriate use in rechargeable lithium batteries.

In the recent years, electrospinning method has been adopted to develop the non-woven fibrous membranes for lithium-ion batteries. Many recent studies reported that the electrospinning technique is a simple and an effective approach to produce highly interconnected ultrafine porous structure with lower average fiber diameter by an optimization of electrospinning parameters, such as polymer concentration, applied voltage and distance between spinneret and collector, solution feed rate, spinneret size and drum collector rotating speed [87, 137]. The schematic diagram of an electrospinning setup for the fabrication of non-woven membrane and morphological images are shown in Fig. 13. Various polymers like poly(vinylidene fluoride) [P(VdF)], poly(vinylidene fluoride-co-hexafluoropropylene) [P(VdF-co-HFP)], polyacrylonitrile [PAN], polyimide (PI), polyurethane (PU), polystyrene (PS) and poly(methyl methacrylate) (PMMA), etc., have been used to develop electrospun non-woven fibrous membranes for lithium based batteries over the past decades [86, 87, 117, 137-145].

Among them, P(VdF) is found to be a suitable candidate for electrospun non-woven membrane as separator because of its attractive features such as, good electrochemical stability, an excellent affinity to electrolyte solution and high dielectric constant ($\varepsilon \approx 8.4$). For example, Liang et al. (2013) [146] prepared and investigated the mechanical properties and electrochemical performances of heat-treated electrospun P(VdF) non-

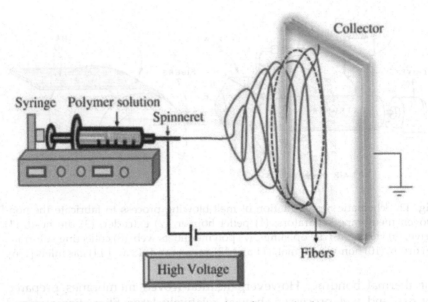

Fig. 13. Schematic diagram of an electrospinning setup.

woven membrane at 160° C for lithium-metal battery applications. The heat-treated P(VdF) electrospun membrane at 160° C exhibited high ionic conductivity (1.3×10^{-3} Scm^{-1}), less interfacial resistance (93 Ω) with lithium electrode, high electrochemical stability window up to ~4.8 V *vs* Li$^+$/Li, than that of commercial Celgard®2400 separator. In addition, the Li/LiFePO$_4$ cell with heat-treated PVdF fibrous membrane showed high charge and discharge capacities are 168.2 and 162.3 mAh g^{-1} at 0.2 C rate, respectively. Gao et al. (2006) [147] investigated the morphology and crystal structure of as-prepared electrospun PVdF non-woven membrane by an effect of applied voltage and average fiber diameter (AFD) to understand the physical properties. The interconnected electrospun PVdF non-woven membrane showed high porosity (80%) with uniform pore distribution, which enables to absorb large amount of liquid electrolyte, suppress the lithium dendrite growth and also exhibited higher discharge capacity (125 mAh g^{-1}) with stable capacity retention (119 mAh g^{-1}) even after 50 cycles at 0.5 C rate than Celgard®2400 separator. Though, the P(VdF) based non-woven membrane has relatively high crystallinity, which could limit the electrolyte wettability and mobility of ions.

Therefore, various approaches have been used to modify the P(VdF) based non-woven membrane, including a suitable P(VdF-co-HFP) copolymer and polymer blends, etc., in order to reduce the crystallinity of P(VdF) polymer and it has been reported. For instance, Cheruvalley et al. [148] developed P(VdF-co-HFP) copolymer based electrospun non-woven membranes and investigated their physical and electrochemical properties after soaking in different LiCF$_3$SO$_3$-TEGDME, LiTFSI-BMITFSI electrolyte solutions for lithium-ion batteries. The as-prepared P(VdF-co-HFP) copolymer based non-woven membrane as separator showed good mechanical strength and also excellent electrochemical performances, owing to the reduction of crystallinity, high electrolyte uptake/retention and ionic conductivity (mS cm^{-1}). Similarly, Kadar et al. developed electrospun P(VdF) grafted tBA copolymer [poly(vinylidene fluoride)-graft-poly(tert-butyl acrylate)] non-woven membrane by electrospinning technique. The P(VdF-g-tBA) based electrospun membrane exhibited high ionic conductivity, better electrochemical stability (>4.5 V *vs* Li$^+$/Li) and stable cyclic performances due to the reduction of P(VdF) crystallinity than that of pure P(VdF) membrane. Accordingly, electrospun polymer blends such as PVdF/PEO [149], P(VdF-co-HFP)/PMMA [86], PVdF/PAN [87], PVdF/PU [150], P(VdF-co-HFP)/PI [139], PAN/PMMA/PS [141], etc., have been reported by the reduction of crystallinity in order to enhance the electrolyte wettability and ionic conductivity without sacrificing their mechanical and thermal stabilities.

Interestingly, electrospun multilayer or sandwiching type of two different polymers based non-woven membranes such as, P(VdF-HFP)/PET/P(VdF-HFP) [151], PI/PVdF/PI [152], PAN/P(VdF-co-HFP)/PAN [153], P(VdF-co-HFP)/PAN/P(VdF-co-HFP) [153] and PVdF/PMMA/PVdF [154], etc., have been reported in order to improve their thermal stability *via* thermal self-shutdown property at high temperature, better electrolyte uptake/retention ability and safety for lithium battery applications. For instance, Wu et al. (2015) [152] fabricated a novel sandwich-type PI/PVdF/PI separator by a simple electrospinning technique with the thermal shutdown function for lithium-ion batteries and also investigated the high temperature treatment above 170 °C. The sandwich type PI/PVdF/PI non-woven membrane showed superior electrolyte uptake (476%) and ionic conductivity (3.46 mS cm^{-1}) compared to the pristine polyolefin separators and also it has lower impedance, high discharge capacity and stable cycle life. Under high-temperature treatments above 170 °C, the self-shutdown function of the trilayer [PI/PVdF/PI] non-woven membrane was observed within 10 minutes, which could be served as the potential separator to defend the thermal runaway issues, associated to lithium-metal batteries. Similarly, Raghavan et al. (2010) [153] developed sandwich type P(VdF-co-HFP) and PAN-based non-woven membrane by electrospinning technique for lithium batteries. The prepared trilayer non-woven membrane showed uniform bead free morphology with an average fiber diameter of 320-490 nm, high electrolyte uptake and ionic conductivity (10^{-3} S cm^{-1}) at room temperature even after soaking in 1 M LiPF6, EC: DEC (1:1, v/v) electrolyte solution. It was found that the activated sandwiched type non-woven membrane showed better anodic stability >4.6 V with stable interfacial resistance, good charge/

discharge capacities and stable cycle performances. Correspondingly, Xiao et al. (2009) [155] prepared a novel sandwiched type PVdF and PMMA polymers based non-woven membrane and investigated the surface morphology, crystallinity, electrolyte uptake, electrical conductivity and electrochemical performances at room temperature for lithium-ion batteries. The electrolyte activated sandwiched type [PVdF/ PMMA/PVdF – 1M $LiClO_4$, EC/PC] non-woven gel polymer electrolyte fibrous membrane exhibited high ionic conductivity and electrochemical potential stability of >4.5 V *vs* Li^+/Li. Despite, electrospun non-woven membranes relatively lose their mechanical strength after soaking in an organic liquid electrolyte solution. Moreover, electrospinning is a slow process, which limits the production rate; as a result, the separator cost is relatively expensive.

4.2.4 Composite Membranes

Composite membrane is one of the most effective modification methods to improve their physical, chemical, thermal and electrochemical properties for rechargeable lithium-based batteries. It can be formed in two different ways, such as (i) coating of inorganic ceramic particles with a polymeric binder onto the surface of microporous polyolefin/non-woven membranes, and (ii) directly filled or dispersed into the polymer host matrix through different methods as shown in Fig. 14. Recent studies reported that the coating of nanosized inorganic particles, such as Al_2O_3, TiO_2, SiO_2, ZrO_2, etc., with or without a polymer binder onto the surface of microporous polyolefin/non-woven membrane separators can significantly improve their wettability, thermal stability, mechanical strength, ionic conductivity and interfacial stability with lithium metal electrode, due to their high dielectric constant, hydrophilicity and high surface active sites [90, 156-162]. In the case of inorganic particles filled or dispersed composite membranes, the crystallinity of polymer host matrix can appreciably reduce through the Lewis acid-base interactions between the inorganic nanoparticles and polymer units, which could be able to absorb a large quantity of liquid electrolyte and promote the migration or mobility of lithium ions between the electrodes without sacrificing their physical properties [91, 93, 163-168]. Further, the inorganic particles coated and filled composite membrane separators are discussed briefly in the forthcoming sections.

4.2.4.1 Inorganic Particles Coated Polyolefin/Non-woven Composite Membranes

It is well known that the safety and stability issues which occur in the rechargeable lithium metal based batteries are mostly related to the dimensional shrinkage owing to thermal instability, poor electrolyte

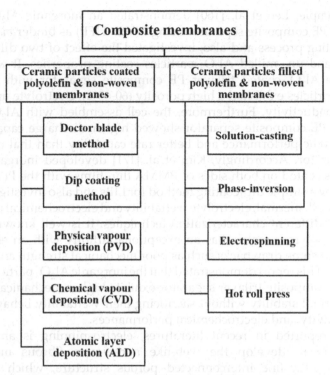

Fig. 14. Classification of composite membranes and their processing methods.

wettability and interfacial stability of the commercial polyolefin separators. In order to overcome the aforementioned issues, nanosized inorganic particles either commercially available or synthesized powders including Al_2O_3, SiO_2, TiO_2, etc., are coated onto the surface of commercial polyolefin (PE, PP or laminated PP/PE/PP) or non-woven membranes via different coating processes. The resultant polyolefin/non-woven composite separators exhibited good thermal stability with zero-dimensional shrinkage even at high temperature up to 150 °C, excellent wettability with an aprotic electrolytes and better interfacial stability with lithium metal anode and it has been reported in the recent years [90, 159, 169, 170]. Additionally, the polymers namely, PVdF [90], P(VdF-HFP) [170], PI [156], PVA [88], etc., have been used as a binder to bond the inorganic particles onto the surface of polyolefin/non-woven substrate and also, it helps to facilitate the electrolyte absorption/retention during electrochemical performances. In another approach to improve the affinity with an electrolyte and thermal shutdown property, novel trilayer composite membranes have been prepared by simple dip-coating or sandwiching type of inorganic Al_2O_3, SiO_2 and ZrO_2 nanoparticles on both sides of PE, PP, PMMA and PI membranes using P(VdF-HFP) as binder.

For example, Lee et al. [160] demonstrated an inorganic Al_2O_3-coated trilayer PE composite separator using P(VdF-HFP) as binder *via* a simple dip-coating process and also, investigated the effect of two different jet-milled and un- milled Al_2O_3 particles coating properties. It was found that the Al_2O_3-coated trilayer PE composite separator with jet-milled Al_2O_3 particles exhibited a high porosity (60-63%), electrolyte uptake and ionic conductivity. Furthermore, the cell assembled with Al_2O_3-coated trilayer PE composite separator showed higher discharge capacity with stable cyclic performance and better rate capability than that of pristine PE separator. Accordingly, Kim et al. [171] developed inorganic Al_2O_3 particles coated on both sides of PMMA thin films with the P(VdF-HFP) binder by a simple dip-coating method for LIBs and also investigated their mechanical, thermal, electrolyte wettability and electrochemical properties through different characterization techniques. It is well known that the PMMA polymer matrix has an exceptional affinity with an electrolyte, but it has some drawbacks such as poor mechanical strength and thermal stability. This report demonstrated that the inorganic Al_2O_3 particles coated PMMA composite trilayer separator exhibits higher mechanical strength and thermal stability without sacrificing their wettability behavior, ionic conductivity, and electrochemical performances.

As reported in recent literatures, electrospinning is an effective technique to develop the web-like non-woven fibrous membranes with an ultra-fine interconnected porous structure, which can offer excellent wettability and promotes the migration of charge carriers, thus, improves the ionic conductivity and capacity retention with stable cyclic performances at different C-rates. However, the as-prepared electrospun fibrous membrane lose their mechanical strength and dimensional stability, when soaked into an organic liquid electrolyte to form the gel-polymer electrolyte, owing to the lower crystallinity and weak bonding between fibers. Therefore, inorganic nanosized particles have been used to develop the composite non-woven membranes by coating on the surface of polymer matrix. For example, Lee et al. [172] demonstrated a non-woven Al_2O_3 inorganic particles coated electrospun PI composite separator using P(VdF-HFP) as binder *via* a simple dip-coating process, for improving the thermal stability and electrochemical properties of rechargeable lithium-ion batteries. The Al_2O_3-coated electrospun PI composite separator showed high thermal stability and electrolyte wettability, owing to the attractive intrinsic properties of both PI membrane and Al_2O_3 inorganic nanoparticles. Besides, the cell with Al_2O_3-coated electrospun PI composite separator exhibits higher capacity retention (95.5% after 200 cycles at 1C) with stable cycling performance and rate capability (78.9% at 10 C) than commercial PP separator.

Compared to the aforementioned coating processes, "grafting" method is a superficial and feasible way to achieve the surface modification of polyolefin/non-woven separators with the negligible interfacial resistance. For example, Jiang et al. [173] reported a novel Al_2O_3 ceramic grafted PE composite separator prepared by using vinylsilane coupling reagent *via* electron-beam irradiation method. It was also demonstrated that the novel Al_2O_3 ceramic-grafted PE separator improved an excellent thermal stability – almost no shrinkage even at 150 °C – and wettability to carbonate based liquid electrolytes, thus enhancing safety and stability of battery performances compared to the bare PE separator. In particular, the cell fabricated with Al_2O_3 ceramic grafted PE separator exhibits better cyclic performance and rate capability than that of pristine PE separator. Accordingly, Zhu et al. [174] demonstrated grafting of vinyl trimethoxysilane and isopropyl triethyltitanate species on the surface of commercial polyolefin PE separator *via* electron-beam irradiation method. Subsequently, the electron- beam treated/modified PE separators were immersed in an appropriate ethanol solutions containing 0.15 mol/L sodium metasilicate and 0.36 mol/L hydrochloric acid for 4 h at 60° C and 0.12 mol/L tetrabutyltitanate, 0.5 mol/L hydrochloric acid, 0.5 mol/L water and 0.03 mol/L acetylacetone at 70 °C for 4 h to obtain SiO_2 and TiO_2-grafted PE separator through hydrolysis reactions. It was found that the surface modification of PE separator endows the excellent thermal stability and electrolyte affinity without sacrificing its ion transport properties. Besides, the $Li/LiFePO_4$ cell assembled with SiO_2 and TiO_2-grafted PE separator exhibits high discharge capacity (139 and 133 mAh g^{-1} at 0.2 C), better rate capability at different C-rate (0.2–5 C) with stable cycle performance. Though, the inorganic ceramic particles coating processes lose their basic requirements, such as separator thickness, porosity and permeability *via* pore blocking, which may lead to hinder the electrochemical performances with stable cycle life of rechargeable lithium-based batteries.

Thus, worldwide many researchers have been focusing to deposit an ultrathin inorganic metal oxides, such as Al_2O_3, SiO_2 and TiO_2, etc., over the surface of commercial microporous polyolefin (PE and PP)/non-woven separators through different thin film technologies, including magnetron sputtering [161, 175], pulsed laser deposition (PLD) [176], chemical vapor deposition (CVD) [177] and atomic layer deposition (ALD) [162, 178-180], etc. These technologies are beneficial for optimizing the coating thickness and surface properties without sacrificing their basic requirements, which could effectively increase the electrolyte wettability, thermal stability, mechanical strength and interfacial stability with the electrodes for improving the safety and stability issues of rechargeable lithium-based batteries for practical applications. For instance, Lee et al. [175] demonstrated the effects of Al_2O_3 nanoparticles coated PE separator by radio-frequency (RF) magnetron sputtering for lithium-ion batteries. The novel Al_2O_3 coated PE separator solved the thermal shrinkage problem even after high-temperature exposure at 140 °C for 30 min than that of commercial PE separators. Besides, the sputtered Al_2O_3 ceramic layer effectively improved its wettability with liquid electrolyte and rate capability (~130%) over high current densities compared to the pristine PE separator. Peng et al. [161] developed a novel TiO_2 nanoparticles coated microporous PP composite separator by magnetron sputtering deposition (MSD) technique for lithium-ion batteries and also investigated the surface morphology of TiO_2-MSD coated layers, thickness, thermal stability, electrolyte affinity and the electrochemical properties including ionic conductivity, discharge capacity, cyclic performances and rate capability. It has been found that the TiO_2-MSD coated PP composite separator suppressed the thermal shrinkage and improvement in wettability with liquid electrolytes and cell performances compared to the pristine PP separator. Chen et al. [162] investigated an effect of atomic layer deposition (ALD) cycles (20, 50, 100, 150, 200 and 500) of inorganic TiO_2 nanostructures on plasma-treated polypropylene (PP) membranes for LIBs. It was observed that the improvement of thermal stability even up to 160° C, wettability towards an electrolyte and capacity retention over different C-rates of the TiO_2-ALD coated PP membrane could be achieved at high ALD cycles.

4.2.4.2 Inorganic Particles Filled Composite Membranes

Alternatively, a simple approach has been employed to develop the nanocomposite membranes by an incorporation or filled of nanosized inorganic metal oxides, such as Al_2O_3, TiO_2, SiO_2, ZnO, ZrO_2, $BaTiO_3$, etc., into the polymer host matrices (PEO, PAN, PI, PVdF, P(VdF-co-HFP), P(VdF-co-HFP)/PMMA, etc.) through different techniques like solvent casting, phase inversion, electrospinning, etc., over the past decades [91, 93-95, 164, 165, 167, 168, 181, 182]. It has been demonstrated that the ionic conductivity (σ) and Li^+ ion transference number was increased by an incorporation of nanosized inorganic particles into the polymer host matrices, owing to the reduction of polymer host crystallinity through Lewis acid-base interactions between the ceramic nanoparticles and polymer units [95, 183-185]. For example, PEO based nanocomposite solid polymer electrolyte was developed by dissolving a different lithium salt (Li^+X^-, $X = ClO_4$, BF_4, CF_3SO_3) and an incorporation of nanosized inorganic particles *via* a solution-casting method for enhancing the room temperature ionic conductivity, Li^+ transference number, wide electrochemical potential window and high interfacial stability with lithium metal anode [91, 93, 94, 186]. Nevertheless, the solvent-free nanocomposite solid-state polymer electrolytes have quite low ionic conductivity of the order of 10^{-4} to 10^{-6} S cm^{-1} at room temperature, owing to the poor dissociation of ions and its mobility, which limits the practical applications of lithium batteries. Therefore, a facile phase inversion method has been used to fabricate the microporous PVdF and P(VdF-co-HFP) co-polymer based composite membranes filled with different inorganic particles such as Al_2O_3, SiO_2, TiO_2, ZrO_2, MMT clay and cellulose, etc., to increase the ionic conductivity at room temperature and high thermal stability without sacrificing their mechanical strength. In fact, the as-prepared microporous nanocomposite polymer membranes were activated into nanocomposite gel polymer electrolyte by soaking in an appropriate organic carbonate, namely EC, PC, DEC, DMC, etc., based electrolytes [92, 165, 182, 187]. Hence, the resultant microporous nanocomposite polymer membrane should be an appropriate porous structure with a uniform pore size and an excellent affinity with an electrolyte to absorb/retain a larger amount of liquid electrolyte, which can lead to increase the room temperature ionic conductivity, interfacial stability, and electrochemical performance.

Electrospinning is an effective technique to develop an inorganic particle filled non-woven nanocomposite polymer fibrous membranes with highly interconnected ultrafine multi-fiber layers, which generates high porous structure and large surface area. For example, Kim et al. [166] prepared the non-woven P(VdF-HFP) based composite membrane filled with 6, 10 wt% of SiO_2 nanoparticles by electrospinning technique. The as-prepared electrospun nanocomposite membranes were activated into gel polymer electrolytes by soaking in a non-volatile room temperature ionic-liquid (RTIL), 1-butyl-3-methylimidazolium bis(trifluoromethane sulfonylimide) (BMITFSI) and investigated their thermal stability, mechanical strength, average fiber diameter, porosity, electrolyte wettability, ionic conductivity, interfacial stability, potential stability window and electrochemical performances. It was found that, the NCGPE membrane with 6 wt% of SiO_2 depicts lower average fiber diameter (<1 μm), high porosity and electrolyte uptake, which lead to exhibit high ionic conductivity (4.3 mS cm^{-1}) at room temperature, better compatibility with lithium metal electrode and also, the Li/LiFePO$_4$ cells with 6 wt% of SiO_2. NCGPE membrane delivered high initial discharge capacity (169 mAh g at 0.1 C) than other composite membranes.

Blending is another approach to address the crystallinity, mechanical strength, thermal and electrochemical stability issues in the rechargeable lithium-based batteries. Recent studies showed that electrospun polymer blend matrices such as PEO/PVdF [186], PVAc/P(VdF-co-HFP) [181], PI/P(VdF-co-HFP) [188] and P(VdF-co-HFP)/PMMA [183, 184], etc., have also been filled with inorganic particles to develop the non-woven composite membranes for improving the ionic conductivity, interfacial stability and electrochemical performances. For example, Chen et al. [188] developed a novel composite non-woven membrane by an incorporation of three different concentrations (1%, 2% and 3%) functionalized-TiO_2 nanoparticles into PI/P(VdF-co-HFP) polymers by co-electrospinning technique for lithium battery applications. The surface morphology, porosity, electrolyte uptake, contact angle and ionic conductivity of all the as-prepared polymer blend nanocomposite fibrous membranes were investigated through different characterization techniques. It was observed that the f-TiO$_2$@PI/P(VdF-co-HFP) composite membrane with 2% of f-TiO$_2$ nanoparticles showed low average fiber diameter (689 nm), higher porosity (65%), smaller contact angle (12°) and high ionic conductivity (2.36 mS cm^{-1}) at room temperature. Furthermore, the Li/LiCoO$_2$ cell assembled with 2% f-TiO$_2$@PI/P(VdF-co-HFP) composite membrane delivered higher initial discharge capacity (170 mAh g^{-1}) with the capacity retention of 95% after 50 cycles at 0.5 C rate than the other two composite membranes. However, in this section, various strategies have been discussed briefly with the sufficient number of reported literatures, related to the modification of commercial polyolefin separators/non-woven separators for improving the electrochemical kinetics and stable cycling performances through an enhancement of electrolyte wettability, interfacial stability and also overcome the safety risks of rechargeable lithium batteries.

4.3 Modification of Electrolytes

In general, liquid electrolyte has a higher reactivity with Li metal anode owing to the instability of SEI film during electrochemical processes, which results in serious Li dendrite growth, low coulombic efficiency and safety hazards. Moreover, it is highly influenced by the chemical nature of different electrolytes including carbonate, ester and ether based liquid electrolytes [189, 190]. Therefore, in the past decades, many research activities have been proposed for constructing a stable SEI film *via* modifying an electrolyte that can adequately prevent the electrolyte decomposition on Li metal anode during subsequent cycles, which are briefly described in this section.

4.3.1 Electrolyte Additives

Electrolyte additive is one of the most convenient method to promote the formation of stable SEI film and control Li-ion diffusion & plating behaviours on Li metal anode through different category of additive species mechanisms during cycling processes. Based on this scenario, it can be classified into two types namely, (i) SEI film forming additives and (ii) Li ion plating additives.

4.3.1.1 SEI Film Forming Additives

The formation of stable SEI film plays an important role in suppressing the dendrite growth by preventing the electrolyte decomposition, which achieves a high coulombic efficiency without safety issues, such as Li-S, Li-O$_2$, formation for practical applications [30, 191, 192]. Therefore, worldwide, many researchers are trying to develop an effective electrolyte additive with the required features to form a stable SEI for Li metal anodes as follows:

- It should have a higher HOMO and lower LUMO to ensure superior reactions with Li metal in preference to other electrolyte components or species to form a stable SEI film between Li metal/electrolyte interface during electrochemical processes.
- The SEI components should remain stable in both chemical as well as electrochemical conditions with a high ionic conductivity and electronically insulating nature.
- The formed SEI film should have a high modulus of elasticity with dense and a uniform structure.

Various electrolyte additives, including fluoroethylene carbonate (FEC) and vinylene carbonate (VC) have been designed and extensively used to form a stable SEI film over the graphite anode in the commercial LIBs [28, 36, 193]. Besides, they were also tried for Li metal batteries containing various electrolyte systems such as, 1 M LITFSI, tetraethylene glycol dimethyl ether (TEGDME), 1 M $LiPF_6$, dimethyl carbonate (DMC), and 1 M $LiPF_6$, ethylene carbonate (EC)/diethyl carbonate (DEC), in order to suppress Li dendrite growth and to achieve a better cyclic performances [194-199]. However, the protective role of FEC strongly depends on its concentration, due to the instability of the formed SEI film against Li metal during long-term cycling processes [196]. Therefore, worldwide, researchers have been putting an effort to find a suitable, dense and stable electrolyte additive specially designed for Li metal anodes. In recent years, multielectron conversion chemistry [synergistic effect of Li polysulfide (LiPS) and lithium nitrate ($LiNO_3$)] have also been adopted as electrolyte additives to protect Li metal anode, especially for Li-Air and Li-S batteries [200]. For example, Li et al. [201], Zhao et al. [202], and Yan et al. [203] demonstrated that the growth of Li dendrites can be significantly suppressed through synergistic effects between $LiNO_3$ and LiPS additives in LITFSI-DOL/DME electrolyte. $LiNO_3$ reacts first to passivate the Li surface and Li polysulfide then reacts to form Li_2S/Li_2S_2 in the upper layer of the SEI to prevent further electrolyte decomposition, which enables stable cycling even at high current density. Interestingly, a controlled trace amount of water (H_2O, 25-50 ppm) molecule in 1 M $LiPF_6$ based electrolyte is widely regarded as a detrimental factor for Li metal batteries [204, 205]. The controlled content of H_2O can protect Li metal anode well by constructing a uniform and dense LiF-rich SEI film, which can effectively achieve a dendrite-free Li metal deposition during electrochemical processes.

4.3.1.2 Li Ion Plating Additives

In recent years, another category of additives has been used to control Li ion diffusion and plating behaviour during cycling processes, which are less often proposed than the SEI-forming additives. So far, only two kinds of additives have been revealed such as, alkali metal ions and halide ions [26]. For instance, Ding et al. [206, 207] demonstrated alkali metal ions [such as caesium (Cs^+) or rubidium (Rb^+) ions] with specific concentrations as the additives and proposed a positively charged electrostatic shielding mechanism during Li plating, owing to lower effective reduction potential than that of Li ions at low concentrations. This strategy can effectively alter the deposition of Li ions around the initial Li dendrite growth tip and improve the safety performance of Li metal anode. Significantly, the concentration of alkali metal ions needs to be carefully selected to avoid electrochemical reduction. Besides, other alkali metal ions (such as sodium), alkaline earth metal ions (such as calcium, strontium and barium), and N-methyl-N-butylpiperidinium can render a similar effect on dendrite growth [208]. Alternatively, Li halides with low diffusing and high surface energies have displayed an awesome status in controlling the diffusion of Li ions during electroplating process, with the result a dendrite-free Li deposition morphology was achieved [209]. The halogenated salt blends in liquid electrolyte as additive exhibits long-term stability for upto thousands of charging/discharging cycles, often with no signs of deposition instabilities at room temperature [210, 211]. Hence, these additives potential for suppressing Li dendrite growth.

4.3.2 Super-concentrated Electrolyte

It was generally accepted that 1 M Li salt concentration in the liquid electrolyte achieved a high ionic conductivity, an optimized viscosity and lower cost for Li-ion batteries. In the case of the high-energy density Li-S and $Li-O_2$ batteries employing Li metal as anode, a new salt concentration should be explored to use Li metal anode safely for practical applications in the near future [212-216]. Therefore, super-concentrated electrolyte system with a concentration of >3 M has been receiving extensive attention for rechargeable Li-metal batteries. Moreover, it displays the bulk properties in the likeness of ionic liquids and reduced free solvent molecules owing to the high concentration of Li salts [217]. For instance, S.K. Jeong et al. reported about super-concentrated electrolytes with Li-metal anode and verified the reversibility of Li ion plating/striping processes in a propylene carbonate (PC) electrolyte containing different concentrations of $LiN(SO_2C_2F_5)_2$ salt [218]. The observed results revealed that a highly concentrated electrolyte rendered a thinner SEI than a lower concentration electrolyte solution and achieved a superior suppression of Li dendrite growth. Similarly, Y.S. Hu group [219] proposed a solvent-in-salt electrolyte concept and systematically investigated the effect of different concentrations ranging from 1 mol to 7 mol of LITFSI salt in DOL:DME (1:1, v/v) solvent for high energy density rechargeable Li-S batteries. An ultrahigh concentrated electrolyte solution contains 4 M LITFSI salt, enhances the cyclic and safety performances by an effective suppression of Li dendrite growth. Further, the advantage of this electrolyte system inhibits lithium polysulfide dissolution during electrochemical processes. Correspondingly, J-G. Zhang group [35] investigated the use of highly concentrated electrolytes composed of ether [1,2-dimethoxyethane (DME)] solvents and 4 M LIFSI [lithium (bisfluorosulfonyl)imide] salt for lithium-metal batteries. The results revealed that the highly concentrated electrolyte solution enables the cycling performances for more than 6000 cycles, very high coulombic efficiencies were achieved: 99.1% for > 1000 cycles at 0.2 mA cm^{-2}, stable SEI layer and dendrite-free morphology than that of routine electrolyte (1 M LIFSI in DME) solution [214].

In addition, the highly concentrated 3.2 mol kg^{-1} LIFSI in ionic liquids (ILs) [217], 4.0 M LiNO$_3$ in DMSO electrolyte system can also suppress Li dendrite growth [220]. Interestingly, it also affords the possibility of aqueous electrolyte for Li-metal batteries in recent years. For example, K. Xu et al. (2015) [221] demonstrated a highly concentrated 21 mol LITFSI salt in water as electrolyte for lithium-ion batteries and also investigated the electrochemical stability window with the formation of an electrode-electrolyte interphase. The obtained results revealed that a highly concentrated aqueous electrolyte expanded the working potential window up to ~3.0 volts and also better electrochemical charge/discharge capacity at different C rates. Because, it can be regarded as an intermediate state between the liquid electrolyte and solid-state electrolyte, which holds both the attractive features attractive features including high ionic conductivity of the liquid electrolyte and high modulus of the solid-state electrolyte. Hence, the highly concentrated electrolyte could be a great potential to suppress Li dendrite growth and achieving a high coulombic efficiency for practical applications.

4.3.3 Solid-state Electrolytes

In most of the commercial lithium-ion batteries, liquid electrolytes have been used as a medium to transfer Li-ions between the electrodes. They contain organic solvents, typically carbonates such as ethylene carbonate (EC), propylene carbonate (PC), dimethyl carbonate (DMC), diethyl carbonate (DEC) and ethylmethyl carbonate (EMC) with different lithium salts including lithium hexafluorophosphate (LiPF$_6$), lithium perchlorate (LiClO$_4$), lithium hexafluoroarsenate (LiAsF$_6$), lithium tetrafluoroborate (LiBF$_4$) and lithium trifluoromethane sulfonate [Li(CF$_3$SO$_3$)] [189, 190]. Although, it has some safety issues such as volatile nature, flammability, electrolyte leakage, electrode corrosion and low operating temperature, which limits for electric vehicle applications. In order to address safety concerns, numerous efforts have been devoted to finding the appropriate solid electrolytes with positive features, such as leak proof construction, wide operating temperatures, reduced flammability, and electrochemical stability. It also provides the possibility of improved safety, energy density and cycling performances, due to an efficient cell packing and better electrochemical stability against high voltage cathodes.

Since the 1970s, Solid State Ionics (SSI) has been emerging as one of the major interdisciplinary fields of physics, chemistry, materials science, engineering and technology. It deals with the ionic transport in solid-state materials, which exhibit a wide range of ionic conductivities (σ = 10^{-14} to 10^{-1} S cm^{-1}). The solids, which exhibit high ionic conductivity (σ = 10^{-6} to 10^{-1} S cm^{-1}) are called "Superionic conductors" (SICs) or "Fast ionic conductors" (FICs) also called Solid electrolytes (SEs) [222]. The general characteristic features of solid electrolytes are as follows:

- High ionic conductivity (10^{-1} – 10^{-5} S cm^{-1})
- Low activation energy (< 0.3 eV)
- High ionic transference number, closer to unity (t_{ion} ~ 1)
- Negligible electronic conductivity

All the above-mentioned features of superionic solids are governed by various physical properties like crystal structure, degree of lattice disorder, size of mobile ions, high concentration of mobile ions, ionic polarizability, ion-ion interactions, bonding characteristics and number of available pathways/vacant sites, etc.

Classification of Solid-state Electrolytes

On the basis of their microstructures and physical properties, solid electrolytes can be classified into four different categories as mentioned in Fig. 15.

4.3.3.1 Crystalline Electrolytes

Crystalline electrolytes are of great interest owing to an attractive feature such as, thermal stability, mechanical strength, wide electrochemical window for high potential cathode materials (>5 V) and wide operating temperature range (>100 °C), which solve most of the safety issues related to Li metal anode. Nevertheless, it has some drawbacks such as, low ionic conductivity at room temperature, high ionic diffusion resistance, poor interfacial and chemical stabilities [223]. Interestingly, Li$^+$ ion conductivity of an inorganic solid-state crystalline electrolyte increases with an increase in temperature in accordance with the Arrhenius equation. So far, various

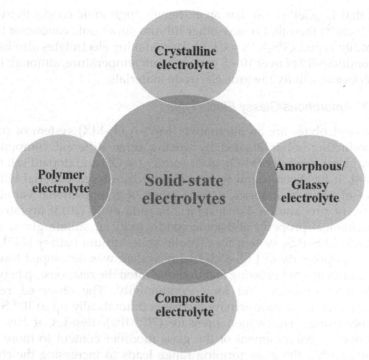

Fig. 15. Classification of solid-state electrolytes.

types of solid-state Li^+ ion conducting crystalline electrolytes, such as LISICON-type $Li_{14}Zn(GeO_4)_4$ [224], NASICON-type $Li_{1+x}A_xTi_{2-x}(PO_4)_3$ (A = Al, Fe, Sc, etc.) (LATP), $Li_{1+x}Al_xGe_{2-x}(PO_4)_3$ (LAGP) [225], perovskite-type $Li_3xLa_{2/3}-xTiO_3$ (LLTO) [226], garnet-type $Li_7La_3Zr_2O_{12}$ (LLZO) [227-232], sulfide based $Li_{10}GeP_2S_{12}$, $Li_7P_3S_{11}$ [223, 233], and argyrodite-type Li_6PS_5X (X = Cl, Br, I) [234-237] have been reported for all solid-state lithium batteries.

Among them, oxides with garnet structure have been attracting much attention because of their high ionic conductivity (10^{-3} to 10^{-4} S cm^{-1}), negligible electronic conductivity and high chemical stability against lithium negative electrode. Despite, it requires an effective sintering to reduce the effect of grain boundary resistance in order to increase the total conductivity. Recently, sulfide based electrolytes have the benefit of high conductivity in the order of 10^{-3} S cm^{-1} at room temperature, due to highly reduced grain-boundary resistance by conventional cold-pressing and also a preferable mechanical property for all solid-state lithium battery applications [238]. A series of lithium superionic conductors $Li_{4-x}Ge_{1-x}P_xS_4$ in x = 0.75 (thio-LISICON) exhibit high ionic conductivity of 2.2×10^{-3} S cm^{-1} at 25 °C [224]. Very recently, N. Kamaya et al. (2011)

found that $Li_{10}GeP_2S_{12}$ SE has an extremely high ionic conductivity (1.2 × 10^{-2} S cm^{-1}) than that of any other lithium superionic conductor [223]. Argyrodite-type Li_6PS_5X (X = Cl, Br, I) crystalline electrolytes also have a high conductivity of over 10^{-3} S cm^{-1} at room temperature, although it has an interface reactivity towards electrode materials.

4.3.3.2 Amorphous/Glassy Electrolytes

Glassy electrolytes are an alternative [M_2O-A_xO_y-MX] system of superion conducting solids created by mixing three different components, namely glassy modifier (M_2O), glass former (A_xO_y) and dopant salt (MX) [238-244]. It has been found to be a highly disordered material with an absence of grain boundaries, a wide range of compositional variability and ease of preparation. For instance, Sakuda et al. (2013) investigated the mechanical property and ionic conductivity of sulfide glassy solid electrolyte Li_2S–P_2S_5 system for all solid-state lithium battery [245]. The mechanical property of Li_2S–P_2S_5 glassy system was developed through densification by cold pressing and demonstrated the relationship between densification pressure and ionic conductivity. The observed results found that the ionic conductivity increases dramatically up to 10^{-4} S cm^{-1} with increasing densification pressure (70 MPa). Besides, it has been found that an enhancement of the glass modifier content to more than 70 mol% within the glass forming range leads to increasing the charge carrier concentration and hence, improves the ionic conductivity to over 10^{-4} S cm^{-1} at room temperature. Very recently, Fukushima et al. (2017) developed high lithium ion conducting Li_2S-P_2S_5-Li_3N glasses and glass–ceramics electrolyte systems for rechargeable lithium batteries [246]. The various composition ranges of (75–1.5x)Li_2S-25P_2S_5-xLi_3N (0 ≤ x ≤ 20 mol%) glassy electrolytes were prepared by mechanical milling. Glassy electrolyte with 20 mol% of Li_3N dopant salt showed highest conductivity of 5.8 × 10^{-4} S cm^{-1} at room temperature owing to decrease in the activation energy for conduction. Correspondingly, lithium iodide salt was added to (100−x)(0.7Li_2S·0.3P_2S_5)-xLiI glassy system and the conductivity increased from the order of 10^{-4} to 10^{-3} S cm^{-1} at room temperature [247].

4.3.3.3 Composite Electrolyte

In composite electrolytes, two or more phases of materials are mixed together in order to enhance the ionic conductivity at room temperature. The composite solid materials are formed by dispersion of electrically insulating and chemically inert particles (Al_2O_3, SiO_2, ZrO_2, Fe_2O_3, etc.) in the parent/host materials (LiI, AgI, AgBr). The dispersion of smaller sized (nano/micro) inert particles into the host ion conducting materials exhibited an improvement in physical properties without sacrificing their structural and chemical properties. For instance, Liang et al. (1973)

reported a remarkable ionic conductivity enhancement upon addition of Al_2O_3 inert particles into LiI conducting host matrix to develop LiI-Al_2O_3 composite solid electrolyte system [248]. The ionic conductivity of LiI-Al_2O_3 system at room temperature is ~10^{-5} S cm^{-1}, which is a remarkable enhancement as compared to pure lithium iodide. A number of interesting studies on dispersed phase composite electrolytes have been investigated over the past decades [249, 250]. Interestingly, a superionic conducting composite electrolyte is often precipitated by a thermodynamically stable crystalline phase from a precursor glass to produce glass-ceramic electrolytes. Consequently, Li_2S–P_2S_5, $Li_7P_3S_{11}$ or $Li_{3.25}P_{0.95}S_4$ systems were crystallized from glasses at the compositions of 70 mol% or 80 mol% of Li_2S, respectively and the prepared glass–ceramics showed high conductivities of over 10^{-3} S cm^{-1} at room temperature, owing to the reduction of grain-boundary resistance among crystal domains [244, 251-256].

4.3.3.4 *Polymer Electrolyte*

Polymer electrolytes are attractive materials in the category of solid electrolytes, because of high ionic conductivity (~10^{-5} to ~10^{-2} S cm^{-1}) compared to other electrolyte systems, and thus have been considered as potential candidates for the development of electrochemical devices [257]. Generally, poly(ethylene oxide) (PEO) based polymer host matrix with different lithium salts (Li$^+$X$^-$) X = ClO_4, AsF_6, PF_6, BF_4, CF_3SO_3 and $(CF_3SO_2)_2N^-$ have been investigated as solid polymer electrolytes for lithium battery applications [258-267]. It possesses many advantages over other types of ion conduction solid electrolytes including: a simple design with desired size and shape, thin-film formability, flexibility, leak proof construction, resistance to shock and vibration, cost-effective, light weight, resistance to pressure and temperature variations, wider electrochemical stability, and better safety. Nevertheless, it has low ionic conductivity (~10^{-8} S cm^{-1}) at room temperature as compared to organic electrolytes and it thus limits their use in conventional lithium batteries. In order to improve the ionic conductivity at room temperature, many efforts have been made to develop a successful polymer electrolyte (PEs) to meet the basic requirements for all electrochemical devices, which are given below [268]:

(a) *High ionic conductivity*: The high ionic conductivity with reduced thickness assures low internal resistance of polymer electrolytes. Therefore, the ionic conductivity of polymer electrolytes should be as high as possible, at room temperature, in the order of ~10^{-4} to ~10^{-2} S cm^{-1}.

(b) *High ionic transference number (t_{ion})*: The polymer electrolyte must be pre-dominantly an ionic conductor and purely an electronic insulator. Most preferably, the movement of specific ions (H$^+$, Li$^+$,

Na^+, Ag^+, Zn^{2+}, Mg^{2+}, etc.) is desired for the better performance of electrolyte in electrochemical devices.

(c) **High electrochemical stability window**: The fast ion conducting polymer electrolyte should be stable for a wider working voltage range as high as ~4 V.

(d) **Durability (chemical and thermal stability)**: The polymer electrolyte should have high chemical and thermal stability over a wide range of temperatures under working conditions. Any undesirable reactions at electrode/electrolyte interface or between electrolyte and other components of the device may lead to inhibit the cell performances.

(e) **Cost**: The cost of the polymer electrolytes should be as low as possible.

Classification of Polymer Electrolytes

Various approaches have been attempted to achieve high electrical conductivity at room temperature. On the basis of approaches and physical properties, the polymer electrolytes have been categorized into four different classes as mentioned below:

(i) **Solid Polymer Electrolyte**

Solid polymer electrolyte (SPE) films were prepared by dissolving suitable ionic salts like $LiClO_4$, $LiAsF_6$, LiI, LiBr, $LiBF_4$, $LiCF_3SO_4$, $LiPF_6$, $LiN(CF_3SO_3)_2$, etc., in PEO, PEG, PU, etc., host matrixes for lithium polymer batteries [259, 261, 267, 269-276]. It has sufficient donor ability and tendency to form polymer-salt complexes with various ionic salts. The formation of polymer-salt complex depends on solvation and lattice energies between polymers and lithium salts. In such a system, the host polymer itself acts like a solvating agent to dissociate the ions and also the ionic conductivity is associated with local relaxation as well as segmental motion of the polymer chains, when the host polymer is in an amorphous nature [265, 277]. The major drawback of this solvent-free solid polymer electrolyte is low ionic conductivity at room temperature, owing to the presence of high crystalline phase in it. Therefore, researchers have been focusing to find an alternative potential electrolyte system by different approaches for improving the room temperature ionic conductivity and it has been described in the forthcoming sections.

(ii) **Gel Polymer Electrolyte**

Gel polymer electrolytes (GPEs) are one of the most important classes of ion conducting intermediate polymeric materials, which have higher values of ionic conductivity at room temperature compared to solid polymer electrolyte. The idea of gel polymer electrolytes was first given by Feuillade and Perche in 1975, in which a suitable amount of liquid electrolyte was incorporated into the polymer

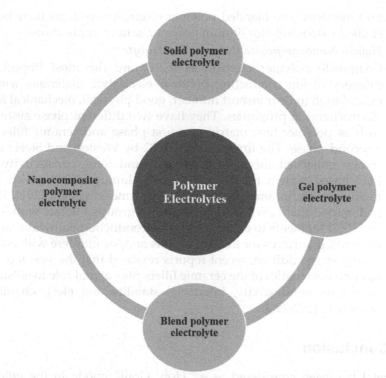

Fig. 16. Classification of polymer electrolytes.

host matrices [278]. Accordingly, researchers put efforts to prepare the GPEs by dipping a variety of porous polymer membranes into an organic liquid electrolyte solution over the past decades [279-287]. The entrapped liquid electrolytes in the polymer matrix, give rise to high ionic conductivity at ambient temperature. Also, the swollen polymer phase takes a tight hold and prevents the leakage of entrapped liquid electrolyte solution. Nevertheless, it has poor mechanical strength and electrochemical instability, which limit its commercialization for practical application.

(iii) Blend Polymer Electrolyte

Polymer blending is one of the most promising and feasible approaches to enhance the ionic conductivity without sacrificing mechanical stability of the blend polymer electrolyte films as reported [86, 288-290]. The polymer blend is a mixture of two or more polymers to form a new distinct material, which is entirely different from that of basic polymers. In the resulting polymer blend matrix, both the polymers should have good compatibility with each other, which improves the mechanical stability and ionic conductivity at room temperature by the reduction of crystallinity of polymer host matrices. The blended polymer electrolyte systems have been studied extensively for lithium polymer battery applications.

(iv) Hybrid Nanocomposite Polymer Electrolyte

Composite polymer electrolytes (CPE) are the most important category of ion conducting polymer electrolyte materials, which exhibit high ionic transport number, good physical, mechanical and electrochemical properties. They have two different phase systems such as polymer host matrix as a first phase and ceramic filler as a second phase. The first report in 1982 by Weston and Steele has shown enhanced mechanical strength and ionic conductivity of PEO based system by the addition of alumina fillers which led to an increase in amorphicity of the polymeric system [291]. It is understood that Lewis acid-base surface groups interact with ions and PEO segments to create additional conduction pathways along the particle surfaces for lithium cations and/or interfere with anion transport. In addition, recent reports revealed that the size and the basic characteristics of the ceramic fillers play a vital role to enhance their ionic conductivity, interfacial stability and electrochemical properties [292-296].

5. Conclusion

Li metal has been considered as a "Holy Grail" anode in the golden age of energy storage, especially rechargeable batteries. Though many breakthroughs have been attained, this field of the high-energy-density LMB is still challenging, but fascinating for practical applications. Hence, in the past four decades, many researchers have made efforts to understand the mechanism *via* qualitative models on Li ion dendrite nucleation, space charge, SEI formation during Li plating/stripping to suppress Li dendrite growth. Based on the theoretical models and experimental results, tremendous progress has been achieved through various strategies, such as Li surface protective layer, modifications of separator and electrolyte, for suppressing Li dendrite growth and addressing safety issues for rechargeable LMBs.

Acknowledgements

Dr. OP gratefully acknowledges SERB/DST and UGC, Govt. of India's financial support through N-PDF (PDF/2015/000502) and D.S.K-PDF (F.4-2/2006 (BSR)/PH/17-18/0081) major research project grants. Also, Prof. NS gratefully acknowledges UGC, Govt. of India for awarding BSR Faculty fellowship (No. F.18-1/2011(BSR), dated: 07-03-2019).

References

1. Winter, M., Barnett, B. and Xu, K. 2018. Before Li ion batteries. Chem. Rev. 118(23), 11433-11456.
2. Wang, Q., Mao, B., Stoliarov, S.I. and Sun, J. 2019. A review of lithium ion battery failure mechanisms and fire prevention strategies. Prog. Energy Combust. Sci. 73, 95-131.
3. Etacheri, V., Marom, R., Elazari, R., Salitra, G. and Aurbach, D. 2011. Challenges in the development of advanced Li-ion batteries: A review. Energy Environ. Sci. 4(9), 3243-3262.
4. Tarascon, J.M. and Armand, M. 2001. Issues and challenges facing rechargeable lithium batteries. Nature 414(6861), 359-367.
5. Goodenough, J.B. and Kim, Y. 2010. Challenges for rechargeable Li batteries. Chem.Mater. 22(3), 587-603.
6. Bruce, P.G., Freunberger, S.A., Hardwick, L.J. and Tarascon, J.M. 2012. LigO$_2$ and LigS batteries with high energy storage. Nat. Mater. 11(1), 19-29.
7. Manthiram, A., Fu, Y., Chung, S.-H., Zu, C. and Su, Y.-S. 2014. Rechargeable lithium-sulfur batteries. Chem. Rev. 114(23), 11751-11787.
8. Cheng, X.-B., Huang, J.-Q. and Zhang, Q. 2018. Review—Li metal anode in working lithium-sulfur batteries. J. Electrochem. Soc. 165(1), A6058-A6072.
9. Chen, L. and Shaw, L.L. 2014. Recent advances in lithium-sulfur batteries. J. Power Sources 267, 770-783.
10. Takeda, Y., Yamamoto, O. and Imanishi, N. 2016. Lithium dendrite formation on a lithium metal anode from liquid, polymer and solid electrolytes. Electrochemistry 84(4), 210-218.
11. Cheng, X.B., Zhang, R., Zhao, C.Z. and Zhang, Q. 2017. Toward safe lithium metal anode in rechargeable batteries: A review. Chem. Rev. 117(15), 10403-10473.
12. Yang, C., Fu, K., Zhang, Y., Hitz, E. and Hu, L. 2017. Protected lithium-metal anodes in batteries: From liquid to solid. Adv. Mater. 29(36), 1-28.
13. Lin, D., Liu, Y. and Cui, Y. 2017. Reviving the lithium metal anode for high-energy batteries. Nat. Nanotechnol. 12(3), 194-206.
14. Barton, J.L. 1962. The electrolytic growth of dendrites from ionic solutions. Proc. R. Soc. London. Ser. A. Math. Phys. Sci. 268(1335), 485-505.
15. Akolkar, R. 2014. Modeling dendrite growth during lithium electrodeposition at sub-ambient temperature. J. Power Sources 246, 84-89.
16. Aryanfar, A., Brooks, D.J., Colussi, A.J. and Hoffmann, M.R. 2014. Quantifying the dependence of dead lithium losses on the cycling period in lithium metal batteries. Phys. Chem. Chem. Phys. 16(45), 24965-24970.
17. Ozhabes, Y., Gunceler, D. and Arias, T.A. 2015. Stability and surface diffusion at lithium-electrolyte interphases with connections to dendrite suppression. Physics 1-7.
18. Chazalviel, J.N. 1990. Electrochemical aspects of the generation of ramified metallic electrodeposits. Phys. Rev. A 42(12), 7355-7367.
19. Nishikawa, K., Mori, T., Nishida, T., Fukunaka, Y. and Rosso, M. 2011. Li dendrite growth and Li$^+$ ionic mass transfer phenomenon. J. Electroanal. Chem. 661(1), 84-89.

20. Steiger, J., Kramer, D. and Mönig, R. 2014. Microscopic observations of the formation, growth and shrinkage of lithium moss during electrodeposition and dissolution. Electrochim. Acta 136, 529-536.

21. Stark, J.K., Ding, Y. and Kohl, P.A. 2013. Nucleation of electrodeposited lithium metal: Dendritic growth and the effect of co-deposited sodium. J. Electrochem. Soc. 160(9), D337-D342.

22. Sano, H., Kitta, M. and Matsumoto, H. 2016. Effect of charge transfer resistance on morphology of lithium electrodeposited in ionic liquid. J. Electrochem. Soc.163(12), D3076-D3079.

23. Sun, F., Moroni, R., Dong, K., Markötter, H., Zhou, D., Hilger, A., Zielke, L., Zengerle, R., Thiele, S., Banhart, J. and Manke, I. 2017. Study of the mechanisms of internal short circuit in a Li/Li cell by synchrotron X-ray phase contrast tomography. ACS Energy Lett. 2(1), 94-104.

24. Park, M.S., Lee, D.J., Im, D., Doo, S. and Yamamoto, O. 2014. A highly reversible lithium metal anode. Scientific Reports 3815(4), 1-8.

25. Dey, A.N. and Sullivan, B.P. 2007. The electrochemical decomposition of propylene carbonate on graphite. J. Electrochem. Soc. 117(2), 222.

26. Peled, E. 2006. The electrochemical behavior of alkali and alkaline earth metals in nonaqueous battery systems—The solid electrolyte interphase model. J. Electrochem. Soc. 126(12), 2047.

27. Goodenough, J.B. and Park, K.S. 2013. The Li-ion rechargeable battery: A perspective. J. Am. Chem. Soc. 135(4), 1167-1176.

28. Kuwata, H., Sonoki, H., Matsui, M., Matsuda, Y. and Imanishi, N. 2016. Surface layer and morphology of lithium metal electrodes. Electrochemistry 84(11), 854-860.

29. Zhang, X.Q., Cheng, X.B. and Zhang, Q. 2018. Advances in interfaces between Li metal anode and electrolyte. Adv. Mater. Interfaces 5(2), 1-19.

30. Tikekar, M.D., Choudhury, S., Tu, Z. and Archer, L.A. 2016. Design principles for electrolytes and interfaces for stable lithium-metal batteries. Nat. Energy 1(9), 1-7.

31. Cheng, X.B., Zhang, R., Zhao, C.Z., Wei, F., Zhang, J.G. and Zhang, Q. 2015. A review of solid electrolyte interphases on lithium metal anode. Adv. Sci. 3(3), 1-20.

32. Peled, E. 1997. Advanced model for solid electrolyte interphase electrodes in liquid and polymer electrolytes. J. Electrochem. Soc. 144(8), L208.

33. Shi, S., Lu, P., Liu, Z., Qi, Y., Hector, L.G., Li, H. and Harris, S.J. 2012. Direct calculation of Li-ion transport in the solid electrolyte interphase. J. Am. Chem. Soc. 134(37), 15476-15487.

34. Uchida, S. and Ishikawa, M. 2017. Lithium bis(fluorosulfonyl)imide based low ethylene carbonate content electrolyte with unusual solvation state. J. Power Sources 359, 480-486.

35. Qian, J., Henderson, W.A., Xu, W., Bhattacharya, P., Engelhard, M., Borodin, O. and Zhang, J-G. 2015. High rate and stable cycling of lithium metal anode. Nat. Commun. 6, 6362.

36. Markevich, E., Salitra, G., Chesneau, F., Schmidt, M. and Aurbach, D. 2017. Very stable lithium metal stripping-plating at a high rate and high areal capacity in fluoroethylene carbonate-based organic electrolyte solution. ACS Energy Lett. 2(6), 1321-1326.

37. Xu, K., Von Cresce, A. and Lee, U. 2010. Differentiating contributions to 'ion transfer' barrier from interphasial resistance and Li$^+$ desolvation at electrolyte/graphite interface. Langmuir 26(13), 11538-11543.

38. Christensen, J. and Newman, J. 2004. A mathematical model for the lithium-ion negative electrode solid electrolyte interphase. J. Electrochem. Soc. 151(11), A1977.

39. Borodin, O., Smith, G.D. and Fan, P. 2006. Molecular dynamics simulations of lithium alkyl carbonates. J. Phys. Chem. B 110(45), 22773-22779.

40. Single, F., Horstmann, B. and Latz, A. 2017. Revealing SEI morphology: In-depth analysis of a modeling approach. J. Electrochem. Soc. 164(11), E3132-E3145.

41. Chen, Y.C., Ouyang, C.Y., Song, L.J. and Sun, Z.L. 2011. Electrical and lithium ion dynamics in three main components of solid electrolyte interphase from density functional theory study. J. Phys. Chem. C 115(14), 7044-7049.

42. Bunde, A., Dieterich, W. and Roman, E. 1985. Dispersed ionic conductors and percolation theory. Phys. Rev. Lett. 55(1), 5-8.

43. Stone, G.M., Mullin, S.A., Teran, A.A., Hallinan, D.T., Minor, A.M., Hexemer, A. and Balsara, N.P. 2011. Resolution of the modulus versus adhesion dilemma in solid polymer electrolytes for rechargeable lithium metal batteries. J. Electrochem. Soc. 59(3), A222-A227.

44. Monroe, C. and Newman, J. 2005. The impact of elastic deformation on deposition kinetics at lithium/polymer interfaces. J. Electrochem. Soc. 152(2), A396.

45. Rong, G., Zhang, X., Zhao, W., Qiu, Y., Liu, M., Ye, F., Xu, Y., Chen, J., Hou, Y., Li, W., Duan, W. and Zhang, Y. 2017. Liquid-phase electrochemical scanning electron microscopy for in situ investigation of lithium dendrite growth and dissolution. Adv. Mater. 29(13), 1606187.

46. Cohen, Y.S., Cohen, Y. and Aurbach, D. 2002. Micromorphological studies of lithium electrodes in alkyl carbonate solutions using in situ atomic force microscopy. J. Phys. Chem. B 104(51), 12282-12291.

47. Liu, X.R., Deng, X., Liu, R.R., Yan, H.J., Guo, Y.G., Wang, D. and Wan, L.J. 2014. Single nanowire electrode electrochemistry of silicon anode by in situ atomic force microscopy: Solid electrolyte interphase growth and mechanical properties. ACS Appl. Mater. Interfaces 6(22), 20317-20323.

48. Zhang, J., Wang, R., Yang, X., Lu, W., Wu, X., Wang, X., Li, H. and Chen, L. 2012. Direct observation of inhomogeneous solid electrolyte interphase on MnO anode with atomic force microscopy and spectroscopy. Nano Lett. 12(4), 2153-2157.

49. Li, N.W., Yin, Y.X., Yang, C.P. and Guo, Y.G. 2016. An artificial solid electrolyte interphase layer for stable lithium metal anodes. Adv. Mater. 28(9), 1853-1858.

50. Liu, Z., Qi, Y., Lin, Y.X., Chen, L., Lu, P. and Chen, L.Q. 2016. Interfacial study on solid electrolyte interphase at Li metal anode: Implication for Li dendrite growth. J. Electrochem. Soc. 163(3), A592-A598.

51. Delaporte, N., Guerfi, A., Demers, H., Lorrmann, H., Paolella, A. and Zaghib, K. 2019. Facile protection of lithium metal for all-solid-state batteries. ChemistryOpen 8(2), 192-195.

52. Zhao, C-Z., Chen, P-Y., Zhang, R., Chen, X., Li, B-Q., Zhang, X-Q., Cheng, X-B. and Zhang, Q. 2018. An ion redistributor for dendrite-free lithium metal anodes. Sci. Adv. 4(11), 3446.

53. Yang, H., Guo, C., Naveed, A., Lei, J., Yang, J., Nuli, Y. and Wang, J. 2018. Recent progress and perspective on lithium metal anode protection. Energy Storage Mater. 14, 199-221.

54. Liu, Y., Lin, D., Liang, Z., Zhao, J., Yan, K. and Cui, Y. 2016. Lithium-coated polymeric matrix as a minimum volume-change and dendrite-free lithium metal anode. Nat. Commun. 7, 1-9.

55. Zhang, Z., Peng, Z., Zheng, J., Wang, S., Liu, Z., Bi, Y., Chen, Y., Wu, G., Li, H., Cui, P., Wen, Z. and Wang, D. 2017. The long life-span of a Li-metal anode enabled by a protective layer based on the pyrolyzed N-doped binder network. J. Mater. Chem. A 5(19), 9339-9349.

56. Liang, Z., Zheng, G., Liu, C., Liu, N., Li, W., Yan, K., Yao, H., Hsu, P.C., Chu, S. and Cui, Y. 2015. Polymer nanofiber-guided uniform lithium deposition for battery electrodes. Nano Lett. 15(5), 2910-2916.

57. Liu, W., Mi, Y., Weng, Z., Zhong, Y., Wu, Z. and Wang, H. 2017. Functional metal-organic framework boosting lithium metal anode performance: Via chemical interactions. Chem. Sci. 8(6), 4285-4291.

58. Zhang, R., Chen, X.R., Chen, X., Cheng, X.B., Zhang, X.Q., Yan, C. and Zhang, Q. 2017. Lithiophilic sites in doped graphene guide uniform lithium nucleation for dendrite-free lithium metal anodes. Angew. Chemie – Int. Ed. 56(27), 7764-7768.

59. Cheng, X.B., Hou, T.Z., Zhang, R., Peng, H.J., Zhao, C.Z., Huang, J.Q. and Zhang, Q. 2016. Dendrite-free lithium deposition induced by uniformly distributed lithium ions for efficient lithium metal batteries. Adv. Mater. 28(15), 2888-2895.

60. Liang, Z., Lin, Di., Zhao, J., Lu, Z., Liu, Y., Liu, C., Lu, Y., Wang, H., Yan, K., Tao, X. and Cui, Y. 2016. Composite lithium metal anode by melt infusion of lithium into a 3D conducting scaffold with lithiophilic coating. Proc. Natl. Acad. Sci. 113(11), 2862-2867.

61. Jin, C., Sheng, O., Luo, J., Yuan, H., Fang, C., Zhang, W., Huang, H., Gan, Y., Xia, Y., Liang, C., Zhang, J. and Tao, X. 2017. 3D lithium metal embedded within lithiophilic porous matrix for stable lithium metal batteries. Nano Energy 37(April), 177-186.

62. Zhang, H., Liao, X., Guan, Y., Xiang, Y., Li, M., Zhang, W., Zhu, X., Ming, H., Lu, L., Qiu, J., Huang, Y., Cao, G., Yang, Y., Mai, L., Zhao, Y. and Zhang, H. 2018. Lithiophilic-lithiophobic gradient interfacial layer for a highly stable lithium metal anode. Nat. Commun. 9(1), 1-11.

63. Lin, D., Liu, Y., Liang, Z., Lee, H.W., Sun, J., Wang, H., Yan, K., Xie, J. and Cui, Y. 2016. Layered reduced graphene oxide with nanoscale interlayer gaps as a stable host for lithium metal anodes. Nat. Nanotechnol. 11(7), 626-632.

64. Zhang, Y., Liu, B., Hitz, E., Luo, W., Yao, Y., Li, Y., Dai, J., Chen, C., Wang, Y., Yang, C., Li, H. and Hu, L. 2017. A carbon-based 3D current collector with surface protection for Li metal anode. Nano Research 10(4), 1356-1365.

65. Kazyak, E., Wood, K.N. and Dasgupta, N.P. 2015. Improved cycle life and

stability of lithium metal anodes through ultrathin atomic layer deposition surface treatments. Chem. Mater. 27(18), 6457-6462.

66. Guan, C. and Wang, J. 2016. Recent development of advanced electrode materials by atomic layer deposition for electrochemical energy storage. Adv. Sci. 3(10), 1-23.

67. Kozen, A.C., Lin, C.F., Pearse, A.J., Schroeder, M.A., Han, X., Hu, L., Lee, S.B., Rubloff, G.W. and Noked, M. 2015. Next-generation lithium metal anode engineering via atomic layer deposition. ACS Nano 9(6), 5884-5892.

68. Chen, L., Connell, J.G., Nie, A., Huang, Z., Zavadil, K.R., Klavetter, K.C., Yuan, Y., Sharifi-Asl, S., Shahbazian-Yassar, R., Libera, J.A., Mane, A.U. and Elam, J.W. 2017. Lithium metal protected by atomic layer deposition metal oxide for high performance anodes. J. Mater. Chem. A 5(24), 12297-12309.

69. Yan, B., Li, X., Bai, Z., Song, X., Xiong, D., Zhao, M., Li, D. and Lu, S. 2017. A review of atomic layer deposition providing high performance lithium sulfur batteries. J. Power Sources 338, 34-48.

70. Li, X., Liu, J., Banis, M.N., Lushington, A., Li, R., Cai, M. and Sun, X. 2014. Atomic layer deposition of solid-state electrolyte coated cathode materials with superior high-voltage cycling behavior for lithium ion battery application. Energy Environ. Sci. 7(2), 768-778.

71. Park, J.S., Mane, A.U., Elam, J.W. and Croy, J.R. 2017. Atomic layer deposition of Al-W-Fluoride on $LiCoO_2$ cathodes: Comparison of particle- and electrode-level coatings. ACS Omega 2(7): 3724-3729.

72. Arora, P. and Zhang, Z. 2004. Battery separators. Chem. Rev. 104(10), 4419-4462.

73. Zhang, S.S. 2007. A review on the separators of liquid electrolyte Li-ion batteries. J. Power Sources 164(1), 351-364.

74. Huang, X. 2011. Separator technologies for lithium-ion batteries. J. Solid State Electrochem. 15(4), 649-662.

75. Xiang, Y., Li, J., Lei, J., Liu, D., Xie, Z., Qu, D., Li, K., Deng, T. and Tang, H. 2016. Advanced separators for lithium-ion and lithium-sulfur batteries: A review of recent progress. ChemSusChem 9(21), 3023-3039.

76. Lee, H., Yanilmaz, M., Toprakci, O. Fu, K. and Zhang, X. 2014. A review of recent developments in membrane separators for rechargeable lithium-ion batteries. Energy Environ. Sci. 7(12), 3857-3886.

77. Kim, J.Y., Lee, Y. and Lim, D.Y. 2009. Plasma-modified polyethylene membrane as a separator for lithium-ion polymer battery. Electrochim. Acta 54(14), 3714-3719.

78. Jin, S.Y., Manuel, J., Zhao, X., Park, W.H. and Ahn, J.H. 2017. Surface-modified polyethylene separator via oxygen plasma treatment for lithium ion battery. J. Ind. Eng. Chem. 45, 15-21.

79. Choi, S.H. and Nho, Y.C. 2000. Radiation-induced graft copolymerization of binary monomer mixture containing acrylonitrile onto polyethylene films. Radiat. Phys. Chem. 58(2), 157-168.

80. Lee, J.Y., Lee, Y.M., Bhattacharya, B., Nho, Y.C. and Park, J.K. 2009. Separator grafted with siloxane by electron beam irradiation for lithium secondary batteries. Electrochim. Acta 54(18), 4312-4315.

81. Lee, J.Y., Bhattacharya, B., Nho, Y.C. and Park, J.K. 2009. New separator prepared by electron beam irradiation for high voltage lithium secondary batteries. Nucl. Instruments Methods Phys. Res. Sect. B Beam Interact. with Mater. Atoms 267(14), 2390-2394.

82. Sohn, J.Y., Im, J.S., Gwon, S.J., Choi, J.H., Shin, J. and Nho, Y.C. 2009. Preparation and characterization of a PVdF-HFP/PEGDMA-coated PE separator for lithium-ion polymer battery by electron beam irradiation. Radiat. Phys. Chem. 78(7-8), 505-508.

83. Choi, S., Lee, K., Lee, J. and Nho, Y.C. 1999. Graft copolymer – Metal complexes obtained by radiation. J. Appl. Polym. Sci. 77, 500-508.

84. Mark H.S.B., Dilip D.D. and Thomas F.G. 1989. Nonwoven Continuously-Bonded Trilaminate. United States Patent, US004863785.

85. Yi, W., Huaiyu, Z., Jian, H., Yun, L. and Shushu, Z. 2009. Wet-laid non-woven fabric for separator of lithium-ion battery. J. Power Sources 189(1), 616-619.

86. Ding, Y., Zhang, P., Long, Z., Jiang, Y., Xu, F. and Di, W. 2009. The ionic conductivity and mechanical property of electrospun P(VdF-HFP)/PMMA membranes for lithium ion batteries. J. Memb. Sci. 329(1-2), 56-59.

87. Gopalan, A.I., Santhosh, P., Manesh, K.M., Nho, J.H., Kim, S.H., Hwang, C.G. and Lee, K.P. 2008. Development of electrospun PVdF-PAN membrane-based polymer electrolytes for lithium batteries. J. Memb. Sci. 325(2), 683-690.

88. Yu, L., Jin, Y. and Lin, Y.S. 2016. Ceramic coated polypropylene separators for lithium-ion batteries with improved safety: Effects of high melting point organic binder. RSC Adv. 6(46), 40002-40009.

89. Shi, C., Zhang, P., Chen, L., Yang, P. and Zhao, J. 2014. Effect of a thin ceramic-coating layer on thermal and electrochemical properties of polyethylene separator for lithium-ion batteries. J. Power Sources 270, 547-553.

90. Liu, K., Zhuo, D., Lee, H-W., Liu, W., Lin, D., Lu, Y. and Cui, Y. 2017. Extending the life of lithium-based rechargeable batteries by reaction of lithium dendrites with a novel silica nanoparticle sandwiched separator. Adv. Mater. 29(4), 1603987.

91. Abdullah, A., Abdullah, S.Z., Ali, A.M.M., Winie, T., Yahya, M.Z.A. and Subban, R.H.Y. 2009. Electrical properties of PEO–LiCF$_3$SO$_3$–SiO$_2$ nanocomposite polymer electrolytes. Mater. Res. Innov. 13(3), 255-258.

92. Deka, M. and Kumar, A. 2011. Electrical and electrochemical studies of poly(vinylidene fluoride)-clay nanocomposite gel polymer electrolytes for Li-ion batteries. J. Power Sources 196(3), 1358-1364.

93. Croce, F., Persi, L., Ronci, F. and Scrosati, B. 2000. Nanocomposite polymer electrolytes and their impact on the lithium battery technology. Solid State Ionics 135(1-4), 47-52.

94. Croce, F., Curini, R., Martinelli, A., Persi, L., Ronci, F., Scrosati, B. and Caminiti, R. 1999. Physical and chemical properties of nanocomposite polymer electrolytes. J. Phys. Chem. B 103(48), 10632-10638.

95. Osińska, M., Walkowiak, M., Zalewska, A. and Jesionowski, T. 2009. Study of the role of ceramic filler in composite gel electrolytes based on microporous polymer membranes. J. Memb. Sci. 326(2), 582-588.

96. Johnson, M.B. and Wilkes, G.L. 2002. Microporous membranes of isotactic poly(4-methyl-1-pentene) from a melt-extrusion process. I: Effects of resin variables and extrusion conditions. J. Appl. Polym. Sci. 83(10), 2095-2113.

97. Johnson, M.B. and Wilkes, G.L. 2002. Microporous membranes of isotactic poly(4-methyl-1-pentene) from a melt-extrusion process. II: Effects of thermal annealing and stretching on porosity. J. Appl. Polym. Sci. 84(5), 1076-1100.

98. Wang, H., Huang, H. and Wunder, S.L. 2002. Novel microporous poly(vinylidene fluoride) blend electrolytes for lithium-ion batteries. J. Electrochem. Soc. 147(8), 2853.

99. Subramania, A., Sundaram, N.T.K. and Kumar, G.V. 2006. Structural and electrochemical properties of micro-porous polymer blend electrolytes based on PVdF-co-HFP-PAN for Li-ion battery applications. J. Power Sources 153(1), 177-182.

100. Hiroyuki, H., Kiichiro, M., Minoru, E. and Toshihiko, S. 1995. Porous Film Process for Producing the Same and Use of the Same. United States Patent, US005385777A.

101. Chaiya, C., Marino, X., Kamalesh, K.S. and Costas, G. 2004. Preparation of Microporous Films from Immiscible Blends via Melt Processing and Stretching. United States Patent, US006824680B2.

102. Martinez-Cisneros, C., Antonelli, C., Levenfeld, B., Varez, A. and Sanchez, J.Y. 2016. Evaluation of polyolefin-based macroporous separators for high temperature Li-ion batteries. Electrochim. Acta 216, 68-78.

103. Deimede, V. and Elmasides, C. 2015. Separators for lithium-ion batteries: A review on the production processes and recent developments. Energy Technol. 3(5), 453-468.

104 Ronald, W.C., J. Robert, D., Shizuo, O., Donald, K.S. and Xiangyun, W. 2005. Battery Separator. United States Patent, US006921608B2.

105. Yang, M. and Hou, J. 2012. Membranes in lithium ion batteries. Membranes 2(3), 367-383.

106. Tian, Z., He, X., Pu, W., Wan, C. and Jiang, C. 2006. Preparation of poly(acrylonitrile-butyl acrylate) gel electrolyte for lithium-ion batteries. Electrochim. Acta 52(2), 688-693.

107. Boudin, F., Andrieu, X., Jehoulet, C. and Olsen, I.I. 1999. Microporous PVdF gel for lithium-ion batteries. J. Power Sources 81-82, 804-807.

108. Gineste, J.L. 1995. Polypropylene separator grafted with hydrophilic monomers for lithium batteries. J. Memb. Sci. 107(1-2), 155-164.

109. Jeong, Y.-B. and Kim, D.-W. 2004. Effect of thickness of coating layer on polymer-coated separator on cycling performance of lithium-ion polymer cells. J. Power Sources 128(2), 256-262.

110. Lee, Y., Ryou, M.H., Seo, M., Choi, J.W. and Lee, Y.M. 2013. Effect of polydopamine surface coating on polyethylene separators as a function of their porosity for high-power Li-ion batteries. Electrochim. Acta 113, 433-438.

111. Li, H., Ma, X.T., Shi, J.L., Yao, Z.K., Zhu, B.K. and Zhu, L.P. 2011. Preparation and properties of poly(ethylene oxide) gel filled polypropylene separators and their corresponding gel polymer electrolytes for Li-ion batteries. Electrochim. Acta 56(6), 2641-2647.

112. Ryou, M.-H., Lee, D.J., Lee, J.-N., Lee, Y.M., Park, J.-K. and Choi, J.W. 2012. Lithium-ion batteries: Excellent cycle life of lithium-metal anodes in lithium-ion batteries with mussel-inspired polydopamine-coated separators. Adv. Energy Mater. 2(6), 645-650.

113. Shi, J.L., Fang, L.F., Li, H., Zhang, H., Zhu, B.K. and Zhu, L.P. 2013. Improved thermal and electrochemical performances of PMMA modified PE separator skeleton prepared via dopamine-initiated ATRP for lithium ion batteries. J. Memb. Sci. 437, 160-168.

114. Choi, S.-H., Kang, H.-J., Ryu, E.-N. and Lee, K.-P. 2001. Electrochemical properties of polyolefin nonwoven fabric modified with carboxylic acid group for battery separator. Radiat. Phys. Chem. 60(4-5), 495-502.

115. Kim, K.J., Kim, Y.H., Song, J.H., Jo, Y.N., Kim, J.S. and Kim, Y.J. 2010. Effect of gamma ray irradiation on thermal and electrochemical properties of polyethylene separator for Li ion batteries. J. Power Sources 195(18), 6075-6080.

116. Gao, K., Hu, X., Yi, T. and Dai, C. 2006. PE-g-MMA polymer electrolyte membrane for lithium polymer battery. Electrochim. Acta, 52(2), 443-449.

117. Cao, C., Tan, L., Liu, W., Ma, J. and Li, L. 2014. Polydopamine coated electrospun poly(vinyldiene fluoride) nanofibrous membrane as separator for lithium-ion batteries. J. Power Sources 248, 224-229.

118. Wang, D., Zhao, Z., Yu, L., Zhang, K., Na, H., Ying, S., Xu, D. and Zhang, G. 2014. Polydopamine hydrophilic modification of polypropylene separator for lithium ion battery. J. Appl. Polym. Sci. 131(15), 1-7.

119. Kang, S.M., Ryou, M.H., Choi, J.W. and Lee, H. 2012. Mussel- and diatom-inspired silica coating on separators yields improved power and safety in Li-ion batteries. Chem. Mater. 24(17), 3481-3485.

120. Man, C., Jiang, P., Wong, K.W., Zhao, Y., Tang, C., Fan, M., Lau, W.M., Mei, J., Li, S., Liu, H. and Hui, D. 2014. Enhanced wetting properties of a polypropylene separator for a lithium-ion battery by hyperthermal hydrogen induced cross-linking of poly(ethylene oxide). J. Mater. Chem. A 2(30), 11980-11986.

121. Park, J.H., Park, W., Kim, J.H., Ryoo, D., Kim, H.S., Jeong, Yeon U., Kim, D.W. and Lee, S.Y. 2011. Close-packed poly(methyl methacrylate) nanoparticle arrays-coated polyethylene separators for high-power lithium-ion polymer batteries. J. Power Sources 196(16), 7035-7038.

122. Huai, Y., Gao, J., Deng, Z. and Suo, J. 2010. Preparation and characterization of a special structural poly(acrylonitrile)-based microporous membrane for lithium-ion batteries. Ionics (Kiel). 16(7), 603-611.

123. Hu, S., Lin, S., Tu, Y., Hu, J., Wu, Y., Liu, G., Li, F., Yu, F. and Jiang, T. 2016. Novel aramid nanofiber-coated polypropylene separators for lithium ion batteries. J. Mater. Chem. A 4(9), 3513-3526.

124. Kim, K.M., Poliquit, B.Z., Lee, Y.G., Won, J., Ko, J.M. and Cho, W.I. 2014. Enhanced separator properties by thermal curing of poly(ethylene glycol)diacrylate-based gel polymer electrolytes for lithium-ion batteries. Electrochim. Acta 120, 159-166.

125. Lee Min, Y., Kim, J.W., Choi, N.S., Lee An, J., Seol, W.H. and Park, J.K. 2005. Novel porous separator based on PVdF and PE non-woven matrix for rechargeable lithium batteries. J. Power Sources 139(1-2), 235-241.

126. Lee, Y.M., Choi, N.S., Lee, Je A., Seol, W.H., Cho, K.Y., Jung, H.Y., Kim, J.W. and Park, J.K. 2005. Electrochemical effect of coating layer on the separator based on PVdF and PE non-woven matrix. J. Power Sources 146(1-2), 431-435.

127. Periasamy, P., Tatsumi, K., Shikano, M., Fujieda, T., Saito, Y., Sakai, T., Mizuhata, M., Kajinami, A. and Deki, S. 2000. Studies on PVdF-based gel polymer electrolytes. J. Power Sources 88(2), 269-273.

128. Fang, L.F., Shi, J.L., Li, H., Zhu, B.K. and Zhu, L.P. 2014. Construction of porous PVDF coating layer and electrochemical performances of the corresponding modified polyethylene separators for lithium ion batteries. J. Appl. Polym. Sci. 131(21), 1-9.

129. Kim, K.J., Kim, J.H., Park, M.S., Kwon, H.K., Kim, H. and Kim, Y.J. 2012. Enhancement of electrochemical and thermal properties of polyethylene separators coated with polyvinylidene fluoride-hexafluoropropylene co-polymer for Li-ion batteries. J. Power Sources 198, 298-302.

130. Kim, H.S., Periasamy, P. and Moon, S.I. 2005. Electrochemical properties of the Li-ion polymer batteries with P(VdF-co-HFP)-based gel polymer electrolyte. J. Power Sources 141(2), 293-297.

131. Robert, R.B., John, W.H., James, P.K. and Vollie, L.M. 1971. Method for Producing a Melt Blown Roving. United States Patent, US003615995.

132. Kui-Chiu Kwok. 1999. Melt-blowing Method and Apparatus. United States Patent, US005902540.

133. Warren, C.M. and Donald V.S. 1977. Wet Laid Laminate and Method of Manufacturing the Same. United States Patent, US004012281.

134. Richard, P.K., Richard, L.G.J. and Joseph, I. 1980. Composite Nonwoven Fabric Comprising Adjacent Microfine Fibers in Layers. United States Patent, US004196245.

135. Kritzer, P. 2006. Nonwoven support material for improved separators in Li-polymer batteries. J. Power Sources 161(2), 1335-1340.

136. Keller, J.P., Prentice, J.S. and Harding, J.W. 1973. Process for Producing Melt-blown Non-Woven Synthetic Polymer Mat Having High Tear Resistance. Unites States Patent, US003755527.

137. Yang, C., Jia, Z., Guan, Z. and Wang, L. 2009. Polyvinylidene fluoride membrane by novel electrospinning system for separator of Li-ion batteries. J. Power Sources 189(1), 716-720.

138. Kim, J.R., Choi, S.W., Jo, S.M., Lee, W.S. and Kim, B.C. 2004. Electrospun PVdF-based fibrous polymer electrolytes for lithium ion polymer batteries. Electrochim. Acta 50(1), 69-75.

139. Chen, W., Liu, Y., Ma, Y., Liu, J. and Liu, X. 2014. Improved performance of PVdF-HFP/PI nanofiber membrane for lithium ion battery separator prepared by a bicomponent cross-electrospinning method. Mater. Lett. 133, 67-70.

140. Ye, W., Zhu, J., Liao, X., Jiang, S., Li, Y., Fang, H. and Hou, H. 2015. Hierarchical three-dimensional micro/nano-architecture of polyaniline nanowires wrapped-on polyimide nanofibers for high performance lithium-ion battery separators. J. Power Sources 299, 417-424.

141. Prasanth, R., Aravindan, V. and Srinivasan, M. 2012. Novel polymer

electrolyte based on cob-web electrospun multi component polymer blend of polyacrylonitrile/poly(methyl methacrylate)/polystyrene for lithium ion batteries – Preparation and electrochemical characterization. J. Power Sources 202, 299-307.

142. Liang, X., Yang, Y., Jin, X. and Cheng, J. 2016. Polyethylene oxide-coated electrospun polyimide fibrous separator for high-performance lithium-ion battery. J. Mater. Sci. Technol. 32(3), 200-206.

143. Cao, L., An, P., Xu, Z. and Huang, J. 2016. Performance evaluation of electrospun polyimide non-woven separators for high power lithium-ion batteries. J. Electroanal. Chem. 767, 34-39.

144. Wang, Q., Song, W.L., Wang, L., Song, Y., Shi, Q. and Fan, L.Z. 2014. Electrospun polyimide-based fiber membranes as polymer electrolytes for lithium-ion batteries. Electrochim. Acta 132, 538-544.

145. Miao, Y.E., Zhu, G.N., Hou, H., Xia, Y.Y. and Liu, T. 2013. Electrospun polyimide nanofiber-based nonwoven separators for lithium-ion batteries. J. Power Sources 226, 82-86.

146. Liang, Y., Cheng, S., Zhao, J., Zhang, C., Sun, S., Zhou, N., Qiu, Y. and Zhang, X. 2013. Heat treatment of electrospun polyvinylidene fluoride fibrous membrane separators for rechargeable lithium-ion batteries. J. Power Sources 240, 204-211.

147. Gao, K., Hu, X., Dai, C. and Yi, T. 2006. Crystal structures of electrospun PVdF membranes and its separator application for rechargeable lithium metal cells. Mater. Sci. Eng. B Solid-State Mater. Adv. Technol. 131(1-3), 100-105.

148. Li, X., Liu, G. and Popov, B.N. 2010. Activity and stability of non-precious metal catalysts for oxygen reduction in acid and alkaline electrolytes. J. Power Sources 195(19), 6373-6378.

149. Prasanth, R., Shubha, N., Hng, H.H. and Srinivasan, M. 2014. Effect of poly(ethylene oxide) on ionic conductivity and electrochemical properties of poly(vinylidenefluoride) based polymer gel electrolytes prepared by electrospinning for lithium ion batteries. J. Power Sources 245, 283-291.

150. Wu, N., Cao, Q., Wang, X., Li, X. and Deng, H. 2011. A novel high-performance gel polymer electrolyte membrane basing on electrospinning technique for lithium rechargeable batteries. J. Power Sources 196(20), 8638-8643.

151. Wu, Y.S., Yang, C.C., Luo, S.P., Chen, Y.L., Wei, C.N. and Lue, S.J. 2017. PVdF- HFP/PET/PVDF-HFP composite membrane for lithium-ion power batteries. Int. J. Hydrogen Energy 42(10), 6862-6875.

152. Wu, D., Shi, C., Huang, S., Qiu, X., Wang, H., Zhan, Z., Zhang, P., Zhao, J., Sun, D. and Lin, L. 2015. Electrospun nanofibers for sandwiched polyimide/poly (vinylidene fluoride)/polyimide separators with the thermal shutdown function. Electrochim. Acta 176, 727-734.

153. Raghavan, P., Zhao, X., Shin, C., Baek, D.H., Choi, J.W., Manuel, J., Heo, M.Y., Ahn, J.H. and Nah, C. 2010. Preparation and electrochemical characterization of polymer electrolytes based on electrospun poly(vinylidene fluoride-co-hexafluoropropylene)/polyacrylonitrile blend/composite membranes for lithium batteries. J. Power Sources 195(18), 6088-6094.

154. Zhai, Y., Wang, N., Mao, X., Si, Y., Yu, J., Al-Deyab, S.S., El-Newehy, M. and Ding, B. 2014. Sandwich-structured PVdF/PMIA/PVdF nanofibrous separators with robust mechanical strength and thermal stability for lithium ion batteries. J. Mater. Chem. A 2(35), 14511-14518.

155. Xiao, Q., Li, Z., Gao, D. and Zhang, H. 2009. A novel sandwiched membrane as polymer electrolyte for application in lithium-ion battery. J. Memb. Sci. 326(2), 260-264.

156. Shi, C., Dai, J., Shen, X., Peng, L., Li, C., Wang, X., Zhang, P. and Zhao, J. 2016. A high-temperature stable ceramic-coated separator prepared with polyimide binder/Al_2O_3 particles for lithium-ion batteries. J. Memb. Sci. 517, 91-99.

157. Jeon, H., Yeon, D., Lee, T., Park, J., Ryou, M.H. and Lee, Y.M. 2016. A water-based Al_2O_3 ceramic coating for polyethylene-based microporous separators for lithium-ion batteries. J. Power Sources 315, 161-168.

158. Li, X., Zhang, M., He, J., Wu, D., Meng, J. and Ni, P. 2014. Effects of fluorinated SiO_2 nanoparticles on the thermal and electrochemical properties of PP nonwoven/PVdF- HFP composite separator for Li-ion batteries. J. Memb. Sci. 455, 368-374.

159. Yang, C., Tong, H., Luo, C., Yuan, S., Chen, G. and Yang, Y. 2017. Boehmite particle coating modified microporous polyethylene membrane: A promising separator for lithium ion batteries. J. Power Sources 348, 80-86.

160. Lee, D.W., Lee, S.H., Kim, Y.N. and Oh, J.M. 2017. Preparation of a high-purity ultrafine α-Al_2O_3 powder and characterization of an Al_2O_3-coated PE separator for lithium-ion batteries. Powder Technol. 320, 125-132.

161. Peng, K., Wang, B., Li, Y. and Ji, C. 2015. Magnetron sputtering deposition of TiO_2 particles on polypropylene separators for lithium-ion batteries. RSC Adv. 5(99), 81468-81473.

162. Chen, H., Lin, Q., Xu, Q., Yang, Y., Shao, Z. and Wang, Y. 2014. Plasma activation and atomic layer deposition of TiO_2 on polypropylene membranes for improved performances of lithium-ion batteries. J. Memb. Sci. 458, 217-224.

163. Raja, M., Sanjeev, G., Prem Kumar, T. and Manuel Stephan, A. 2015. Lithium aluminate-based ceramic membranes as separators for lithium-ion batteries. Ceram. Int. 41(2), 3045-3050.

164. Padmaraj, O., Venkateswarlu, M. and Satyanarayana, N. 2013. Effect of ZnO filler concentration on the conductivity, structure and morphology of PVdF-HFP nanocomposite solid polymer electrolyte for lithium battery application. Ionics. 19(12), 1835-1842.

165. Ali, S., Tan, C., Waqas, M., Lv, W., Wei, Z., Wu, S., Boateng, B., Liu, J., Ahmed, J., Xiong, J., Goodenough, J.B. and He, W. 2018. Highly efficient PVDF-HFP/colloidal alumina composite separator for high-temperature lithium-ion batteries. Adv. Mater. Interfaces 5(5), 1-10.

166. Kim, J.K., Cheruvally, G., Li, X., Ahn, J.H., Kim, K.W. and Ahn, H.J. 2008. Preparation and electrochemical characterization of electrospun, microporous membrane-based composite polymer electrolytes for lithium batteries. J. Power Sources 178(2), 815-820.

167. Zhang, X., Liu, T., Zhang, S., Huang, X., Xu, B., Lin, Y., Xu, B., Li, L., Nan, C.W. and Shen, Y. 2017. Synergistic coupling between $Li_{6.75}La_3Zr_{1.75}Ta_{0.25}O_{12}$

and poly(vinylidene fluoride) induces high ionic conductivity, mechanical strength, and thermal stability of solid composite electrolytes. J. Am. Chem. Soc. 139(39), 13779-13785.

168. Zhang, F., Ma, X., Cao, C., Li, J. and Zhu, Y. 2014. Poly(vinylidene fluoride)/ SiO_2 composite membranes prepared by electrospinning and their excellent properties for nonwoven separators for lithium-ion batteries. J. Power Sources 251, 423-431.

169. Jeon, H., Jin, S.Y., Park, W.H., Lee, H., Kim, H.T., Ryou, M.H. and Lee, Y.M. 2016. Plasma-assisted water-based Al_2O_3 ceramic coating for polyethylene-based microporous separators for lithium metal secondary batteries. Electrochim. Acta 212, 649-656.

170. Li, X., He, J., Wu, D., Zhang, M., Meng, J. and Ni, P. 2015. Development of plasma-treated polypropylene nonwoven-based composites for high-performance lithium-ion battery separators. Electrochim. Acta 167, 396-403.

171. Kim, M., Han, G.Y., Yoon, K.J. and Park, J.H. 2010. Preparation of a trilayer separator and its application to lithium-ion batteries. J. Power Sources 195(24), 8302-8305.

172. Lee, J., Lee, C.L., Park, K. and Kim, I.D. 2014. Synthesis of an Al_2O_3-coated polyimide nanofiber mat and its electrochemical characteristics as a separator for lithium ion batteries. J. Power Sources 248, 1211-1217.

173. Jiang, X., Zhu, X., Ai, X., Yang, H. and Cao, Y. 2017. Novel ceramic-grafted separator with highly thermal stability for safe lithium-ion batteries. ACS Appl. Mater. Interfaces 9, 25970-25975.

174. Zhu, X., Jiang, X., Ai, X., Yang, H. and Cao, Y. 2015. A highly thermostable ceramic-grafted microporous polyethylene separator for safer lithium-ion batteries. ACS Appl. Mater. Interfaces 7(43), 24119-24126.

175. Lee, T., Kim, W.K., Lee, Y., Ryou, M.H. and Lee, Y.M. 2014. Effect of Al_2O_3 coatings prepared by RF sputtering on polyethylene separators for high-power lithium ion batteries. Macromol. Res. 22(11), 1190-1195.

176. Kumar, J., Kichambare, P., Rai, A.K., Bhattacharya, R., Rodrigues, S. and Subramanyam, G. 2016. A high performance ceramic-polymer separator for lithium batteries. J. Power Sources 301, 194-198.

177. Kim, M. and Park, J.H. 2012. Inorganic thin layer coated porous separator with high thermal stability for safety reinforced Li-ion battery. J. Power Sources 212, 22-27.

178. Jung, Y.S., Cavanagh, A.S., Gedvilas, L., Widjonarko, N.E., Scott, I.D., Lee, S.H., Kim, G.H., George, S.M. and Dillon, A.C. 2012. Improved functionality of lithium-ion batteries enabled by atomic layer deposition on the porous microstructure of polymer separators and coating electrodes. Adv. Energy Mater. 2(8), 1022-1027.

179. Wilson, C.A., Grubbs, R.K. and George, S.M. 2005. Nucleation and growth during Al_2O_3 atomic layer deposition on polymers. Chem. Mater. 17(23), 5625-5634.

180. Xu, Q., Yang, J., Dai, J., Yang, Y., Chen, X. and Wang, Y. 2013. Hydrophilization of porous polypropylene membranes by atomic layer deposition of TiO_2 for simultaneously improved permeability and selectivity. J. Memb. Sci. 448, 215-222.

181. Ulaganathan, M., Nithya, R., Rajendran, S. and Raghu, S. 2012. Li-ion conduction on nanofiller incorporated PVdF-co-HFP based composite polymer blend electrolytes for flexible battery applications. Solid State Ionics 218, 7-12.

182. Subramania, A., Kalyana Sundaram, N.T., Sathiya Priya, A.R. and Vijaya Kumar, G. 2007. Preparation of a novel composite micro-porous polymer electrolyte membrane for high performance Li-ion battery. J. Memb. Sci. 294(1-2), 8-15.

183. Padmaraj, O., Venkateswarlu, M. and Satyanarayana, N. 2016. Effect of PMMA blend and $ZnAl_2O_4$ fillers on ionic conductivity and electrochemical performance of electrospun nanocomposite polymer blend fibrous electrolyte membranes for lithium batteries. RSC Adv. 6(8), 6486-6495.

184. Padmaraj, O., Rao, B.N., Venkateswarlu, M. and Satyanarayana, N. 2015. Electrochemical characterization of electrospun nanocomposite polymer blend electrolyte fibrous membrane for lithium battery. J. Phys. Chem. B 119(16), 5299-5308.

185. Padmaraj, O., Nageswara Rao, B., Jena, P., Venkateswarlu, M. and Satyanarayana, N. 2014. Electrochemical studies of electrospun organic/inorganic hybrid nanocomposite fibrous polymer electrolyte for lithium battery. Polym. (United Kingdom) 55(5), 1136-1142.

186. Lee, L., Park, S.J. and Kim, S. 2013. Effect of nano-sized barium titanate addition on PEO/PVdF blend-based composite polymer electrolytes. Solid State Ionics 234, 19-24.

187. Raja, M., Angulakshmi, N., Thomas, S., Kumar, T.P. and Stephan, A.M. 2014. Thin, flexible and thermally stable ceramic membranes as separator for lithium-ion batteries. J. Memb. Sci. 471, 103-109.

188. Chen, W., Liu, Y., Ma, Y. and Yang, W. 2015. Improved performance of lithium ion battery separator enabled by co-electrospinnig polyimide/poly(vinylidene fluoride-co-hexafluoropropylene) and the incorporation of TiO_2-(2-hydroxyethyl methacrylate). J. Power Sources 273, 1127-1135.

189. Xu, K. 2014. Electrolytes and interphases in Li-ion batteries and beyond. Chem. Rev. 114(23), 11503-11618.

190. Xu, K. 2004. Nonaqueous liquid electrolytes for lithium-based rechargeable batteries. Chem. Rev. 104(10), 4303-4418.

191. Chen, X., Hou, T.Z., Li, B., Yan, C., Zhu, L., Guan, C., Cheng, X.B., Peng, H.J., Huang, J.Q. and Zhang, Q. 2017. Towards stable lithium-sulfur batteries: Mechanistic insights into electrolyte decomposition on lithium metal anode. Energy Storage Mater. 8, 194-201.

192. Peled, E. and Menkin, S. 2017. Review—SEI: Past, present and future. J. Electrochem. Soc. 164(7), A1703-A1719.

193. Markevich, E., Salitra, G. and Aurbach, D. 2017. Fluoroethylene carbonate as an important component for the formation of an effective solid electrolyte interphase on anodes and cathodes for advanced Li-ion batteries. ACS Energy Lett. 2(6), 1337-1345.

194. Cheng, X.B. and Zhang, Q. 2015. Dendrite-free lithium metal anodes: Stable solid electrolyte interphases for high-efficiency batteries. J. Mater. Chem. A 3(14), 7207-7209.

195. Zhang, X.Q., Cheng, X.B., Chen, X., Yan, C. and Zhang, Q. 2017. Fluoroethylene carbonate additives to render uniform Li deposits in lithium metal batteries. Adv. Funct. Mater. 27(10), 1-8.

196. Jung, R., Metzger, M., Haering, D., Solchenbach, S., Marino, C., Tsiouvaras, N., Stinner, C. and Gasteiger, H.A. 2016. Consumption of fluoroethylene carbonate (FEC) on Si-C composite electrodes for Li-ion batteries. J. Electrochem. Soc. 163(8), A1705-A1716.

197. Heine, J., Hilbig, P., Qi, X., Niehoff, P., Winter, M. and Bieker, P. 2015. Fluoroethylene carbonate as electrolyte additive in tetraethylene glycol dimethyl ether based electrolytes for application in lithium ion and lithium metal batteries. J. Electrochem. Soc. 162(6), A1094-A1101.

198. Su, Y.S., Fu, Y., Cochell, T. and Manthiram, A. 2013. A strategic approach to recharging lithium-sulphur batteries for long cycle life. Nat. Commun. 4, 1-8.

199. Guo, J., Wen, Z., Wu, M., Jin, J. and Liu, Y. 2015. Vinylene carbonate-LiNO$_3$: A hybrid additive in carbonic ester electrolytes for SEI modification on Li metal anode. Electrochem. Commun. 51, 59-63.

200. Zhang, L., Ling, M., Feng, J., Mai, L., Liu, G. and Guo, J. 2018. The synergetic interaction between LiNO$_3$ and lithium polysulfides for suppressing shuttle effect of lithium-sulfur batteries. Energy Storage Mater. 11, 24-29.

201. Li, W., Yao, H., Yan, K., Zheng, G., Liang, Z., Chiang, Y.M. and Cui, Y. 2015. The synergetic effect of lithium polysulfide and lithium nitrate to prevent lithium dendrite growth. Nat. Commun. 6, 1-8.

202. Zhao, C.Z., Cheng, X.B., Zhang, R., Peng, H.J., Huang, J.Q., Ran, R., Huang, Z.H., Wei, F. and Zhang, Q. 2016. Li$_2$S$_5$-based ternary-salt electrolyte for robust lithium metal anode. Energy Storage Mater. 3, 77-84.

203. Yan, C., Cheng, X.B., Zhao, C.Z., Huang, J.Q., Yang, S.T. and Zhang, Q. 2016. Lithium metal protection through in-situ formed solid electrolyte interphase in lithium-sulfur batteries: The role of polysulfides on lithium anode. J. Power Sources 327, 212-220.

204. Koshikawa, H., Matsuda, S., Kamiya, K., Kubo, Y., Uosaki, K., Hashimoto, K. and Nakanishi, S. 2017. Effects of contaminant water on coulombic efficiency of lithium deposition/dissolution reactions in tetraglyme-based electrolytes. J. Power Sources 350, 73-79.

205. Qian, J., Xu, W., Bhattacharya, P., Engelhard, M., Henderson, W.A., Zhang, Y. and Zhang, J.G. 2015. Dendrite-free Li deposition using trace-amounts of water as an electrolyte additive. Nano Energy 15, 135-144.

206. Ding, F., Xu, W., Graff, G.L., Zhang, J., Sushko, M.L., Chen, X., Shao, Y., Engelhard, M.H., Nie, Z., Xiao, J., Liu, X., Sushko, P.V., Liu, J. and Zhang, J.G. 2013. Dendrite-free lithium deposition via self-healing electrostatic shield mechanism. J. Am. Chem. Soc. 135(11), 4450-4456.

207. Ding, F., Xu, W., Chen, X., Zhang, J., Shao, Y., Engelhard, M.H., Zhang, Y., Blake, T.A., Graff, G.L., Liu, X. and Zhang, J.G. 2014. Effects of cesium cations in lithium deposition via self-healing electrostatic shield mechanism. J. Phys. Chem. C 118(8), 4043-4049.

208. Goodman, J.K.S. and Kohl, P.A. 2014. Effect of alkali and alkaline earth metal salts on suppression of lithium dendrites. J. Electrochem. Soc. 161(9), D418-D424.

209. Xiao, L., Chen, X., Cao, R., Qian, J., Xiang, H., Zheng, J., Zhang, J.G. and Xu, W. 2015. Enhanced performance of Li | LiFePO$_4$ cells using CsPF$_6$ as an electrolyte additive. J. Power Sources 293, 1062-1067.

210. Lu, Y., Tu, Z., Shu, J. and Archer, L.A. 2015. Stable lithium electrodeposition in salt-reinforced electrolytes. J. Power Sources 279, 413-418.

211. Lu, Y., Tu, Z. and Archer, L.A. 2014. Stable lithium electrodeposition in liquid and nanoporous solid electrolytes. Nat. Mater. 13(10), 961-969.

212. Yamada, Y. and Yamada, A. 2015. Review – Superconcentrated electrolytes for lithium batteries. J. Electrochem. Soc. 162(14), A2406-A2423.

213. Wang, J., Yamada, Y., Sodeyama, K., Chiang, C.H., Tateyama, Y. and Yamada, A. 2016. Superconcentrated electrolytes for a high-voltage lithium-ion battery. Nat. Commun. 7(May), 1-9.

214. Liu, B., Xu, W., Yan, P., Sun, X., Bowden, M.E., Read, J., Qian, J., Mei, D., Wang, C.M. and Zhang, J.G. 2016. Enhanced cycling stability of rechargeable Li-O$_2$ batteries using high-concentration electrolytes. Adv. Funct. Mater. 26(4), 605-613.

215. Camacho-Forero, L.E., Smith, T.W. and Balbuena, P.B. 2017. Effects of high and low salt concentration in electrolytes at lithium-metal anode surfaces. J. Phys. Chem. C 121(1), 182-194.

216. Qian, J., Adams, B.D., Zheng, J., Xu, W., Henderson, W.A., Wang, J., Bowden, M.E., Xu, S., Hu, J. and Zhang, J.G. 2016. Anode-free rechargeable lithium metal batteries. Adv. Funct. Mater. 26(39), 7094-7102.

217. Yoon, H., Howlett, P.C., Best, A.S., Forsyth, M. and MacFarlane, D.R. 2013. Fast charge/discharge of Li metal batteries using an ionic liquid electrolyte. J. Electrochem. Soc. 160(10), A1629-A1637.

218. Jeong, S.K., Seo, H.Y., Kim, D.H., Han, H.K., Kim, J.G., Lee, Y.B., Iriyama, Y., Abe, T. and Ogumi, Z. 2008. Suppression of dendritic lithium formation by using concentrated electrolyte solutions. Electrochem. Commun. 10(4), 635-638.

219. Suo, L., Hu, Y.S., Li, H., Armand, M. and Chen, L. 2013. A new class of solvent-in-salt electrolyte for high-energy rechargeable metallic lithium batteries. Nat. Commun. 4, 1-9.

220. Togasaki, N., Momma, T. and Osaka, T. 2016. Enhanced cycling performance of a Li metal anode in a dimethylsulfoxide-based electrolyte using highly concentrated lithium salt for a lithium-oxygen battery. J. Power Sources 307, 98-104.

221. Suo, L., Borodin, O., Gao, T., Olguin, M., Ho, J., Fan, X., Luo, C., Wang, C. and Xu, K. 2015. 'Water-in-Salt' electrolyte enables high-voltage aqueous lithium-ion chemistries. Science 350, 938-943.

222. Weppner, W. 1981. Trends in new materials for solid electrolytes and electrodes. Solid State Ionics 5, 3-8.

223. Kamaya, N., Homma, K., Yamakawa, Y., Hirayama, M., Kanno, R., Yonemura, M., Kamiyama, T., Kato, Y., Hama, S., Kawamoto, K. and Mitsui, A. 2011. A lithium superionic conductor. Nat. Mater. 10(9), 682-686.

224. Kanno, R. and Murayama, M. 2001. Lithium ionic conductor Thio-LISICON: The Li$_2$S-GeS$_2$-P$_2$S$_5$ system. J. Electrochem. Soc. 148(7), A742.

225. Hartmann, P., Leichtweiss, T., Busche, M.R., Schneider, M., Reich, M., Sann, J., Adelhelm, P. and Rgen Janek, J., 2013. Degradation of NASICON-type

materials in contact with lithium metal: Formation of mixed conducting interphases (MCI) on solid electrolytes. J. Phys. Chem. C 117, 21064-21074.

226. Wenzel, S., Leichtweiss, T., Krüger, D., Sann, J. and Janek, J. 2015. Interphase formation on lithium solid electrolytes—An in situ approach to study interfacial reactions by photoelectron spectroscopy. Solid State Ionics 278, 98-105.

227. Suzuki, Y., Kami, K., Watanabe, K., Watanabe, A., Saito, N., Ohnishi, T., Takada, K., Sudo, R. and Imanishi, N. 2015. Transparent cubic garnet-type solid electrolyte of Al_2O_3 doped $Li_7La_3Zr_2O_{12}$. Solid State Ionics 278, 172-176.

228. Murugan, R., Thangadurai, V. and Weppner, W. 2007. Fast lithium ion conduction in garnet-type $Li_7La_3Zr_2O_{12}$. Angew. Chemie - Int. Ed. 46(41), 7778-7781.

229. Murugan, R., Ramakumar, S. and Janani, N. 2011. High conductive yttrium doped $Li_7La_3Zr_2O_{12}$ cubic lithium garnet. Electrochem. Commun. 13(12), 1373-1375.

230. Shao, C., Liu, H., Yu, Z., Zheng, Z., Sun, N. and Diao, C. 2016. Structure and ionic conductivity of cubic $Li_7La_3Zr_2O_{12}$ solid electrolyte prepared by chemical co-precipitation method. Solid State Ionics 287, 13-16.

231. Liu, Q., Geng, Z., Han, C., Fu, Y., Li, S., He, Y-B., Kang, F. and Li, B. 2018. Challenges and perspectives of garnet solid electrolytes for all solid-state lithium batteries. J. Power Sources 389, 120-134.

232. Chan, C.K., Yang, T. and Mark Weller, J. 2017. Nanostructured garnet-type $Li_7La_3Zr_2O_{12}$: Synthesis, properties, and opportunities as electrolytes for Li-ion batteries. Electrochim. Acta 253, 268-280.

233. Wenzel, S., Weber, D.A., Leichtweiss, T., Busche, M.R., Sann, J. and Janek, J. 2016. Interphase formation and degradation of charge transfer kinetics between a lithium metal anode and highly crystalline $Li_7P_3S_{11}$ solid electrolyte. Solid State Ionics 286, 24-33.

234. Yu, C., van Eijck, L., Ganapathy, S. and Wagemaker, M. 2016. Synthesis, structure and electrochemical performance of the argyrodite Li_6PS_5Cl solid electrolyte for Li-ion solid state batteries. Electrochim. Acta 215, 93-99.

235. Wang, H., Yu, C., Ganapathy, S., van Eck, E.R.H., van Eijck, L. and Wagemaker, M. 2019. A lithium argyrodite $Li_6PS_5Cl_{0.5}Br_{0.5}$ electrolyte with improved bulk and interfacial conductivity. J. Power Sources 412(August 2018), 29-36.

236. Auvergniot, J., Cassel, A., Foix, D., Viallet, V., Seznec, V. and Dedryvère, R. 2017. Redox activity of argyrodite Li_6PS_5Cl electrolyte in all-solid-state Li-ion battery: An XPS study. Solid State Ionics 300, 78-85.

237. Boulineau, S., Courty, M., Tarascon, J.-M. and Viallet, V. 2012. Mechanochemical synthesis of Li-argyrodite Li_6PS_5X (X = Cl, Br, I) as sulfur-based solid electrolytes for all solid state batteries application. Solid State Ionics 221, 1-5.

238. Tatsumisago, M., Nagao, M. and Hayashi, A. 2013. Recent development of sulfide solid electrolytes and interfacial modification for all-solid-state rechargeable lithium batteries. J. Asian Ceram. Soc. 1(1), 17-25.

239. Ujiie, S., Inagaki, T., Hayashi, A. and Tatsumisago, M. 2014. Conductivity of $70Li_2S\cdot30P_2S_5$ glasses and glass–ceramics added with lithium halides. Solid State Ionics 263, 57-61.

240. Minami, K., Hayashi, A., Ujiie, S. and Tatsumisago, M. 2011. Electrical and electrochemical properties of glass–ceramic electrolytes in the systems $Li_2S-P_2S_5-$ P_2S_3 and $Li_2S-P_2S_5-P_2O_5$. Solid State Ionics 192(1), 122-125.

241. Nagao, M., Hayashi, A. and Tatsumisago, M. 2011. Sulfur–carbon composite electrode for all-solid-state Li/S battery with $Li_2S-P_2S_5$ solid electrolyte. Electrochim. Acta 56(17), 6055-6059.

242. Takada, K., Aotani, N., Iwamoto, K. and Kondo, S. 1996. Solid state lithium battery with oxysulfide glass. Solid State Ionics 86-88, 877-882.

243. Tatsumisago, M., Mizuno, F. and Hayashi, A. 2006. All-solid-state lithium secondary batteries using sulfide-based glass–ceramic electrolytes. J. Power Sources 159(1), 193-199.

244. Minami, T., Hayashi, A. and Tatsumisago, M. 2006. Recent progress of glass and glass-ceramics as solid electrolytes for lithium secondary batteries. Solid State Ionics 177(26-32), 2715-2720.

245. Sakuda, A., Hayashi, A. and Tatsumisago, M. 2013. Sulfide solid electrolyte with favorable mechanical property for all-solid-state lithium battery. Scientific Reports 3(2261), 1-5.

246. Fukushima, A., Hayashi, A., Yamamura, H. and Tatsumisago, M. 2017. Mechanochemical synthesis of high lithium ion conducting solid electrolytes in a $Li_2S-P_2S_5-Li_3N$ system. Solid State Ionics 304, 85-89.

247. Ujiie, S., Hayashi, A. and Tatsumisago, M. 2012. Structure, ionic conductivity and electrochemical stability of $Li_2S-P_2S_5-LiI$ glass and glass–ceramic electrolytes. Solid State Ionics 211, 42-45.

248. Liang, C.C. 1973. Conduction characteristics of the lithium iodide-aluminum oxide solid electrolytes. J. Electrochem. Soc. 120(10), 1289.

249. Yamada, H., Bhattacharyya, A.J. and Maier, J. 2006. Extremely high silver ionic conductivity in composites of silver halide (AgBr, AgI) and mesoporous alumina. Adv. Funct. Mater. 16(4), 525-530.

250. Asai, T., Hu, C.H. and Kawai, S. 1987. 7Li NMR study on the $LiIAl_2O_3$ composite electrolyte. Mater. Res. Bull. 22(2), 269-274.

251. Ujiie, S., Inagaki, T., Hayashi, A. and Tatsumisago, M. 2014. Conductivity of $70Li_2S\cdot30P_2S_5$ glasses and glass-ceramics added with lithium halides. Solid State Ionics 263, 57-61.

252. Trevey, J., Jang, J.S., Jung, Y.S., Stoldt, C.R. and Lee, S.H. 2009. Glass-ceramic $Li_2S-P_2S_5$ electrolytes prepared by a single step ball billing process and their application for all-solid-state lithium-ion batteries. Electrochem. Commun. 11(9), 1830-1833.

253. Minami, K., Hayashi, A., Ujiie, S. and Tatsumisago, M. 2011. Electrical and electrochemical properties of glass-ceramic electrolytes in the systems $Li_2S-P_2S_5-$ P_2S_3 and $Li_2S-2S_5-P_2O_5$. Solid State Ionics 192(1), 122-125.

254. Rangasamy, E., Liu, Z., Gobet, M., Pilar, K., Sahu, G., Zhou, W., Wu, H., Greenbaum, S. and Liang, C. 2015. An iodide-based $Li_7P_2S_8I$ superionic conductor. J. Am.Chem. Soc. 137(4), 1384-1387.

255. Tadanaga, K., Takano, R., Ichinose, T., Mori, S., Hayashi, A. and Tatsumisago, M. 2013. Low temperature synthesis of highly ion conductive $Li_7La_3Zr_2O_{12}$-Li_3BO_3 composites. Electrochem. Commun. 33(3), 51-54.

256. Tatsumisago, M. and Hayashi, A. 2012. Superionic glasses and glass-ceramics in the Li_2S-P_2S_5 system for all-solid-state lithium secondary batteries. Solid State Ionics 225, 342-345.

257. De Paoli, M.A., Casalbore-Miceli, G., Girotto, E.M. and Gazotti, W.A. 1999. All polymeric solid state electrochromic devices. Electrochim. Acta 44(18), 2983-2991.

258. Bouridah, A., Dalard, F., Deroo, D. and Armand, M.B. 1986. Potentiometric measurements of ionic mobilities in poly(ethyleneoxide) electrolytes. Solid State Ionics 18-19(Part 1), 287-290.

259. Gray, F.M., MacCallum, J.R. and Vincent, C.A. 1986. Poly(ethylene oxide) – $LiCF_3SO_3$ – polystyrene electrolyte systems. Solid State Ionics 18-19(Part 1), 282-286.

260. Staunton, E., Andreev, Y.G. and Bruce, P.G. 2005. Structure and conductivity of the crystalline polymer electrolyte β-PEO_6: $LiAsF_6$. J. Am. Chem. Soc. 127(35), 12176-12177.

261. Marzantowicz, M., Dygas, J.R., Krok, F., Tomaszewska, A., Florjańczyk, Z., Zygadło-Monikowska, E. and Lapienis, G. 2009. Star-branched poly(ethylene oxide) $LiN(CF_3SO_2)_2$: A promising polymer electrolyte. J. Power Sources 194(1), 51-57.

262. Wright, P.V. 1975. Electrical conductivity in ionic complexes of poly(ethylene oxide). Br. Polym. J. 7, 319-327.

263. Wright, P.V. 1998. Polymer electrolytes—The early days.pdf. Electrochim. Acta 43, 1137-1143.

264. Armand, M.B., Duclot, M.J. and Rigaud, P. 1981. Polymer solid electrolytes: Stability domain. Solid State Ionics 3-4(C), 429-430.

265. Mustarelli, P., Quartarone, E., Tomasi, C. and Magistris, A. 2000. New materials for polymer electrolytes. Solid State Ionics 135(1-4), 81-86.

266. Fenton, D.E., Parker, J.M. and Wright, P.V. 1973. Complexes of alkali metal ions with poly(ethylene oxide). Polymer 14(11), 589.

267. Karan, N.K., Pradhan, O.K., Thomas, R., Natesan, B. and Katiyar, R.S. 2008. Solid polymer electrolytes based on polyethylene oxide and lithium trifluoro-methane sulfonate (PEO-$LiCF_3SO_3$): Ionic conductivity and dielectric relaxation. Solid State Ionics 179(19-20), 689-696.

268. Agrawal, R.C. and Pandey, G.P. 2008. Solid polymer electrolytes: Materials designing and all-solid-state battery applications: An overview. J. Phys. D. Appl. Phys. 41(22), 223001.

269. Zygadło-Monikowska, E., Florjańczyk, Z., Rogalska-Jońska, E., Werbanowska, A., Tomaszewska, A., Langwald, N., Golodnitsky, D., Peled, E., Kovarsky, R., Chung, S.H. and Greenbaum, S.G. 2007. Lithium ion transport of solid electrolytes based on PEO/CF_3SO_3Li and aluminum carboxylate. J. Power Sources 173, 734-742.

270. Bandyopadhyay, S., Marzke, R.F., Singh, R.K. and Newman, N. 2010. Electrical conductivities and Li ion concentration-dependent diffusivities, in polyurethane polymers doped with lithium trifluoromethanesulfonimide

(LiTFSI) or lithium perchlorate (LiClO$_4$). Solid State Ionics 181(39-40), 1727-1731.

271. Jeon, J.-D., Kwak, S.-Y. and Cho, B.-W. 2005. Solvent-free polymer electrolytes. J. Electrochem. Soc. 152(8), A1583.

272. Stoeva, Z., Martin-Litas, I., Staunton, E., Andreev, Y.G. and Bruce, P.G. 2003. Ionic conductivity in the crystalline polymer electrolytes PEO$_6$:LiXF$_6$, X = P, As, Sb. J. Am. Chem. Soc. 125(15), 4619-4626.

273. Singh, T.J. and Bhat, S.V. 2003. Morphology and conductivity studies of a new solid polymer electrolyte: (PEG) x LiClO$_4$. Bull. Mater. Sci., 26(7), 707-714.

274. Zhang, H., Liu, C., Zheng, L., Xu, F., Feng, W., Li, H., Huang, X., Armand, M., Nie, J. and Zhou, Z. 2014. Lithium bis(fluorosulfonyl)imide/poly(ethylene oxide) polymer electrolyte. Electrochim. Acta 133, 529-538.

275. Mindemark, J., Lacey, M.J., Bowden, T. and Brandell, D. 2018. Beyond PEO—Alternative host materials for Li$^+$-conducting solid polymer electrolytes. Prog. Polym. Sci. 81, 114-143.

276. Płocharski, J., Wieczorek, W., Przyłuski, J. and Such, K. 1989. Mixed solid electrolytes based on poly(ethylene oxide). Appl. Phys. A Solids Surfaces 49(1), 55-60.

277. Aziz, S.B., Woo, T.J., Kadir, M.F.Z. and Ahmed, H.M. 2018. A conceptual review on polymer electrolytes and ion transport models. J. Sci. Adv. Mater. Devices 3(1), 1-17.

278. Feuillade, G. and Perche, P. 1975. Ion-conductive macromolecular gels and membranes for solid lithium cells. J. Appl. Electrochem. 5, 63-69.

279. Stephan, A.M. 2006. Review on gel polymer electrolytes for lithium batteries. Eur. Polym. J. 42(1), 21-42.

280. Li, G., Li, Z., Zhang, P., Zhang, H. and Wu, Y. 2008. Research on a gel polymer electrolyte for Li-ion batteries. Pure Appl. Chem. 80(11), 2553-2563.

281. Pu, W., He, X., Wang, L., Tian, Z., Jiang, C. and Wan, C. 2006. Preparation of P(AN-MMA) microporous membrane for Li-ion batteries by phase inversion. J. Memb. Sci. 280(1-2), 6-9.

282. Deka, M., Nath, A.K. and Kumar, A. 2009. Effect of dedoped (insulating) polyaniline anofibers on the ionic transport and interfacial stability of poly(vinylidene fluoride-hexafluoropropylene) based composite polymer electrolyte membranes. J. Memb. Sci. 327(1-2), 188-194.

283. Lin, D.J., Chang, C.L., Lee, C.K. and Cheng, L.P. 2006. Preparation and characterization of microporous PVdF/PMMA composite membranes by phase inversion in water/DMSO solutions. Eur. Polym. J. 42(10), 2407-2418.

284. Kumar, A., Deka, M. and Banerjee, S. 2010. Enhanced ionic conductivity in oxygen ion irradiated poly(vinylidene fluoride-hexafluoropropylene) based nanocomposite gel polymer electrolytes. Solid State Ionics 181(13-14), 609-615.

285. Liao, Y.H., Rao, M.M., Li, W.S., Yang, L.T., Zhu, B.K., Xu, R. and Fu, C.H. 2010. Fumed silica-doped poly(butyl methacrylate-styrene)-based gel polymer electrolyte for lithium ion battery. J. Memb. Sci. 352(1-2), 95-99.

286. Kumar, D., Suleman, M. and Hashmi, S.A. 2011. Studies on poly(vinylidene fluoride-co-hexafluoropropylene) based gel electrolyte nanocomposite for sodium-sulfur batteries. Solid State Ionics 202(1), 45-53.

287. Sarnowska, A., Polska, I., Niedzicki, L., Marcinek, M. and Zalewskał, A. 2011. Properties of poly(vinylidene fluoride-co-hexafluoropropylene) gel electrolytes containing modified inorganic Al_2O_2 and TiO_2 filler, complexed with different lithium salts. Electrochim. Acta 57(1), 180-186.

288. Gebreyesus, M.A., Purushotham, Y. and Kumar, J.S. 2016. Preparation and characterization of lithium ion conducting polymer electrolytes based on a blend of poly(vinylidene fluoride-co-hexafluoropropylene) and poly(methyl methacrylate). Heliyon 2(7), e00134.

289. Cui, Z.Y., Xu, Y.Y., Zhu, L.P., Wang, J.Y., Xi, Z.Y. and Zhu, B.K. 2008. Preparation of PVdF/PEO-PPO-PEO blend microporous membranes for lithium ion batteries via thermally induced phase separation process. J. Memb. Sci. 325(2), 957-963.

290. Xu, J.J. and Ye, H. 2005. Polymer gel electrolytes based on oligomeric polyether/cross-linked PMMA blends prepared via in situ polymerization. Electrochem. Commun. 7(8), 829-835.

291. Weston, J. and Steele, B. 1982. Effects of inert fillers on the mechanical and electrochemical properties of lithium salt-poly(ethylene oxide) polymer electrolytes. Solid State Ionics 7(1), 75-79.

292. Chen, L., Li, Y., Li, S.-P., Fan, L.-Z., Nan, C.-W. and Goodenough, J.B. 2018. PEO/garnet composite electrolytes for solid-state lithium batteries: From 'ceramic-in-polymer' to 'polymer-in-ceramic'. Nano Energy 46, 176-184.

293. Langer, F., Bardenhagen, I., Glenneberg, J. and Kun, R. 2016. Microstructure and temperature dependent lithium ion transport of ceramic–polymer composite electrolyte for solid-state lithium ion batteries based on garnet-type $Li_7La_3Zr_2O_{12}$. Solid State Ionics 291, 8-13.

294. Zhang, W., Nie, J., Li, F., Wang, Z.L. and Sun, C. 2018. A durable and safe solid-state lithium battery with a hybrid electrolyte membrane. Nano Energy 45, 413-419.

295. Zhao, Y., Wu, C., Peng, G., Chen, X., Yao, X., Bai, Y., Wu, F., Chen, S. and Xu, X. 2016. A new solid polymer electrolyte incorporating $Li_{10}GeP_2S_{12}$ into a polyethylene oxide matrix for all-solid-state lithium batteries. J. Power Sources 301, 47-53.

296. Jung, Y.-C., Park, M.-S., Doh, C.-H. and Kim, D.-W. 2016. Organic-inorganic hybrid solid electrolytes for solid-state lithium cells operating at room temperature. Electrochim. Acta 218, 271-277.

Advances in Rechargeable Lithium Ion Batteries and their Systems for Electric and Hybrid Electric Vehicles

Shashank Arora[1] and Weixiang Shen[2*]

[1] School of Engineering, Aalto University, Espoo, Finland
[2] Faculty of Science, Engineering and Technology, Swinburne University of Technology, Hawthorn, Australia

1. Introduction

Rising demand for off-grid homes and long-range electric vehicles (EVs) powered with smart electric systems has created a surge in efforts to develop new generation Li-ion battery based energy storage systems (EESs) [1, 2]. In their present form and configuration, Li-ion batteries offer energy storage capacities ranging from 110-180 Whkg^{-1}. They can be up to four-times higher on material's scale. Volumetric energy density at pack level is also two times lower than energy density available at cellular level, highlighting overwhelming impact of inactive components like current collectors, binders and conductive agents, electric wiring and power electronics, cooling/heating system and packaging on battery weight [3, 4].

The new generation EESs should be extremely compact and lightweight, i.e. possess high energy and power densities to maximize the usable space in next generation EVs, HEVs and smart homes [5-7]. Since electrical energy stored in an electrochemical cell or a Li-ion battery based EES is a function of its nominal capacity and operating voltage, focus is on developing a Li-ion battery system with high terminal voltage. One method of achieving this is through replacement of traditional cathodes such as spinel lithium manganese oxide and layered lithium cobalt

*Corresponding author: wshen@swin.edu.au

oxide with active materials that exhibit higher redox potentials than lithium reference electrodes (Li^+/Li) [8]. As far as anodes are concerned, commercially implemented capacity for graphite, the most common carbon-based anodic material, is fast approaching its theoretical maximum limit. Silicon (Si), a naturally abundant mineral with a low operating potential (< 0.5 V vs. Li/Li^+) and the highest known theoretical capacity (4200 $mAhg^{-1}$), is considered a promising replacement for carbon-based anodes in high-energy Li-ion batteries. However, Si-anodes experience 100-320% volumetric change during the lithiation and delithiation cycle [9]. In addition, Li-ion batteries are categorized as temperature-sensitive systems. Sluggish charge transfer kinetics and reduced conductivities make charging and use of regenerative braking difficult in sub-zero temperatures. On the other hand, heat accumulation inside battery packs during operation at high ambient temperatures is a serious safety concern.

This chapter presents a brief overview of various research efforts made to improve energy density and power density as well as performances of Li-ion batteries in different operating conditions, thereby solving the aforementioned problems. Section two of the chapter summarises developments made in materials domain. Subsequent sections describe strategies to improve battery performances by combining batteries with other energy storage technologies to achieve balance between energy density and power density or by integrating batteries with thermal management to attain stringent control over battery temperature.

2. Materials for Advanced Lithium Ion Batteries

2.1 Anode

Anodic reactions can be divided in three categories:

(a) intercalation;
(b) conversion; and
(c) alloying

Correspondingly, three types of anode materials facilitate these anodic reactions.

2.1.1 Intercalation Anode Materials

Graphite and lithium titanate oxide (LTO) are commonly used examples of intercalation anodic materials. Present-day Li-ion batteries are mainly designed using carbonaceous anodes due to their low cost, favorable electrochemical properties, and easy availability. However, graphitic anodes offer a relatively low theoretical lithium intercalation capacity of 372 $mAhg^{-1}$ and a poor Li-ion diffusion coefficient of the order of 10^{-9} cm^2s^{-1}, which limit the scope for their application in next generation Li-ion batteries. Amorphous carbon, on the other hand, has a malleable microstructure and can support a much higher lithium uptake than graphitic anodes. Through pyrolysis of organic precursors, amorphous carbons can be fashioned into a planar hexagonal frame of carbon without any crystallographic ordering. This disordering is linked to their high reversible lithium insertion capacities. Efforts at synthesizing artificial varieties of graphite have also been made but the process involves heat treatment at over 3000 °C and management of hot gases. Modifying the structure of natural graphite by blending it with other forms such as kish graphite is, thus, preferred. Such surface and structural modifications have resulted in a 10% average increase in energy density of Li-ion batteries comparing with carbonaceous anodes for successive years [10]. Nevertheless, further improvements are needed to compete with increasing choices of alternate materials.

Graphene with its high electrical conductivity, superior mechanical strength, and a surface to mass ratio greater than 2600 m^2g^{-1} has been tested as a probable substitute for graphite anodes. In practice, a single layer of graphene is deposited onto a current collector foil by methods such as chemical vapor deposition. This creates repulsive forces between Li-ions, deposited on either side of the thin graphene layer, making them dominant in the structure limiting the Li-ion uptake capability. For this reason, chemically modified graphene, e.g. graphene oxide is preferred over a single layer graphene. Specific capacity of 100 mAg^{-1} at current rate of 29 mAg^{-1} has been achieved already at full-cell scale with it [11].

Graphene can also be utilized in nanoflakes form to manufacture advanced Li-ion anodes. Graphene nanoflakes possess high crystallinity and can be mass-produced through liquid exfoliation method from pristine graphite. With respect to graphene basal plane, edges of graphene nanoflakes are regarded as highly active Li-ion intercalation sites. They offer up to 50% higher binding energy (1.70 eV to 2.27 eV vs. 1.55 eV) and an energy barrier smaller by up to 0.15 eV for Li-ion diffusion. Furthermore, lateral size of nanoflakes used in anode manufacturing is a crucial parameter affecting battery performances. Graphene nanoflakes with lateral size smaller than 100 nm provide a larger edge to bulk ratio of carbon atoms. Theoretically, nanostructured anodes can have a significantly higher gravimetric density as opposed to a conventional graphitic anode [10].

Among the recently developed anode materials, $Li_4Ti_5O_{12}$ has attracted the most attention. It operates at a potential of 1.55 V, which inhibits dendritic formation during low temperature as well as high discharge operations. $Li_4Ti_5O_{12}$ exists in a spinel-shaped crystal structure with Fd3m space grouping. This means that the three-dimensional structure remains unchanged during the electrochemical reactions and the resultant volume changes are less than 0.1%. Crystal structure allows further modifications to enhance the charge transfer kinetics by doping it with conductive materials. Decreasing particle size is also a viable option. The chemistry is also inherently safe with good cyclic stability. Stability of $Li_4Ti_5O_{12}$ anodes does not depend on presence of a passive surface film; operating potential of 1.55 V is significantly higher than redox voltages of a majority of commonly used electrolytes. As a result, they exhibit remarkable low temperature performances. $Li_4Ti_5O_{12}$ can also be employed in the form of a nanopowder, which is partly responsible for its excellent charge transfer kinetics and enables fast charging of electrodes. Nevertheless, small theoretical capacity (175 mAhg^{-1}) restricts their large-scale adoption by EV industries. Thus, quest for better anode materials, i.e. materials that offer comparable benefits to $Li_4Ti_5O_{12}$ along with a larger specific capacity, continues.

Han et al. reported an intercalation-type compound ($TiNb_2O_7$) featuring a layered monoclinic structure. Its crystal structure falls under C2/m space category. Edges and corners of the $TiNb_2O_7$ monoclinic crystal are shared by three ReO_3 blocks containing octahedral sites that are occupied by Nb_5 and Ti_4 ions in random order. Consequently, $TiNb_2O_7$ exhibits a more open crystal framework than LTO materials. It also facilitates safer operation than LTO anodes due to a comparatively higher operating potential (1.60 V vs. 1.55 V). More importantly, energy capacities over two times greater than that of the conventional LTO anodes can be realized by replacing them with anodes made from $TiNb_2O_7$ material (388 mAhg^{-1} vs 175 mAhg^{-1}), essentially due to availability of five free electrons per formula unit. Limited charge transfer capability of $TiNb_2O_7$ material is a problem though. Vacant 4d/3d orbitals render insulator-like characteristics to the $TiNb_2O_7$ crystal, and is noticeable in the form of its poor electronic conductivity and ionic conductivity [12]. This problem can be addressed through doping method. Wen et al. used V_5 as doping agent to improve Li-ion diffusion in the host material. They demonstrated a comparable capacity of 172 mAhg^{-1} in $TiV_{0.02}Nb_{1.98}O_7$ at a 10C-rate [13]. Lin et al. managed to achieve a capacity of 181 mAhg^{-1} at 5C using $Ru_{0.01}Ti_{0.99}Nb_2O_7$ [14]. On the other hand, Song and Kim preferred Mo_6 as a doping material and demonstrated an impressive energy storage capacity (190 mAhg^{-1}) of $Mo_{0.05}Ti_{0.95}Nb_2O_7$ at 10C-rate [15]. Although the research is promising, issues related to toxicity, high cost, and large molecular weight of the three aforementioned doping agents, respectively, limits applicability of the doped variants. New doping materials and methods are, therefore, needed.

Carbon nanotubes (CNTs) have also received significant attention due to their large specific surface area, high conductivity and a unique one-dimensional microstructure. Multi-walled CNTs, in particular, provide a large number of Li-intercalation sites owing to local turbostatic disorder and a central core. In fact, Li-uptake capacities in excess of 1400 mAhg^{-1} are regularly reported in published literature. Moreover, CNTs are chemically stable and have high electrical conductivity and thermal conductivity, 10^{-4} Scm^{-1} and 2000–4000 Wm^{-1}K^{-1}, respectively. They also possess high shear strength (~500 MPa) in addition to excellent tensile strength (~50 GPa). Short Li-ion diffusion pathways created by virtue of the CNT microstructure enhance charge transfer dynamics for the system and enable charging/discharging at high rates. Durability, high-strength and significantly improved power density and energy density are anticipated to be key features of CNT-based anodes. However, similar to the most other novel anode materials, poor cyclic stability and a large irreversible capacity loss over time affect CNT-based anodes as well. Combining CNTs with Li-alloying metals has been proposed as a possible solution to the problem. For example, Kumar et al. demonstrated Sn-filled multi-walled CNT-based high-capacity Li-insertion anodes. De-insertion capacity of 889 mAhg^{-1} was realized for the first-cycle [16]. Chen et al. used aligned-CNT arrays to design anode for a Li-ion battery. Discharge capacity of the battery (over 265 mAhg^{-1}) remained stable for more than 50 cycles. An additional benefit of aligned-CNT based anodes was that they made the copper current collector redundant. This resulted in a weight saving of more than 85% [17]. Risk of potential battery failure due to copper pitting and dissolution in electrolyte was also eliminated using copper less anode.

2.1.2 Alloying Anode Materials

Alloy anodes are fabricated either from pure metals, or from alloys or from intermetallic compounds. Both amorphous and crystalline states are acceptable in form of fine powders and as thin films. Primary benefit of alloy anodes is that they provide 2 – 10 times and 4 – 20 times higher theoretical specific capacities than graphite anodes and LTO anodes, respectively [18]. A moderate onset potential of 0.3–0.6 V above Li/Li$^+$ mitigates the risk of lithium deposition and offers another crucial advantage over graphite anodes. Energy penalty is also not as large as in the case of LTO anodes (1.5 V above Li/Li$^+$). They have thus been researched heavily in last few years. The key elements are silicon (Si) and tin (Sn). Lithiation up to a stoichiometric ratio of 4.4 can be readily performed with both Si and Sn, which corresponds to specific capacities of 4200 mAhg^{-1} and 994 mAhg^{-1}, respectively. Despite demonstrating better capacity retainability than plain Sn/Si electrodes, intermetallic anodes, e.g. Ni$_3$Sn$_4$ are not considered suitable for prolonged cycling. Cycle life of these anodes is compromised by large volume changes (can be up to 320% in Si anodes) accompanying the alloying process. Mechanical stress created due to volume expansion of anode particles results in failure of active mass integrity, which causes capacity fading in batteries. In addition, formation of Li-alloys in an electrochemical cell with Si-anode is followed by pulverization of electrode and, subsequently, loss of electrical contact between the Si-layer and the adjoining current collector [9]. Figure 1 depicts different failure mechanisms of Si-electrode.

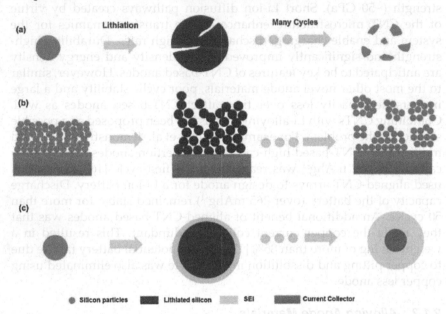

Fig. 1. Three different failure mechanisms of Si electrode: (a) electrode pulverization, (b) collapse of the entire electrode, and (c) continuous breaking and re-growth of the SEI layer [19].

Volumetric charge density of alloy anodes (at fully expanded state) is also 2–5 times greater than LTO and graphite anodes. Therefore, modification of intermetallic space and optimizing morphology are proposed to reduce the observed capacity decrease. Adopted methodologies include use of: (a) thin-films and amorphous alloys, (b) macro- and nano-powders – nanowires in a brush-type morphology have been found to be the most effective in accommodating volumetric strain, (c) different binders and electrolyte formulations – carboxymethyl cellulose (CMC) is the binder of choice for composite Si-anodes, and (d) dispersion in active and inactive matrices. A comparison of various alloy anode materials is presented in Table 1.

2.1.3 *Conversion Anode Materials*

Insertion reaction requires an electronically conductive active material and cystrallographic voids to facilitate Li-ion diffusion. However, Poizot et al. discovered that presence of Li-ions can cause various interstitial-

Table 1. Comparison of theoretical specific capacity, charge density, volume change, and onset potential of various anode materials [18]

Materials	Li	C	LTO	Si	Sn	Sb	Al	Mg	Bi
Density (gcm⁻³)	0.53	2.25	3.5	2.33	7.29	6.7	2.7	1.3	9.78
Lithiated Phase	Li	LiC_6	$Li_7Ti_5O_{12}$	$Li_{4.4}Si$	$Li_{4.4}Sn$	Li_3Sb	LiAl	Li_3Mg	Li_3Bi
Theoretical Specific Capacity (mAhg⁻¹)	3682	372	175	4200	994	660	993	3350	385
Theoretical Charge Density (mAhcm⁻³)	2047	837	613	9786	7246	4422	2681	4355	3765
Volume Change (%)	100	12	1	320	260	200	96	100	215
Potential vs. L (~ V)	0	0.05	1.6	0.4	0.6	0.9	0.3	0.1	0.8

free transition metal oxides, e.g. Fe_2O_3, CuO and CoO, which have been classified unsuitable for the intercalation reaction to go through reversible reduction reactions. Some of these metal oxides exhibit reversible capacities up to 1000 mAhg^{-1} [20]. However, the nanostructured reactants must undergo total chemical and structural transformation as a part of this reaction, represented by equation below [21]:

$$nano - MO + 2 Li^+ + 2e^- = nano - M + Li_2O$$

High capacity reversible conversion reactions involving as many as four electrons per $3d$-orbital have been reported for transitional nitrides, fluorides, phosphides, borides and sulphides as well. Conversion anodes are not considered practical mainly because:

1. development of stable passivation film is unlikely in particular at elevated temperatures since conversion reactions occur at potentials below the threshold reduction potential of standard Li-ion electrolytes
2. energetics of the electrochemical conversion reaction are affected by large hysteresis losses over cyclic discharge process.

2.2 Cathodes

Energy density, working voltage and rate capability of a Li-ion battery are all dependent on thermodynamics and theoretical capacity of cathode materials. A cathode material is, therefore, considered the key to the development of advanced Li-ion batteries. For the same reason, cathode materials in a Li-ion battery are priced twice as much as anode materials. Cathode materials can also facilitate either intercalation or conversion reactions.

2.2.1 Intercalation Cathode Materials

Intercalation cathodes consist of a solid network of host material in which guest ions are cyclically inserted and removed. Traditionally, transitional metal compounds, or complex oxides with a spinel, layered or an olivine crystal structure are used as cathode materials. Automotive grade battery cathodes are mainly represented by the oxide family, LiFePO$_4$ being the sole exception. For the advanced Li-ion batteries though, cathode materials with an upper cut-off voltage greater than 4.2 V, e.g. Ni-rich layered oxides (LiNi$_{1-x}$M$_x$O$_2$ where M = metal $\sim$ 4.5 V), Li-rich layered oxides (Li$_{1+x}$M$_{1-x}$O$_2$ with $\sim$ 4.7 V), high voltage spinel oxides (LiNi$_{0.5}$Mn$_{1.5}$O$_4$ with $\sim$ 4.8 V) and high voltage polyanionic compounds (sulphates and phosphates, up to $\sim$ 5.2 V), are recommended [22]. With the help of these materials, overall contribution of inactive components to battery energy density would be reduced by approximately 20%. Still, cathode materials with practical energy densities close to 800 Whkg^{-1} will be needed to design a battery that can offer 250 Whkg^{-1} [3].

During the electrochemical reaction, Li-ions either take up the vacant tetrahedral/octahedral sites or diffuse into the space between adjacent layers of the host crystal. For charge stability, the same number of electrons as the number of migrating Li-ions must occupy the vacant *d*-orbitals in the crystal structure of the host transition metal. Electrode potential and the Gibbs free energy charge are determined by electronegativity, ionic radius, valence state and local environment of the cations. Figure 2 illustrates relationship between voltage and number of electrons in *d*-orbital of transition metal ions.

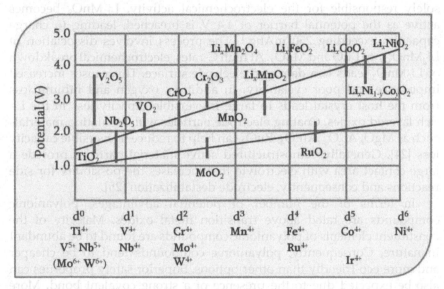

Fig. 2. Voltage range of compounds consisting of transition metal ions. This schematic exhibits the relationship between electrode potential and the number of electrons in *d*-type electronic orbitals of transition metal ions; in general, the potential increases with increasing number of electrons in *d*-type electronic orbitals (courtesy of [23]).

Majority of the next-generation cathodes belong to the oxide category as it offers the right balance between energy density, working potential, theoretical capacity and rate capabilities. Ni-rich NMC oxides with general composition $Li[Ni_{1-x}Co_yMn_z]O_2$, where $x \leq 0.4$ are preferred over others by EV sectors because the merits of this class of novel materials increase with increasing Ni-content. For example, reversible capacity of the oxide increases from 163 mAhg^{-1} to over 200 mAhg^{-1} as the composition changes from NMC-111 to NMC-811. Similarly, cathodes using NMC-811 instead of NMC-111 benefit from a higher Li-ion diffusivity (10^{-8}–10^{-9} cm^{2}s^{-1} vs. 10^{-10}–10^{-11} cm^{2}s^{-1}). Electronic conductivity also increases from 5.2 ×10^{-8} Scm^{-1} to 2.8 × 10^{-5} Scm^{-1} upon replacing NMC-111 with NMC-811.

Reactivity with electrolyte and structural instability are the issues that need to be resolved before actual implementation.

High specific capacities of high-energy NMCs layered composite materials, general formula: $x\text{Li}_2\text{MnO}_3(1-x)\text{LiMO}_2$ where, M = Ni, Co or Mn, is attractive to EV industries as well. Li_2MnO_3 serves dual purposes in the composite structure. Firstly, it provides Li^+ to the LiMO_2 component and acts as the structural stabilizing member within the composite. Secondly, it enables a higher degree of lithiation than that is normally achievable with a pure layered oxide. Between 2.0 and 4.4 V, LiMO_2 is solely responsible for the electrochemical activity. Li_2MnO_3 becomes active as the potential barrier of 4.4 V is breached, leading to charge capacities exceeding 250 mAhg^{-1}. The process involves dissociation of Li_2MnO_3 into Li_2O and MnO_2. At high C-rates, electrochemical breakdown of Li_2MnO_3 leads to a damaged electrode surface. This causes increased impedance and poor cyclability. In addition, oxygen and lithium loss from the host crystal leads to large irreversible capacity loss in the Li-rich layered oxides. Coating electrode particles with insulating materials such as MgO, Al_2O_3, AlPO_4, RuO_2 can help to reduce irreversible capacity loss [24]. Generally, nanostructured active material particles provide a large contact area with electrolyte that increases the possibility for side reactions and consequently, electrode destabilization [25].

In terms of the number of potential advantages, polyanionic compounds are rated above transition metal oxides. Majority of the constituent elements of polyanionic compounds are found to be abundant in nature. Consequently, polyanionic compounds tend to be cheaper and more eco-friendly than other options. Superior safety properties can also be expected due to the presence of a strong covalent bond. More importantly, theoretical potential and ionic character of the M-O bond can be easily tuned by selecting appropriate non-transition metal, e.g. B, S, P, Si [23]. Polyanionic compounds make it possible to design customized or "on-demand" cathode materials. It means that a large pool of available polyanionic compounds with varied crystallographic structures enable designing an energy dense system as well as reducing the working potential if the risk of electrolyte decomposition becomes too high. Selection pool is limited to an extent due to incompatibility of a high-potential (4.7–5.5 V) group of sulphates and phosphate compounds comprising Co, Ni and Mn with the current generation of electrolytes. In certain other compounds, adequate amounts of conductive carbon needs to be added to counter high electronic energy gaps and poor Li^+ diffusion coefficients. Lastly, high molecular weight of compounds like pyrophosphates in comparison to oxides makes it difficult for them to meet the 2025 volumetric energy density targets for batteries [3]. Consequently, they too may be dropped from the selection pool in future.

Some of the technologically mature members of this family are $LiFeBO_3$, $LiVPO_4F$ and $LiMnPO_4$. De-intercalation voltage of $LiFeBO_3$ is nearly 3 V. It offers a theoretical energy density of 660 $Whkg^{-1}$. In addition, excellent cyclic stability is ensured by a volume change in the order of less than 2% during cycling. In contrast, inductive effect of fluorine causes $LiVPO_4F$ to operate at a potential close to 4.26 V. However, the energy density of $LiVPO_4F$ is quite similar to that of $LiFeBO_3$ (650 $Whkg^{-1}$ (or 2000 Whl^{-1}) versus 660 $Whkg^{-1}$. It also demonstrates fast Li-ion diffusion rates and a good long-term reversible capacity retainability. $LiMnPO_4$ too provides energy storage capacity above 600 $Whkg^{-1}$. A few other manganese based compounds, e.g. Li_2MnSiO_4 and $LiMnPO_4(OH)$ operate at 4.5 V. They provide gravimetric and volumetric capacities close to 1300 $Whkg^{-1}$ and 4000 Whl^{-1}, respectively. Electron transfer process initiates multi-step electrode reactions. In most cases, only one-electron transfer is detected since it occurs at a potential that is quite close to the breakdown potential of standard electrolyte solution. However, two distinct electron transfer steps are encountered in certain molecules, mainly with two identical electroactive groups. Two-electron transfer process is responsible for their impressive performance. Performance can be further improved by modifying transport properties, where low Li-ion diffusion coefficients and low electronic conductivities are reported for silicates (5×10^{-16} Scm^{-1} at 25 °C).

2.2.2 Conversion Cathode Materials

Intercalation reactions typically involve one-electron transfer for each metal atom. In contrast, another category of cathodic materials, known as conversion materials, that supports multi-electron reactions and offer theoretical gravimetric densities up to 700 $mAhg^{-1}$ also exist. Reaction mechanism is quite similar to anodic conversion reactions in which metal atoms are reduced to their metallic states. Reaction product is a Li_nX matrix containing metal nanoparticles in a dispersed state, where X is any anion, e.g. F, N, P, S etc. and n is the oxidation state of X in the conversion compound M_aX_b.

Conversion compounds are known for their low molar weight, low cost and high intrinsic thermal stability. It should however be noted that operating potential of majority of the conversion compounds is lower than the operating potential of oxides and polyanionic compounds. Further, it strongly depends on the reaction potential of the negative ion forming the compound. They are also characterized by low ionic conductivities, large hysteresis and poor cycling stability caused by particle delamination due to large volume changes during lithiation and de-lithiation processes. Some of the compounds, e.g. $CuCl_2$, are soluble in organic solvents or exhibit polysulphide dissolution. Armstrong et al. pointed that consistent particle morphology and systematic material preparation is the key to reliable performance of conversion cathodes as conversion materials prepared through different synthetic methods demonstrate significantly different behaviors [26].

Material loading in the present generation EV batteries is between 15-25 mgcm^{-2}. However, most of the novel cathode materials are tested with a loading as low as 1 mgcm^{-2} [3]. Rigorous testing is, therefore, needed before any of the novel materials can be applied in traction batteries.

2.3 Electrolytes

A polar aprotic Li salt, typically lithium hexafluorophosphate (LiPF$_6$) in a binary non-aqueous solvent mixture including either ethyl carbonate (EC) or ethyl methyl carbonate (EMC), diethyl carbonate (DEC) or dimethyl carbonate (DMC), is used as standard electrolyte in currently available commercial Li-ion batteries. High anodic stability, high ionic conductivity even at a temperature as low as −15 °C and high solubility of LiPF$_6$ in all alkyl carbonates are the key reasons for the preferred selection. Usually, anodic stability and oxidation potential of electrolytes increases as the concentration of atoms in high oxidation states increases. However, presence of Li-ions alters the reaction thermodynamics completely. Salt anions of the type MX$_y$ − and polar aprotic solutions are reduced to insoluble ionic Li compounds. Reduction compounds precipitate over carbon electrodes and non-active metal electrodes to form a thin surface film at the electrode and electrolyte interphase. This thin surface film allows Li-ion transfer but starts blocking electron flow after attaining certain thickness. The observed passivating effect is related to apparent stability of commercial electrodes. DMC and EC solutions are known to form highly stable passivating surface films over graphite anodes upon reduction.

Viscosity of these solutions is relatively high. Consequently, wetting process for composite electrodes and separators becomes tricky. In addition, alternative electrolyte solutions are required for low temperature applications since ionic conductivity of the EC/DMC solutions decreases remarkably below −15 °C. Decomposition of LiPF$_6$ to PF$_5$ and LiF at elevated temperatures is also an issue. More importantly, the standard electrolyte solvents pose significant safety challenges during handling, usage and events such as thermal runaway due to their high flammability and high toxicity. Compatible electrolytes are crucial for enhanced performance, cycle life and safety of advanced Li-ion batteries.

Organic solvents such as cyclic phosphates, and halogenated cyclic carbonates exhibit better fire retardancy than EC/DMC solution. Water sensitivity, high viscosity and poor stability are some issues that need to be addressed 27-29]. Use of aqueous electrolytes, developed by dissolving salts such as $LiNO_3$, Li_2SO_4 and LiTFSI into water such that molality exceeds 20 m, is a promising approach to increase fire retardancy of Li-ion batteries 30]. Non-volatile electrolytes based on ionic liquids are also being investigated for next generation batteries. Ionic liquids demonstrate non-flammability. Nonetheless, their practical applications are restricted owing to their high costs and incompatibility with carbonaceous anodes. Formation of solid electrolyte interphase film on graph cyclic phosphate is difficult in Li-ion batteries using ionic-liquids. Li-ion migration is also reduced as cations of ionic liquids compete with Li-ions and occupy the intercalation sites in graphite layers 31-33]. It must be noted that average working voltages of organic electrolyte solvents lie between 1.5 V and 4.5 V. Corresponding energy density values are in the range of 150 Whkg^{-1} to 300 Whkg^{-1}. In contrast, solid-state electrolytes are more stable, safe and compatible with new-age high voltage cathode materials. They can achieve energy densities above 500 Whkg^{-1} 34].

3. Self-heating Lithium Ion Batteries

Poor performance at low temperatures is a major issue affecting wide adoption of the current and next generation Li-ion batteries. Sluggish charge transfer kinetics and reduced ionic and electronic conductivities at subzero environmental temperatures results in a significant power loss in Li-ion batteries [35]. Under such circumstances, fast charging becomes difficult and regenerative braking is not easy either. Mainstream research is thus focused on developing novel electrode materials and electrolyte formulations for use in low temperature applications. However, automotive batteries are expected to deliver performance goals in a variety of operating conditions [36]. Electrolytes and electrodes targeted specifically at low temperatures may provide marginalized performances at elevated temperatures. Various external and internal heating strategies including pulse heating and alternating current heating have thus far been developed as feasible solutions. A detailed discussion on the pre-heating methods along with a comprehensive review of the existing as well as emerging battery thermal management systems is available [37]. Nonetheless, majority of the pre-heating methods take several minutes and approximately 10% of the battery capacity to preheat battery packs. Moreover, complex heating/electrical circuits are needed with certain techniques. A research group led by Prof Chao-Yang Wang from Pennsylvania State University, USA proposed a novel self-heating battery architecture to alleviate the aforementioned concerns [38].

The disclosed battery architecture, shown in Fig. 3, features an embedded nickel foil that heats the battery structure from –20 °C to 0 °C in less than 20 seconds using only 3.8% of battery capacity. The batteries

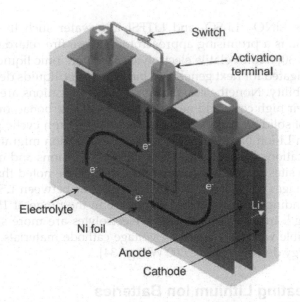

Fig. 3. Architecture of self-heated battery (courtesy of [38]).

are fabricated by placing a 3-layer assembly of a 28 μm thick Ni-foil sandwiched between two single side coated anodes in the middle of the electrode stack. Polyethylene tetraphthalate is coated on the Ni-foil to make an electrical insulation. One end of the Ni-foil is then electrically connected in parallel to the anode whereas the opposite end, referred to as the activation terminal, protrudes outside the battery as a third terminal.

A power switch is connected between the cathode and the activation terminal. The switch is temperature sensitive and turns on automatically at low temperatures. Ni-foil heats up quickly due to the resultant electric current flow, which serves the purpose of heating the battery internally. Current flow through the Ni-foil shuts down when the battery temperature matches a pre-set value. From this moment onwards, battery functions like a typical battery with two external terminals.

An improved version of this design includes two Ni-foils instead of a single sheet. The first foil is placed at ¼th while the second is placed at ¾th of the battery thickness. Temperature and foil resistance follow a linear relationship, as shown in Fig. 4. Ni-foil (resistance) is carefully selected in accordance with battery nominal capacity. The two foils are welded together in parallel. Consequently, the two-sheet design does not differ much from the original single-sheet design in external appearance. Nonetheless, addition of an extra Ni-foil results in a three-times more uniform heat distribution throughout the battery and makes it possible to heat the battery internally from –20 °C to 0 °C in 12.5 seconds with just 2.9% of battery capacity [39].

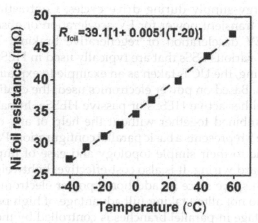

Fig. 4. Relationship between temperature of self-heated battery and Ni-foil resistance [39].

4. Lithium Ion Batteries Integrated with Other Energy Storage Technologies

Rated capacity of battery packs should be large enough to meet the energy/power requirements of EV drive cycles. From the cost perspective, peak power demand is an important design variable in battery pack design since high power batteries cost much more than other variants. When employing low power batteries, power and energy demands of a battery system are met by increasing the number of parallel strings, each of which consists of the number of cells in series. This approach too has a negative effect on the cost of a battery system. In addition, thermal management and cell balancing becomes increasingly difficult as the total number of cells in the system increases. Lack of efficient thermal management system and proper cell balancing system causes the capacity of the battery system to degrade faster. Degradation rate is further exacerbated in high C-rate and extreme weather applications. Current surges triggered by unexpected regenerative braking may accelerate battery pack ageing as well [5]. As a convenient solution to the aforementioned problems, lithium ion batteries integrated with other energy storage technologies have thus been proposed through hybridization.

Hybrid energy storage systems (HESSs) are physical systems that combine two or more different energy storage technologies such as batteries, fuel cells and ultra-capacitors (UCs) through electronic circuitry to form a single on-board unit and benefits from their combined advantages. Configured unit contains a right mix of high-energy (HE) storage elements and high-power (HP) storage elements. Configuration ratio is pre-decided by the end application. The HE elements ensure steady state energy supply during drive cycles. Contrastingly, the HP elements supply transient power for EV acceleration or absorb transient power during EV deceleration or regenerative braking. Table 2 lists characteristics of various ESSs that are typically used in HESSs.

In the following, the UC is taken as an example to explain the ways of designing HESSs. Based on power electronics used, the configuration can be classified as either active HESSs or passive HESSs. In passive HESSs, two ESSs are combined together without the help of any dc/dc power converter. Figure 5 represents a basic parallel configuration. Passive HESSs are preferred due to their simple topology and ease of implementation and straightforward wiring. It is also cost-effective, lightweight and does not require much space since no additional power electronics is needed. However, they do not allow taking full advantage of high power density of UCs since voltage in parallel branches is controlled by the weakest, in this case, voltage of the battery. Consequently, load or the dc-link voltage must match with voltage characteristics of the constituting energy sources.

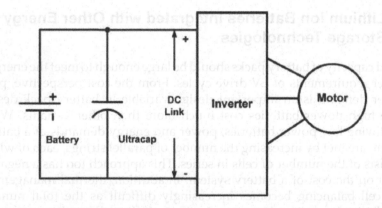

Fig. 5. Basic parallel configuration of battery and UC.

Dougal et al. demonstrated reduction in internal losses of batteries in HESSs. In this hybrid system, batteries supply relatively constant and only one-tenth of the peak current demand which decreases the internal heat generation in batteries because majority of the current comes from UCs. Degree of the benefit is of course dependent on capacity of a UC and pulse width ratio and pulse amplitude of currents required [40].

On the other hand, active HESSs use one or more dc/dc converters as the interface between two energy sources and the dc link. De-coupling between two energy sources makes this hybrid system control more accurate and easier. Interfacing UCs via a bi-directional dc/dc converter, as shown in Fig. 6(a), facilitates operation of the UC in a considerably wide voltage range (from 40% to 100%), enabling utilization of up to 75%

Table 2. Comparison of various energy storage systems typically used to design HESSs

ESS	Rated power (MW)	Typical discharge Time	Power density (W/kg)	Energy density (Wh/kg)	Self-discharge rate per day (%)	Response time	Efficiency (%)	Lifetime (years)	Lifetime (cycles)
Compressed air	100-300	1 – day		30-60	0	minutes	40-70	20—0	
Flywheel	0-0.25	s – h	400-1600	5-130	20-100	ms – s	80-90	15-20	10^4-10^7
Ultra-capacitor	0-0.3	ms – 1 h	0.1-10	0.1-15	2-40	ms	85-98	5-12	10^5-10^6
Li-ion batteries	0-0.1	min – h	200-340	130-250	0.1-0.3	ms	65-95	5-8	600-1200
Fuel cells	0-50	s – days	>500 (W/L)	500-3000	0.5-2	ms – minutes	20-66	5-30	10^3-10^4

of the UC capacity. Tight control over dc-link voltage is necessary in this configuration though due to a direct connection with the battery. Large and frequent fluctuations in voltage of the dc-link affect battery cycle life adversely. Energy consumption can be reduced by 7% through swapping positions of the UC and the battery and connecting the UC to the dc-link directly. Example of a battery/UC topology is shown in Fig. 6(b) [41]. The UC functions as a low-pass filter in this topology which absorbs peak and high frequency currents generated due to power fluctuations. Larger UCs provide better filtering effects [42]. Improved system efficiency is a noticeable feature of this arrangement. Moreover, battery voltage can differ from voltage of the UC in both the aforementioned configurations.

Working range and energy utilization for the UC bank can be further improved by adding a second bi-directional dc/dc convertor between the UC and the dc-link. This topology is referred to as the cascaded network and is schematically represented by Fig. 7(a). Voltage stabilization across the dc-link and the batteries is also easy to attain in cascaded systems.

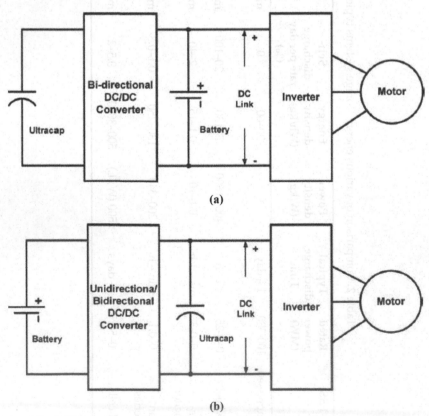

Fig. 6. Illustration of (a) UC/Battery configuration and (b) Battery/UC configuration [43].

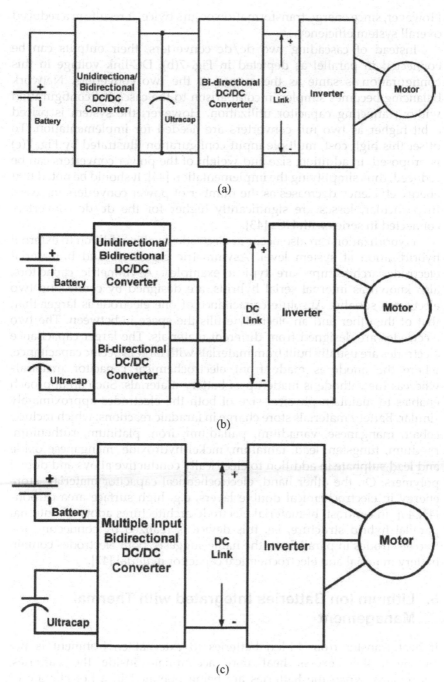

(a)

(b)

(c)

Fig. 7. Illustration of (a) Cascaded configuration, (b) Multiple converter configuration and (c) Multiple input configuration [43].

However, since energy transformation occurs twice, it results in a reduced overall system efficiency.

Instead of cascading two dc/dc converters, their outputs can be connected in parallel as depicted in Fig. 7(b). DC-link voltage in this configuration is same as the output of the two converters. Network balancing becomes simpler in comparison to the cascaded configuration without affecting capacitor utilization. However, the system is priced a bit higher as two full converters are needed for implementation. To offset this high cost, multiple input configuration illustrated by Fig. 7(c) is proposed. In addition, size and weight of the power converters can be reduced, thus simplifying the implementation [44]. It should be noted that energy efficiency decreases as the number of power converters increase. In particular, losses are significantly higher for the dc/dc converters connected in series with UCs [43].

Hybridization can also occur at electrode level in addition to external hybridization at system level. Asymmetric capacitors and bi-material electrode architectures are typical examples. Asymmetric capacitors, also known as internal serial hybrids, are designed by connecting two electrodes serially. Absolute capacitance of one electrode is larger than that of the other and an electrolyte fills the space in-between. The two electrodes are designed from different materials. The larger capacitance electrodes are usually built from materials with larger specific capacitance, where the anode is made from electrochemical capacitor materials whereas the cathode is made up of battery materials. Such an approach enables to maintain physical size of both the electrodes approximately similar. Battery materials store charge in faradaic reactions, which include cobalt, manganese, vanadium, palladium, iron, platinum, ruthenium, rhodium, tungsten, lead, tantalum, nickel hydroxide, manganese oxide and lead sulphate in addition to electrically conductive alloys and doped polymers. On the other hand, electrochemical capacitor materials store energy in electrochemical double layers, e.g. high surface area carbons [45, 46]. In contrast, bi-material electrode architectures adopt an internal parallel hybrid structure, i.e. this device is realized by connecting the two electrodes in parallel. As the name suggests, both electrodes contain battery material and electrochemical capacitor material [47].

5. Lithium Ion Batteries Integrated with Thermal Management

If heat transfer from Li-ion batteries to external environment is not sufficient, then excess heat may accumulate inside the batteries, particularly when the batteries are being operated in a hot climate or under an insulating environment. Hot spots can also develop, leading to an uneven temperature distribution across the batteries, which can alter charging and discharging characteristics of the batteries. More importantly, battery temperature may rise beyond the safety limits of 60 °C for Li-ion batteries using $LiBF4$ as electrolyte, risking battery failure. Studies indicate that battery temperature must be regulated within a predefined operating range to sustain a rate of reaction considered healthy for the efficient operation of batteries. The recommended operational range for Li-ion batteries is generally between 25 °C and 40 °C. Managing large temperature spikes and non-uniform thermal gradients across Li-ion battery packs is, therefore, a major concern in the design of large Li-ion battery packs essential for supporting EV driveline. For these reasons, a battery thermal management system (TMS) needs to be integrated with the EV battery pack.

In recent years, different research groups have tested a number of techniques including reciprocating airflow systems, cold plates, heat pipes and phase change materials etc. for readily removing excess heat from battery packs. A comprehensive review of the latest developments made in this area is available [37], where some of emerging alternatives, such as thermo-acoustic systems, magnetic refrigeration method and traditional cooling techniques, are also discussed. It is understood that power consumed by auxiliaries in actively cooled systems can vary significantly based on ambient temperature; parasitic load can be as high as 40% in some cases. Passive cooling involving phase change materials is, therefore, recommended for EV battery packs. In addition, plumbing and the auxiliary systems used in liquid cooled devices limit their scalability, which is essential for retaining modularity of battery packs. Through a careful qualitative analysis based on the current technological state and associated risks, phase change materials and thermoelectric devices were identified as the most suitable candidates for a modular TMS for EV battery packs. These two methods will be discussed in more detail in the following section.

5.1 Phase Change Materials

A simple passive solution comprising battery cells placed in a matrix of phase change materials with zero maintenance requirements has piqued interest of several research groups. For example, Khateeb et al. studied a Li-ion battery pack made of eighteen 18650 Li-ion cells and filled with a mixture of PCM and aluminium foam. They examined the thermal behavior of this pack with the help of a numerical model that was validated through experiments. It demonstrated through their experiments that the temperature rise in a battery pack can be reduced to half by using a TMS with PCM and aluminum foam, as opposed to the case where no TMS is applied [48]. Mills et al. simulated a laptop battery pack with six 2.2

Ah Li-ion cells and attained a uniform thermal distribution after using expanded graphite saturated with PCM as the thermal management solution [49]. Sabbah et al. compared the performance of a TMS with PCM to that of an air-cooled system using numerical methods and experiments. They demonstrated that the former could keep the temperature of a Li-ion battery cell below 55 °C, even at a constant discharge rate of 6.67C [50]. Kizilel et al. experimented with PCM-filled high-energy Li-ion battery packs and achieved a uniform thermal distribution under both normal and abusive test conditions [51]. Rao et al. also tested eutectic PCMs for 8 Ah prismatic $LiFePO_4$ battery cells. The results of numerous experiments and simulations indicate that PCMs may be a practical solution to the thermal issues affecting EV battery packs [52]. Li et al. tested the effectiveness of a PCM-filled copper foam sandwich panel as a cooling system for prismatic power batteries. The tests reflected a lower surface temperature and a better thermal uniformity in the battery pack after the integration of the PCM-filled panels [53]. Arora et al. also demonstrated that the high capacity PCM (RT28HC) could maintain 20 Ah $LiFePO_4$ pouch cell in a near-isothermal state during a 3C discharge process [8].

However, phase-change materials have relatively low thermal conductivities and slow regeneration rates. Consequently, they cannot be effective in applications where quick regeneration of cooling media may be needed, e.g. fast charging followed by a quick discharge of a battery pack in a short time. Several heat transfer enhancement techniques have been investigated for PCMs, which include:

1. Use of fixed and non-moving surfaces such as fins and honeycombs.
2. Employment of composite PCMs
3. Impregnation of porous material
4. Dispersion of high-conductivity particles in PCMs

In [35], it is shown that it is possible to improve effective thermal conductivity of PCM-systems without compromising their latent heat capacity. This solution is based on using inverted cell layout to design battery packs. Inverting battery cells introduces convective currents in the battery pack, which increases the heat and mass transfer within the pack resulting in an improved system.

5.2 Thermoelectric Coolers

Thermoelectric coolers (TECs) are maintenance-free solid-state heat pumps with no moving parts. They use doped semiconductor elements, comprising a series of p-type and n-type thermo-elements, sandwiched in thermally-conductive but electrically-insulating substrates to transfer heat across a junction of two dissimilar materials via the Peltier effect. The p-type material has excess positive charge carriers called holes, whereas n-type material carries more negative charge carriers called electrons. The direction of heat transfer depends on the polarity of voltage applied to the TE modules. An illustration of a TEC module is provided in Fig. 8.

As voltage is applied to a TEC, electrons jump from a lower energy level of the p-type thermo-element to a higher energy state in the n-type thermo-element by absorbing thermal energy from one side of the module, in effect cooling it. The electrons drop to a stable energy level by rejecting this heat on the other side of the TEC module. Accordingly, the same TEC module can be made to function both as a cooler and as a heater by reversing the direction of current flow across the junction.

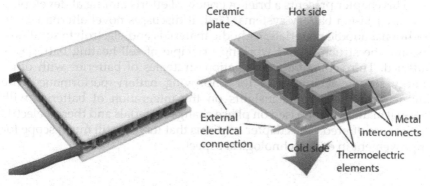

Fig. 8. Illustration of a thermo-electric cooling module (courtesy of [4]).

Other advantages of TECs are:

- They are compact and lightweight
- They are acoustically silent and operate without any vibrations
- They facilitate precise temperature control to within ±0.1 °C
- They have low manufacturing costs and a wide operating temperature range
- They can cool below ambient temperature
- They are location independent, and can operate in any spatial orientation, at high G-levels or in zero gravity

More importantly, the cooling elements of a TEC module are easily scalable [54-56]. For these reasons, TECs have been applied for climate control in EVs [57-60]. However, the figure of merit (ZT) for the currently available bellerium telluride based TECs is approximately one. Consequently, the maximum coefficient of performance obtainable with these devices is limited to 10%. It has been suggested that a ZT close to 4 is required to achieve cooling performance comparable to other thermal management techniques [61].

6. Conclusion

Customer expectations related to available energy capacity and deliverable performance-level of Li-ion battery based energy storage systems are higher than the capacity of the current state-of-the-art systems. In this context, advancements in Li-ion batteries are critical to their continued proliferation in electric vehicles and hybrid electric vehicles. Such advancements can be made either by developing novel active materials for anodes, cathodes and electrolytes of Li-ion batteries or combining batteries with other energy storage technologies or integrating batteries with thermal management.

This chapter presents a brief overview of efforts aiming at developing advanced Li-ion battery systems. First, it discusses novel alternatives to traditional anodic materials, cathodic materials and electrolyte solutions. Second, the structure and working principle of self-heating batteries is outlined. Third, possible combination strategies of batteries with other energy storage technologies for improving battery performances are summarized. Last, some insights on the integration of batteries with thermal management based on phase change materials and thermoelectric devices are offered. The chapter confirms that there is still much scope for improvement in each technological aspect.

References

1. Grimm, M., Lenz, L., Peters, J. and Sievert, M. 2017. Demand for off-grid solar electricity – Experimental evidence from Rwanda. Discussion Paper Series, Environment for Development. EfD DP 17-15.
2. Solar PEG. 2016. Off-Grid Homes in High Demand. PEG ALTERNATIVE ENERGY Inc.
3. Andre, D., Kim, S.-J., Lamp, P., Lux, S.F., Maglia, F. and Paschos, O. 2015. Future generations of cathode materials: An automotive industry perspective. Journal of Materials Chemistry A 3, 6709-6732.
4. Arora, S. 2017. Design of a modular battery pack for electric vehicles. (Doctoral Thesis). Australia: Swinburne University of Technology, Melbourne.
5. Arora, S., Kapoor, A. and Shen, W.X. 2018. Application of robust design methodology to battery packs for electric vehicles: Identification of critical technical requirements for modular architecture. Batteries 4, 30.
6. Arora, S., Shen, W.X. and Kapoor, A. 2015. Designing a Robust Battery Pack for Electric Vehicles Using a Modified Parameter Diagram. SAE Technical Paper 2015-01-0041. https://doi.org/10.4271/2015-01-0041.
7. Arora, S. and Kapoor, A. 2018. Mechanical design and packaging of battery packs for electric vehicles. pp. 175-200. In: Pistoia, G., Liaw, B. (eds). Behaviour of Lithium-Ion Batteries in Electric Vehicles: Battery Health, Performance, Safety, and Cost. Cham: Springer International Publishing.

8. Arora, S., Shen, W.X. and Kapoor, A. 2017. Critical analysis of open circuit voltage and its effect on estimation of irreversible heat for Li-ion pouch cells. Journal of Power Sources 350, 117-126.

9. Xing, Y., Shen, T., Guo, T., Wang, X., Xia, X. and Gu, C. 2018. A novel durable double-conductive core-shell structure applying to the synthesis of silicon anode for lithium ion batteries. Journal of Power Sources 384, 207-213.

10. Shukla, A. and Kumar, T.P. 2008. Materials for next-generation lithium batteries. Current Science 94, 314-331.

11. Hassoun, J., Bonaccorso, F., Agostini, M., Angelucci, M., Betti, M.G. and Cingolani, R. 2014. An advanced lithium-ion battery based on a graphene anode and a lithium iron phosphate cathode. Nano Letters 14, 4901-4906.

12. Han, J.-T. and Goodenough, J.B. 2011. 3-V full cell performance of anode framework $TiNb_2O_7$/Spinel $LiNi_{0.5}Mn_{1.5}O_4$. Chemistry of Materials 23, 3404-3407.

13. Wen, X., Ma, C., Du, C., Liu, J., Zhang, X. and Qu, D. 2015. Enhanced electrochemical properties of vanadium-doped titanium niobate as a new anode material for lithium-ion batteries. Electrochimica Acta 186, 58-63.

14. Lin, C., Yu, S., Wu, S., Lin, S., Zhu, Z.-Z. and Li, J. 2015. $Ru_{0.01}Ti_{0.99}Nb_2O_7$ as an intercalation-type anode material with a large capacity and high rate performance for lithium-ion batteries. Journal of Materials Chemistry A 3, 8627-8635.

15. Song, H. and Kim, Y.-T. 2015. A Mo-doped $TiNb_2O_7$ anode for lithium-ion batteries with high rate capability due to charge redistribution. Chemical Communications 51, 9849-9852.

16. Kumar, T.P., Ramesh, R., Lin, Y. and Fey, GT.-K. 2004. Tin-filled carbon nanotubes as insertion anode materials for lithium-ion batteries. Electrochemistry Communications 6, 520-525.

17. Chen, J., Liu, Y., Minett, A.I., Lynam, C., Wang, J. and Wallace, G.G. 2007. Flexible, aligned carbon nanotube/conducting polymer electrodes for a lithium-ion battery. Chemistry of Materials 19, 3595-3597.

18. Zhang, W.-J. 2011. A review of the electrochemical performance of alloy anodes for lithium-ion batteries. Journal of Power Sources 196, 13-24.

19. Wu, H. and Cui, Y. 2012. Designing nanostructured Si anodes for high energy lithium ion batteries. Nano Today 7, 414-429.

20. Poizot, P., Laruelle, S., Grugeon, S., Dupont, L. and Tarascon, J. 2000. Nano-sized transition-metal oxides as negative-electrode materials for lithium-ion batteries. Nature 407, 496.

21. Etacheri, V., Marom, R., Elazari, R., Salitra, G. and Aurbach, D. 2011. Challenges in the development of advanced Li-ion batteries: A review. Energy & Environmental Science 4, 3243-3262.

22. Li, W., Dolocan, A., Oh, P., Celio, H., Park, S. and Cho, J. 2017. Dynamic behaviour of interphases and its implication on high-energy-density cathode materials in lithium-ion batteries. Nature Communications 8, 14589.

23. Liu, C., Neale, Z.G. and Cao, G. 2016. Understanding electrochemical potentials of cathode materials in rechargeable batteries. Materials Today 19, 109-123.

24. Manthiram, A. 2011. Materials challenges and opportunities of lithium ion batteries. The Journal of Physical Chemistry Letters 2, 176-184.
25. Xu, J., Dou, S., Liu, H. and Dai, L. 2013. Cathode materials for next generation lithium ion batteries. Nano Energy 2, 439-442.
26. Armstrong, M.J., Panneerselvam, A., O'Regan, C., Morris, M.A. and Holmes, J.D. 2013. Supercritical-fluid synthesis of FeF 2 and CoF 2 Li-ion conversion materials. Journal of Materials Chemistry A 1, 10667-10676.
27. Tsujikawa, T., Yabuta, K., Matsushita, T., Matsushima, T., Hayashi, K. and Arakawa, M. 2009. Characteristics of lithium-ion battery with non-flammable electrolyte. Journal of Power Sources 189, 429-434.
28. Ota, H., Kominato, A., Chun, W.-J., Yasukawa, E. and Kasuya, S. 2003. Effect of cyclic phosphate additive in non-flammable electrolyte. Journal of Power Sources 119, 393-398.
29. Nakagawa, H., Shibata, Y., Fujino, Y., Tabuchi, T., Inamasu, T. and Murata, T. 2010. Application of nonflammable electrolytes to high performance lithium-ion cells. Electrochemistry 78, 406-408.
30. Li, Q., Chen, J., Fan, L., Kong, X. and Lu, Y. 2016. Progress in electrolytes for rechargeable Li-based batteries and beyond. Green Energy & Environment 1, 18-42.
31. Ishaq, M., Jabeen, M., Song, W., Xu, L. and Deng, Q. 2017. 3D hierarchical $Ni^{2+}/Mn^{2+}/Al^{3+}$ layered triple hydroxide@ nitrogen-doped graphene wrapped hybrids on nickel foam for supercapacitor applications. Journal of Electroanalytical Chemistry 804, 220-231.
32. Nakagawa, H., Fujino, Y., Kozono, S., Katayama, Y., Nukuda, T. and Sakaebe, H. 2007. Application of nonflammable electrolyte with room temperature ionic liquids (RTILs) for lithium-ion cells. Journal of Power Sources 174, 1021-1026.
33. Xing, W. 2018. High energy/power density, safe lithium battery with nonflammable electrolyte. ECS Transactions 85, 109-114.
34. Xu, G.L., Amine, R., Abouimrane, A., Che, H., Dahbi, M. and Ma, Z.F. 2018. Challenges in developing electrodes, electrolytes, and diagnostics tools to understand and advance sodium-ion batteries. Advanced Energy Materials 8, 1702403.
35. Arora, S., Kapoor, A. and Shen, W.X. 2018. A novel thermal management system for improving discharge/charge performance of Li-ion battery packs under abuse. Journal of Power Sources 378, 759-775.
36. Arora, S., Shen, W.X. and Kapoor, A. 2017. Neural network based computational model for estimation of heat generation in $LiFePO_4$ pouch cells of different nominal capacities. Computers & Chemical Engineering 101, 81-94.
37. Arora, S. 2018. Selection of thermal management system for modular battery packs of electric vehicles: A review of existing and emerging technologies. Journal of Power Sources 400, 621-640.
38. Wang, C.-Y., Zhang, G., Ge, S., Xu, T., Ji, Y., Yang, X.-G. 2016. Lithium-ion battery structure that self-heats at low temperatures. Nature 529, 515.
39. Zhang, G., Ge, S., Xu, T., Yang, X.-G., Tian, H. and Wang, C.-Y. 2016. Rapid self-heating and internal temperature sensing of lithium-ion batteries at low temperatures. Electrochimica Acta 218, 149-155.

40. Dougal, R.A., Liu, S. and White, R.E. 2002. Power and life extension of battery-ultracapacitor hybrids. IEEE Transactions on Components and Packaging Technologies 25, 120-131.
41. Min, H., Lai, C., Yu, Y., Zhu, T. and Zhang, C. 2017. Comparison study of two semi-active hybrid energy storage systems for hybrid electric vehicle applications and their experimental validation. Energies 10, 279.
42. Shin, D., Kim, Y., Wang, Y., Chang, N. and Pedram, M. 2012. Constant-current regulator-based battery-supercapacitor hybrid architecture for high-rate pulsed load applications. Journal of Power Sources 205, 516-524.
43. Cao, J. and Emadi, A. 2012. A new battery/ultracapacitor hybrid energy storage system for electric, hybrid, and plug-in hybrid electric vehicles. IEEE Transactions on Power Electronics 27, 122-132.
44. Li, Z., Onar, O., Khaligh, A. and Schaltz, E. 2009. Design and control of a multiple input DC/DC converter for battery/ultra-capacitor based electric vehicle power system. Applied Power Electronics Conference and Exposition, 2009. APEC 2009 Twenty-Fourth Annual IEEE: IEEE p. 591-596.
45. Zhu, W.H. and Tatarchuk, B.J. 2016. Characterization of asymmetric ultracapacitors as hybrid pulse power devices for efficient energy storage and power delivery applications. Applied Energy 169, 460-468.
46. Razoumov, S., Klementov, A., Litvinenko, S. and Beliakov, A. 2001. Asymmetric electrochemical capacitor and method of making. Google Patents.
47. Cericola, D. and Kötz, R. 2012. Hybridization of rechargeable batteries and electrochemical capacitors: Principles and limits. Electrochimica Acta 72, 1-17.
48. Khateeb, S.A., Farid, M.M., Selman, J.R. and Al-Hallaj, S. 2004. Design and simulation of a lithium-ion battery with a phase change material thermal management system for an electric scooter. Journal of Power Sources 128, 292-307.
49. Mills, A. and Al-Hallaj, S. 2005. Simulation of passive thermal management system for lithium-ion battery packs. Journal of Power Sources 141, 307-315.
50. Sabbah, R., Kizilel, R., Selman, J.R. and Al-Hallaj, S. 2008. Active (air-cooled) vs. passive (phase change material) thermal management of high power lithium-ion packs: Limitation of temperature rise and uniformity of temperature distribution. Journal of Power Sources 182, 630-638.
51. Kizilel, R., Lateef, A., Sabbah, R., Farid, M.M., Selman, J.R. and Al-Hallaj, S. 2008. Passive control of temperature excursion and uniformity in high-energy Li-ion battery packs at high current and ambient temperature. Journal of Power Sources 183, 370-375.
52. Rao, Z., Wang, S. and Zhang, G. 2011. Simulation and experiment of thermal energy management with phase change material for ageing LiFePO$_4$ power battery. Energy Conversion and Management 52, 3408-3414.
53. Li, W., Qu, Z., He, Y. and Tao, Y. 2014. Experimental study of a passive thermal management system for high-powered lithium ion batteries using porous metal foam saturated with phase change materials. Journal of Power Sources 255, 9-15.
54. Thielmann, J. 2013. Thermoelectric Cooling Technology. Will Peltier Modules Supersede the Compressor. SelectScience

55. Corporation, T. 2017. Thermoelectric Cooling America Corporation. Accessed from: https://www.thermoelectric.com/technology/

56. Corporation, F.-N. 2015. Peltier Device: A Single Device Used for Cooling and Heating. WordPress.com.

57. Alaoui, C. and Salameh, Z.M. 2005. A novel thermal management for electric and hybrid vehicles. IEEE Transactions on Vehicular Technology 54, 468-476.

58. Cosnier, M., Fraisse, G. and Luo, L. 2008. An experimental and numerical study of a thermoelectric air-cooling and air-heating system. International Journal of Refrigeration 31, 1051-1062.

59. Miranda, A., Chen, T. and Hong, C. 2013. Feasibility study of a green energy powered thermoelectric chip based air conditioner for electric vehicles. Energy 59, 633-641.

60. Suh, I.-S., Cho, H. and Lee, M. 2014. Feasibility study on thermoelectric device to energy storage system of an electric vehicle. Energy 76, 436-444.

61. Rowe, D. and Goldsmid, H. 2005. A new upper limit to the thermoelectric figure-of-merit. Thermoelectrics Handbook: Macro to Nano. CRC Press.

Advanced Lithium Ion Batteries for Electric Vehicles: Promises and Challenges of Nanostructured Electrode Materials

Venkata Sai Avvaru[1,2], Mewin Vincent[1,2] and Vinodkumar Etacheri[2*]

[1] Faculty of Science, Universidad Autónoma de Madrid,
  C/ Francisco Tomás y Valiente, 7, Madrid 28049, Spain
[2] IMDEA Materials Institute, C/ Eric Kandel 2, Getafe, Madrid 28906, Spain

1. Introduction

In the recent years, energy consumption and depleting fossil fuel resources along with CO_2 emissions, has been increasing exponentially. There is a huge concern about the detrimental impacts of global warming caused by greenhouse gases from primary sources like fossil fuels, transportation vehicles etc. [1]. The increasing fuel consumption that affects environment has made it essential to develop electric vehicles (EVs), hybrid electric vehicles (HEVs) and/or plug-in hybrid electric vehicles (PHEVs), replacing fossil fuel vehicles, with low greenhouse gas emissions [1-6]. Therefore, it has become more important to develop clean and advanced renewable energy storage systems with an efficient energy saving technology like batteries.

Nickel metal hydride (Ni-MH) and Lithium-ion batteries (LIB) are the two main battery technologies used in current marketing systems [7]. Figure 1 show the comparison of various battery technologies in terms of volumetric and gravimetric energy density, where Li-ion battery offers the great potential for developing for EV/HEVs. Due to the limited

*Corresponding author: vinodkumar.etacheri@imdea.org

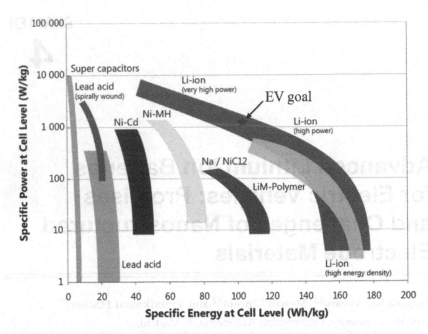

Fig. 1. Ragone Plot illustrating the range of specific energy and power levels by current generation battery cell technology.

specific energy and high self-discharge rate (~50% >NiCd) for Ni-MH batteries, the adoption of LIBs is growing faster particularly in EVs [7, 8]. Rechargeable LIBs are well established as a reliable energy device for portable electronics and also have the potential to power EVs, since its first market introduction from 1991. High energy density, lightweight, very low self-discharge rate, and long service life with high coulombic efficiency for LIBs make them inimitable for electric powered transportation vehicles [2-6]. Fast lithium-ion diffusion enables LIBs to charge and discharge faster than commercial lead-acid and NiMH batteries. For further boosting up the electrification in the transport systems, the main research has been focused on improving the LIB performance by reducing the cost with maximum safety and calendar life. According to United States Advanced Battery Consortium (USABC), for HEV applications the requirements defined for LIBs are to have an energy density of 400 Wh/kg with 90% efficiency, calendar life of 5 years, and operating temperature spanning −20 to 55 °C [9].

To meet the requirements of advanced battery systems for EVs, the recent research and development is mostly focused on high performance electrode materials. Developing electrodes with high charge capacity and high voltage (for cathodes) can meet the specific energy and power densities of battery systems in EVs. With the rise of nanotechnology from the 20th century, researchers initiated their focus more on the morphological aspects of the electrode materials [10-12]. Integrating the nano-engineered materials in the framework of the battery electrodes can improve the reaction kinetics for achieving the desired energy and power densities. Nanostructured materials can lead to new Li storage mechanisms, enabling higher capacities than conventional intercalation mechanisms. Various efforts for incorporating the nanostructured materials, their challenges and state-of-the-art anodes and cathodes are summarised.

2. Negative Electrodes

Selection of anode material is very crucial for Li batteries to maintain and improve their electrochemical performances. In the history of LIBs, graphite is the most extensively studied anode for its excellent behaviour towards a practical Li host. However, bottle necks in the electrochemical performance of graphite restrict its application for the EV industry [10]. At present most EV batteries use graphite based anodes with continues improvements to attain higher practical capacities and high energy densities. At this instant, development of carbon based materials to function as high capacity as well as safe anodes for lithium batteries has huge economical and commercial importance. Owing to the high demand for EV's, an extensive investigation was carried out in the recent past focused on material properties for enhancing specific capacity, coulombic efficiency and redox potential. This research led to the invention of numerous promising anode materials exhibiting exceptional performance of LIBs. In addition, it was recognised that macro-sized materials with poor Li[+] diffusion rates, smaller surface area, drastic morphological variations between charge/discharge states and higher electronic resistivity are not suitable for high energy applications. Hence, tuning size and shape in the nano scaled materials, could provide better structural stability and boost the electrochemical behaviour [11].

In general, according to the lithiation/delithiation mechanism, the anode materials can be classified into three main classes, namely Intercalation, alloying and conversion type anodes. Figure (2) represents the schematic of these reactions.

3. Intercalation Type Anodes

Intercalation anodes are one of the most exploited materials for the LIB technology. High reversibility as well as the ease of ion storage/extraction mechanism make intercalation anodes more stable [13, 14]. Intercalation mechanism is an electrochemical redox reaction between the host materials

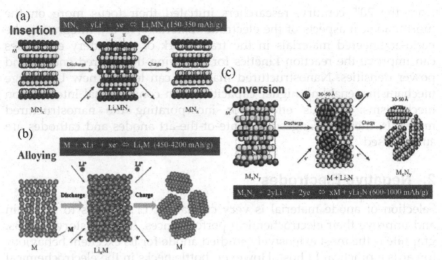

Fig. 2. Different types of anodes and anode reactions in Li-ion batteries (a) Insertion (b) Alloying and (c) Conversion reactions.

(anode) and guest ion (Li^+) where chemical charge transfer is associated with the movement of mobile Li^+ from an electrolyte [12, 15].

An intercalation lithiation /DE lithiation reaction can be represented as follows:

$$MN_x + yLi^+ + ye^- \Leftrightarrow Li_yMN_x$$

in which MN_x represents intercalation lithium host [14]. Some of the anode intercalation materials are carbon, graphite, TiO_2, LTO, Nb_2O_5 etc.

3.1 Carbon Based Anodes

Research on carbon anodes was considered most promising for rechargeable batteries for their reversibility; high capacity (372 Ah/kg) and lithiation at low potential. However, micrometre sized graphite was a stable anode, but in the nano regime is proven to impart adverse effects due to the constraints on the intercalation potentials [16]. Hence, several allotropes of carbon with one dimensional (1-D), two dimensional (2D) and porous morphologies were investigated for improving the battery performance (Fig. 3) [17-20]. Most significant 1-D material used as anode was carbon nanotubes (CNT). Recently reported CNT anodes exhibited reversible capacities of 1116 mAh/g at current density 50 mAh/g [21-22]. In 2D domain, graphene, the novel monolayer carbon, gained much attention as an anode due to its high Li storage capacity on additional sites like defects, edges, disorders etc. [13, 18].

Owing to its ultrathin 2D structure, it shows unique properties of high electrical conductivity, large surface area, and high mechanical

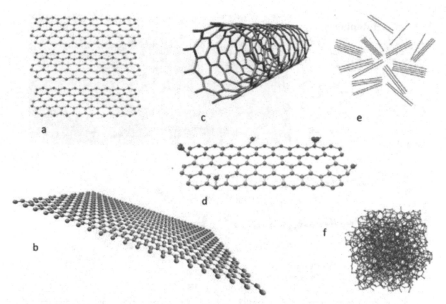

Fig. 3. Different carbon structures studied as LIB anodes (a) Graphite, (b) Graphene, (c) CNT, (d) RGO, (e) Hard carbon, (f) Amorphous carbon.

stabilities [13, 23-25]. These characteristics enable the graphene to exhibit better specific capacities (700 mAh/g) and good cycling stability. Hybrid graphene together with metal, metal oxide or other dopants like nitrogen was found to form very stable structures with a better performance. For example, boron doped graphene sheet as an anode developed by Wu et al. showed (Fig. 4) very high reversible capacity of nearly 1400 mAh/g at current density of 50 mAh/g [26]. Moreover, this anode was compatible with fast charge and discharge rates, a 235 mAh/g capacity was obtained at a high current rate of 20 A/g with a very short charging span of 30 s. Apart from constantly higher capacities, the graphene anodes also possess superior electronic conductivity, chemical tolerance, and a broad electrochemical window as inherent properties. Practically aligned 2D framework of graphene with semi-metallic nature combines both the advantages of semiconductor as well as metallic nature imparting very huge electronic mobility. Recent investigations reveal that the electronic mobility of graphene can go high as 20,000 cm^2/V [27, 28]. Owing to the extremely high charge transfer capabilities, graphene is also used as composites, coating on TMO or structural matrix for performance enhancement of anode materials or vice versa [29-31]. Different graphene derivative compounds, found attractive based on outstanding performance, were also studied among which reduced graphene oxide (RGO) was found to be one of the good choices [32]. Heterogeneous RGO

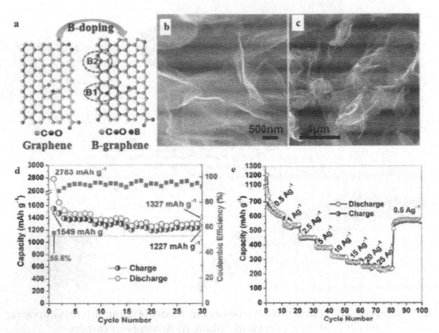

Fig. 4. (a) Schematic of graphene, (b-c) SEM images of undoped and doped graphene, (d) Cycle performance and Coulombic efficiency of the graphene electrode at current rate of 50 mA/g between 3.0 and 0.01 V, (e) Rate performance of graphene electrodes. (Copyright 2011 American Chemical Society)

surface with functional groups like carboxyl or hydroxyl groups have imparted relatively high interlayer spacing making this material more suitable for effective storage. Many investigations also demonstrated that the surface of the RGO provides effective sites for electrochemical reactions owing to the high surface area and porosity [33]. Fu et al. reported RGO as an anode material with a layered irregular structure. This intrinsic structure performed better lithium capacity and good cycling stability [34].

Another form of carbon investigated to a large extent is amorphous carbon. Amorphous carbon gained more attention for its synthesis at lower temperatures, with superior lithium storage capacities. Main attraction of amorphous carbon is its high surface area which is nearly 8.6 times higher than that of graphite, which is mainly attributed to the mixed morphology of graphitic nature and microporous structures. Graphitic crystallites embedded in the matrix ensure excellent conductivity of the material [35]. On the other hand, the amorphous phase leads the lithium storage function. Defective surfaces and closed voids of the amorphous region have a direct relation to the storage mechanism. An optimized combination of defects and void sites with priority on the former is usually more favourable for better ion transport and capacity retention. Owing to the morphological and electrochemical advantages amorphous carbon is widely used to coat on other anode materials like TiO_2, graphite, Si, Sn etc. to increase conductivity as well as to buffer the volume change during cycling [36-39]. Carbon allotrope which is of more interest in the recent years is the hard carbon or disordered carbon [40]. The main advantage of hard carbon as an anode is the absence of exfoliation induced by intercalation due to available active sites on graphitic structures. This effect leads to a lower sensitivity towards the electrolyte compositions. Hence, it could be used with any electrolyte even without the addition of a film forming agent [41, 42]. In addition, these hard carbons can also exhibit competent specific capacities and stable cycling using LiTFSI electrolytes [41].

3.2 Lithium Titanium Oxide (LTO)

Among the available anode materials, only LTO has been successfully commercialised into EVs such as Honda Fit. LTO was known for its 'zero-strain' lithium insertion and extraction mechanism. It has been commercialised in various applications due to combined properties like superior thermal stability, high rate, relatively high volumetric capacity and high cycle life. The lithiation/delithiation in LTO is associated with the structural reorientation of the LTO (Fig. 5) between the $Li_4Ti_5O_{12}$ and $Li_7Ti_5O_{12}$ [43-44]. However, in spite of this structural reorientation, both belong to $Fd\overline{3}m$ crystallographic phase group with almost a negligible volumetric variation (~ 2%) of the spinel lattice upon an exchange of 3 Li^+ in each phase of operation [45-46].

In general, the significant technical characteristics which made LTO-Li batteries more viable and competent for commercial applications,were enhanced safety, long life, fast charging, and improved low temperature performance. LTOs have the limitations of lower capacity and higher intercalation potential. However, they do not form any SEI layer during cycling which is the foremost reason for the very low irreversible capacity loss. In addition, volume expansion of LTOs on lithiation is also extremely negligible. The near zero volume change and the absence of SEI layer together makes the anodes stable against the electrode degradation occurring in each charge/discharge cycle. Practically, LTOs exhibit cell cycle life of 20,000 cycles while the available commercial batteries like Tesla have a maximum of up to 5000 cycles only. Meanwhile, high surface area (100 m^2/g) of nanosized LTOs in effect reduces the diffusion time and charging time tremendously, from several hours to a few minutes. Again, the higher intercalation potential of LTO anodes reduces the chances of Li plating and hence dendrite growth related safety challenges. In addition, the LTO anodes are highly resistant to the thermal run away or overheating that makes the battery highly resistant to fire explosion.

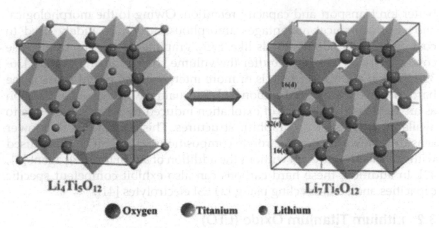

Recent investigations showed that LTO anodes can retain 82% of the initial capacity even at –47 °C. These characteristics made the LTO anodes the first priority for aerospace or avionics applications where the safety measures are non-negotiable.

3.3 Titanium Dioxide (TiO_2)

The foremost successful class of materials next to carbon is TiO_2 based materials. Properties like low cost, non-toxicity and stability against Li deposition are the main attractions of TiO_2 [47]. All the polymorphs of TiO_2, namely anatase, rutile, bronze and anionic or cationic deficient designs, defect structures etc. were investigated as possible anodes [47]. Irrespective of the phase TiO_2 materials demonstrate a theoretical capacity of 335 mAh/g [48]. In addition to these, TiO_2 phases usually share all characteristic features of LTO anodes like higher intercalation potential, near zero SEI layer, thermally stable and long cycle life [12].

Among the various polymorphs of TiO_2, the anatase phase is assumed to be the most electro active against the Li insertion. It is the more preferable phase in the group for its relatively faster Li insertion/extraction kinetics [49]. Owing to the advantages, TiO_2 anodes on further investigations revealed that they can enhance their capacities from theoretical capacity by utilizing the advantages of nano structuring. Hierarchical patterns of mesoporous anatase phase combined with 3D network improved the capacity to more than 240 mAh/g. Texturing in to nanotubes with mesoporous walls further enhanced the capacity of the anodes to 303 mAh/g against Li^+/Li at 1 A/g current density [10, 50]. Very recently TiO_2-B nanotubes and nanowires also gained much attention due to their near LTO characteristics. Crystal structure of TiO_2-B having perovskite-like windows formed of edge and corner sharing TiO_6 octahedra allows a fast Li-ion intercalation. The experimental studies of TiO_2-B anodes showed a high percentage of reversibility and very stable cycling up to 100 cycles with almost no capacity retardation [10].

Some of the intercalation anodes like TiO_2 when sized down to nano exhibit an additional Li storage mechanism called as pseudo capacitance, a surface charge storage phenomenon, besides the usual intercalation of Li on the bulk. Usually, pseudo capacitance is comparatively less in the lower scan rates but its contribution increases linearly with the scan rate and becomes superior to the bulk storage contribution. Additional surface charge storage mechanism helps to obtain practical capacities nearly close to theoretical predictions [51]. Since it is a surface phenomenon, it is faster than the pure intercalation based charging/discharging. Intensive studies suggest that, this surface storage accumulation could be the formation of a highly reversible crystal phase within the vicinity of 3-4 nm from the anode surfaces. Increasing contribution from the pseudo capacitance with increasing porosity justifies this observation.

3.4 Other Intercalation Materials

Black phosphorus is another anode material with its high electrochemical performance as well as solvent stability. Theoretically, it can store 3 Li ions per phosphorous atom with a very high capacity of 2569 mAh/g. Structurally it is very close to the graphite morphology with its double layer structure along (0k0) plane [52]. Owing to the high demand for high capacity anodes, recently there is an increasing interest for the realization of phosphorus based anodes. Black phosphorous prepared by a high temperature and high pressure method showed relatively stable electrochemical cycling more than 60 cycles with capacities ~ 600 mAh/g [52]. Phosphorous as a nanocomposite together with carbon, resulted in anodes exhibiting better cycling stability up to 100 cycles. More noticeably, the specific capacity degradation was too negligible from the initial cycles [53]. Owing to the advantages obtained from graphite to graphene, the studies are now focused on the 2D graphene like phosphorene anodes.

In the vicinity of new anode materials for LIB, Nb_2O_5 is one of the best available anode materials. Though its theoretical capacities are limited to some 200 mAh/g, near 2 V working potential and long cycling durability made it a favourite choice of material where safety concerns are important [54]. In spite of the poor intrinsic electric conductivity of the bulk form, the nanostructures exhibit excellent pseudo capacitive lithium storage. For example, the hollow sphere, core shell, nitrogen doped porous structures etc. exhibited stable rate performances close to the theoretical predictions [55, 56].

4. Alloying Type Anodes

Owing to the very huge specific capacity values, alloying anodes are supposed to be the anodes for the future EV battery technology. Industry is exhibiting great interest in the realization of alloying anodes. LIBs working on partial alloying anodes are already commercialized for portable electronics. From the very first report of the possibility of electrochemical lithium alloying at room temperature, many metallic and semi-metallic materials were investigated as anodes for Li battery. Lithium alloying mechanism could be represented as

$$M + xLi^+ + xe^- \Leftrightarrow Li_xM$$

in which M represents the alloying material. In some of alloying oxide materials there could be a possibility of an *in situ* conversion of oxide phase into metallic phase which then cyclically functions as the alloying reaction. For example, in SnO_2 anodes it is usually observed that an initial irreversible conversion of the oxide to metallic phase further undergoes a cyclic alloying reaction with Lithium which would be further described in the following sections.

4.1 Silicon

As per the theoretical evaluations, the second most abundant and nontoxic material Silicon is the most suitable material for LIBs in terms of its highest theoretical capacity of 4200 mAh/g [57]. Fracture mechanical studies showed the mechanical stability of the nano sized silicon against the strain constraints confirming the feasibility of nanostructuring in realisation of practical Si anodes. Under normal temperatures the performance of an alloying silicon anode is highly dependent on the size and morphology of the nanoparticles and also related to the rate of operation.

Attempts to develop a reliable Si anode have been investigated and a number of nanostructures, have been found to be suitable candidates for rechargeable battery systems. Various structures like nanoparticles, nanotubes, nanowires, nanofibres were grown directly on the current collectors, hollow nanospheres, and composites with carbon allotropes to enhance structural properties [58-61]. Many of these designs put forward stable anodes having reliable capacity, stable cycling and other mechanical properties. Structuring of the anodes with spherical hollow spheres is found to be a reliable approach. Hollow structures effectively buffer the volume strain caused by 240% of volume expansion and reduce the SEI layer formation [52]. Interconnected sphere based anodes showed initial reversible capacities close to 2500 mAh/g maintaining stable capacities to more than 700 cycles. A pomegranate-inspired carbon silicon hybrid showed cycling stability more than 1000 cycles with a volumetric capacity of 1270 mAh/cm^3 and a capacity retention of more than 97% [62]. Moreover, the carbon coating over the matrix improved the lithiation/delithiation state of battery aiding high-energy density. Morphology and the galvanostatic rate performance of the anode are shown in Fig. 6.

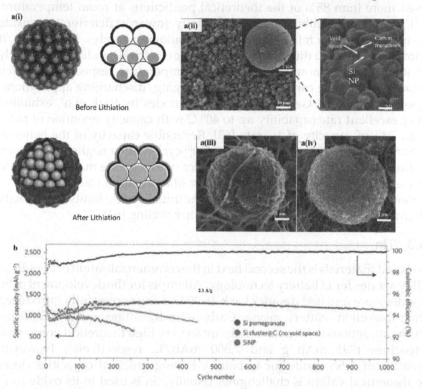

Fig. 6. (a) (i) schematic of morphological changes of Si-C composite on lithiation, (ii) SEM images of prepared composite (iii) SEM images of silicon composite after 100 cycles, (iv) SEI removed, (b) Galvanostatic cycling performance of silicon alone and composite anodes. (Copyright 2014 Springer Nature)

In spite of an extensive ongoing investigations, so far Si anodes are now available to replace the commercial anode in the very near future. EV industries like giant Tesla already announced their project with high energy density Si-based anodes. Few start-ups like Angstron Materials and Sila nanotechnologies are claiming that they would commercialize Si-LIBs in a time span of nearly 2 years with 20-40% of capacity hype over the traditional graphite-LIB.

4.2 Germanium

Germanium is one of the most competing alloying material with promising features to be a good performing LIB anode. According to the Li-Ge phase diagram, a germanium atom can alloy with Li by accommodating 1 to 4.4 Li. In the highest lithiated state, theoretically, Ge can have a capacity of 1620 mAh/g and an average volumetric expansion of 370%. Interestingly, the practical investigations verified that Li storage capacity of Ge can go even more than 85% of the theoretical predictions at room temperature. All the above, exhibits negligible affinity towards deteriorating oxide layer making Ge a reference material for alloying anodes. In addition, it shows very huge Li diffusivity and intrinsic electrical conductivity (nearly 15 and 10^4 times more than Si at room temperature respectively) which makes it a suitable candidate for fast charging/discharging applications. Specially designed Ge nanotubes based anodes by Park et al, exhibited very excellent rate capability up to $40°$ C with capacity retention of more than 70% of capacity of 1C rate [63]. Reversible capacity of the batteries was more than 1000 mAh/g even at 400th cycle with a negligible capacity fading at 1C rate. Superior characteristics of this anode make it excellent in electrochemical performance. Even *in situ* structural transformation of nanowire arrays to porous network in the initial cycles, limits the capacity decay to 0.01% of each cycle during further cycling [64].

4.3 Tin

Sn based materials is the second next in the commercialisation of the silicon alloy anodes for Li battery technology. Attempts for the development of Sn anodes were initiated decades back. In 2005, Sony successfully introduced their Nexelion battery using Carbon-Tin-transition metal-composite. Prime attractions of Sn as an anode are its very high theoretical volumetric capacities (993 mAh/g and 7,300 mAh/L, respectively). However, creation of a Sn anode for Li battery having practical capacities closer to theoretical values is challenging. Usually, Sn is used in its oxide form for the electrode applications which performs as an *in situ* conversion to metallic Sn followed by alloying reaction with Li. This *in situ* reduction of SnO_2 accompanied with the formation of Li_2O, causes a large irreversible capacity lose (711 mAh/g). However, Li_2O matrix effectively helps to buffer the high volume strain (260%) of Sn-Li alloy formation.

Till now various Sn anode formulations were reported but Sn-hollow carbon nanofiber composite containing nearly 70% Sn proved to be a stable design. Large void volume of the nanofiber is designed to buffer the volume changes of the alloying reaction. Its irreversible capacity of 810 mAh/g and more than 80% of capacity retention after 200 cycles verifies the suitability of the anode for commercial applications [65, 66].

4.4 Aluminium

Owing to the low cost and relatively uniform distribution on the earth, Al gained much attention as a possible Li anode material. Similar to other alloying anodes, Al can also form different states upon lithiation. Out of which $AlLi_{2.5}$ has the highest capacity of 2234 mAh/g. But more interestingly AlLi state has a theoretical capacity of 993 mAh/g with a volume expansion of less than 100%. Practical investigations also agree to the suitability of Al as a good performing anode material. An early stage study reported by Ming et al., used free standing Al nanorods grown directly on the current collectors as anodes of Li half cells. The average dimensions of the rods were confined to 80*200 nm (diameter * length) [67]. These structures were capable of giving a practical capacity of 1243 mAh/g at a current density 350 mA/g. Later Park et al. reported Al-C60 nano cluster hybrid anodes which are highly stable electrodes that showed cycling stability up to 100 cycles without much capacity retardation [68]. More noticeably when the cycling current density was increased from 6 A/g to 20 A/g the capacities maintained almost 50% of the initial (a drop from nearly 600 to 300 mAh/g). Very recently, commonly accepted yolk shell nanoparticle design demonstrated promising behaviour. Al nanoparticles covered with TiO_2 shells gave a very stable performance of up to 500 cycles with capacity retention of 99.2%. Capacities obtained were 1200 and 661 mAh/g at current rates of 1C (=1410 mA/g) and 10C, respectively [69].

5. Conversion Type Anodes

The third class of materials suitable for LIB anodes are conversion materials proposed by Tarascon et al. nearly two decades back. In these mechanisms, active material undergoes continues reduction to form the metallic phase reversibly. They exhibit very high reversible capacities and energy densities which are due to the efficient utilization of the oxidation states of the materials. Moreover, the high lithiation potentials ($\geq$ 0.5 V) makes these anodes resistant to the Li dendrite growth to a great extent. Nano porous structures or nanocomposites with carbon or similar materials is found to be very effective in reducing the undesirable characteristics to nominal limits. Usually oxides or sulphides are found to be effective conversion materials. In general, the conversion mechanism can be represented as

$$M_xN_y + 2yLi^+ + 2ye^- \Leftrightarrow xM + yLi_2N$$

Till now the reported conversion materials belong to transition metal oxides, nitrides, fluorides and sulphides. In addition to these, some poly-anionic systems and mixed transition metal compounds are also identified as possible conversion materials.

Co_3O_4 is one of the most promising conversion type materials which is being investigated by researchers all over the world and is soon to be commercialized. It is a highly stable anode which could deliver 8 electrons per mole on reaction with Li and achieve practical capacities very close to its theoretical capacity of 890 mAh/g. So far Co_3O_4 anodes were reported with different morphologies like nanosheets, nanoneedles, mesoporous structures, nano wires, nanorods etc. All these specific morphologies exhibited distinctive electrochemical performances. A deep microscopic investigation during lithiation/delithiation process of the Co_3O_4 anodes revealed a conversion mechanism to metallic Co upon lithiation accompanied with Li_2O formation. In simple words, the complete reactions can be represented as

$$Co_3O_4 + 8Li^+ + 8e^- \Leftrightarrow 3Co + 4Li_2O$$

A mesoporous cubical Co_3O_4 structure design by Guogon et al., is an effective method which can give very stable anodic performance. Individual cubes composed of irregular nanoparticles of size ranging from 20 to 200 nm and an average surface area of 5-10 m^2/g per cube were identified [70]. Polycrystalline mesoporous cubical anodes (Fig. 7) exhibited excellent discharge capacities of 1041 mAh/g and better columbic efficiency. Specific characteristics of the facile structure prevent the agglomeration of active material. The orderly nanoparticles increased the surface area promoting better electrolyte diffusion that enhanced the electrode-electrolyte interaction. Specially designed multi-shelled Co_3O_4 hollow microspheres also showed very distinguishable performances with reversible capacities more than 1500 mAh/g over 30 cycles of operation at a current density of 50 mA/g [71].

The performance of nano Co_3O_4 anodes has strong dependence with the preparation methods and parameters. Ge et al. synthesized Co_3O_4 from the hydrothermally prepared α-$Co(OH)_2$ precursor, the morphology and hence the Li storage capacity of the anodes varied considerably with the hydrothermal synthesis temperature [72]. Reaction mixtures prepared at 150 °C give rise to incomplete 2D structures consisting of collapsed nanoparticles. Whereas the reaction temperature increased to 210 °C, the Co_3O_4 matrix formed a mesoporous 2D sheet that showed better electrochemical performance and structural stability. Moreover, the structures effectively accommodated the volumetric changes related to strain effects.

In spite of the promising characteristics, the volumetric expansion and conductivity constraints hinder the recognition of Co_3O_4 anodes as good performers. Nano structuring to carbon incorporated composites is the most suitable and effective solution to overcome these drawbacks. In a unique synthesis approach, Yao et al. prepared a graphene-Co_3O_4 hybrid with integration of 3D graphene and mesoporous Co_3O_4 nano wires as anodes [73]. Hybrid material is distinguishable with its substantially higher electrical as well as electrochemical properties. Specific anode design helped in retaining the initial reversible capacities (812 mAh/g)

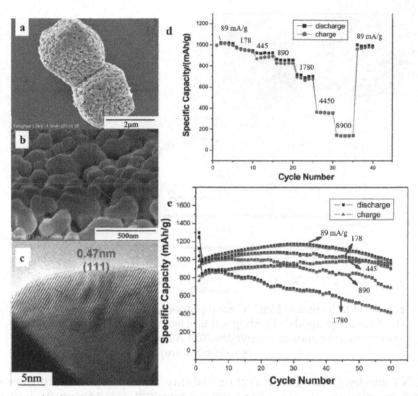

Fig. 7. (a-b) SEM images, (c) HRTEM image of mesoporous cubical Co_3O_4, (d) Rate performance of Co_3O_4 electrodes at various current densities (e) Cycling performances of Co_3O_4 electrodes at different rates. (Copyright 2014 American Chemical Society)

after 200 cycles with close to 100% of coulombic efficiency. These performances attributed to the four intertwining modes of the hybrid structure namely – seaming, bridging, hooping and bundling – revealed from the high resolution microscopy of the structures.

Fe_2O_3 having a theoretical capacity of 1007 mAh/g is a consistent conversion anode. It is already identified that the practical capacity and capacity retention of Fe_2O_3 anodes is highly dependent on the morphology of the nanoparticles design. For example, the hollow sphere nanoparticle anodes cycled at a current density of 200 mA/g retained a capacity of 710 mAh/g only. However, the spindle-like porous architecture could retain a capacity of 911 mAh/g after 50 cycles at a current density of 1 A/g [74, 75]. Similarly, a free standing membrane based Fe_2O_3 anode provided very stable cycling capacities of 600 mAh/g for over 3000 cycles with a capacity retention of more than 92% even at a current rate of 6C [76]. Carbon coating morphologies, though did not exhibit a hike in the capacity, showed excellent cycling and rate performances. Fe_2O_3 filled

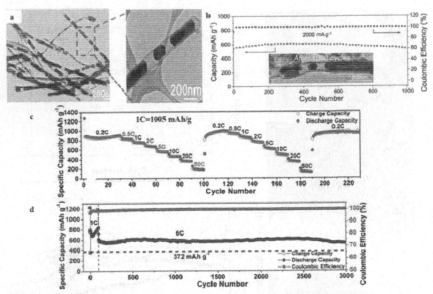

Fig. 8. (a) TEM images of MnO/C nano peapods, (b) cycling performance of the MnO/C nanopeapods (c) cycling stability and (d) rate performance of Fe_2O_3 nanomembrane anodes. (Copyright 2014 American Chemical Society, 2014 Springer Nature)

CNT anodes showed high cycling stability at very high current rates of 3 A/g with specific capacity of 400 mAh/g [77]. In addition, the carbon coating provided consistent electrochemical performances even without additional carbon additives to the electrode which opened a new channel for more active sites. In addition to Fe_2O_3, Fe_3O_4 is also a good performing anode material for LIB with its theoretical capacity of 926 mAh/g. Similar to Fe_2O_3, many attempts are reported to stabilise Fe_3O_4 anodes which are mostly used with modified forms of carbon additive. These attempts resulted in Fe_3O_4 anodes of reversible capacities nearly up to 750 mAh/g with sufficient cycle stability [78].

Another class of materials having similar capacity ranges and the non-toxicity attribute of iron oxide materials are the manganese oxides. Various polymorphs of manganese oxides have theoretical capacities ranging from 700 to 1200 mAh/g in which MnO_2 is dominant over others materials at least by 200 mAh/g [79]. Similar to the other conversion materials, to felicitate the conductivity issues and to buffer the volumetric strain, carbon based hybrids are found to have the most effective architecture for MnO_x anodes. Many of these structures exhibited outstanding performances in terms of both reversible capacity and cycling stability. One of the prominent hybrid MnO_x design is the nano peapod structure in which hollow MnO is coated with a carbon layer. Specially designed structures demonstrated non-degrading capacity of nearly 850 mAh/g for over 1000 cycles at a current density of 2 A/g [80]. Recently, Yuan et al. proposed another type of anode in which MnO is incorporated between the RGO layers [81]. These designs exhibited outstanding performance with a reversible capacity of 379 mAh/g after 4000 cycles when cycled at 15 A/g. With the increase in current rate to 40 A/g the capacities retained were 331 mAh/g. Other than these many alternative structures like mini hollow- thick wall Mn_2O_3 polyhedron, MnO nano crystallites, Mn_3O_4 nano fibres interconnected MnO_x nano wires etc. have also been investigated [82-84].

6. Positive Electrodes

In the past decades, many efforts have been devoted to develop high capacity and high voltage stable cathode materials for electric vehicle batteries. Cathode material generally should have a low diffusion barrier, good electrical conductivity, high specific capacity, and excellent structural stability at higher potentials during lithiation/delithiation [85-87]. Goodenough et al. revolutionised LIB research by introducing $LiCoO_2$ (LCO), one of the commercially successful layered transition metal oxide cathodes. Generally, the cathode materials for commercial rechargeable batteries mainly involve transition metal oxides or phosphate based active materials. However, chemical and structural instability in the charged state accompanied with high cost, toxicity and limited capacities are the issues to be addressed for commercial application of LCOs in EVs. $LiMn_{1.5}Ni_{0.5}O_4$ with capacity comparable to that of $LiCoO_2$ turns out to be a promising candidate for commercialisation due to the high working potential of ~4.7 V (~4.1 V for $LiCoO_2$). $LiMnO_2$, and $LiMn_2O_4$, $LiNi_{0.3}Mn_{0.3}Co_{0.3}O_2$ has also attracted serious attention due to its excellent Li insertion/extraction properties [8, 87, 88]. As discussed earlier, for strengthening the LIBs for EVs the specific capacity of cathodes must be further improved to increase the power and energy densities. Various strategies like layered, spinel and olivine structured materials have been employed to overcome these limitations of conventional cathode material [88-91]. This section focuses on state-of-the-art cathodes used in current LIBs, future high voltage cathode materials and energy storage models beyond the LIB.

6.1 Layered Oxides

Layered oxides with the formula $LiMO_2$ (M=Co, Ni, Mn) have been the most widely deployed cathode materials for commercial LIBs. As discussed, $LiCoO_2$ is the parent compound of this family which was reported as intercalation material by Goodenough et al. for rechargeable LIBs in 1976 [92]. $LiCoO_2$ forms a distorted rock-salt structure in which cations align themselves in alternative [111] planes resulting in a trigonal structure. Crystal structure of layered $LiCoO_2$ forms space group *R-3m* with lithium and cobalt ion along with cubic close-packed oxygen ions. In this structure, the interstitial layers of Li separate into three slabs for edge sharing with CoO_6 octahedral as shown in Fig. 9 [8]. During the discharge state, LCO remains in the layered structures with hexagonal unit cells. Whereas in the charged state, Li^+ from the layered crystal lattice forms a nonstoichiometric $Li_{1-x}CoO_2$ with Co^{3+} oxidation for charge compensation.

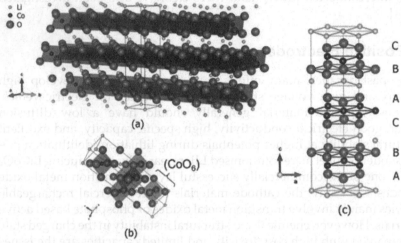

Fig. 9. (a) Layered crystalline structure of the rhombohedra $LiCoO_2$, (b) representation of the octahedral CoO_6 structure; (c) stacking arrangement of the layers (ABCABC). (Copyright 2017 SciELO)

A LCO with a theoretical capacity of 280 mAh/g can also be operated at high potential windows of up to 4 V limiting its practical capacity to 140 mAh/g (Fig. 10). Therefore, $LiNiO_2$, $LiMnO_2$ and $LiMn_2O_4$ layered structures have replaced earlier cathodes as effective replacements due to their Li insertion/extraction properties [8, 93]. $LiNiO_2$ (LNO), isostructural to LCO with the ideal *R-3m* space groups and layered structure proved to be more stable while charging at low lithium loadings with high reversible capacities. Coupling behaviour of $Ni^{3+/4+}$ with high chemical potential towards Li enables extraordinary cell potential of around 4 V [94].

Lithium manganese oxides ($LiMnO_2$) with lower cost, less toxic, safer on overcharging, and environmentally benign material combined the merits of MnO_2 along with layered oxides [95-101]. Layered oxide $LiMnO_2$ existed in zigzag-type orthorhombic and monoclinic crystal structures with *Pmmm* and *C2/m* symmetry space groups respectively [102]. $LiMnO_2$ proved to be electrochemically active material by delivering a very high theoretical capacity of 285 mAh/g. Recently, few reports exhibited a

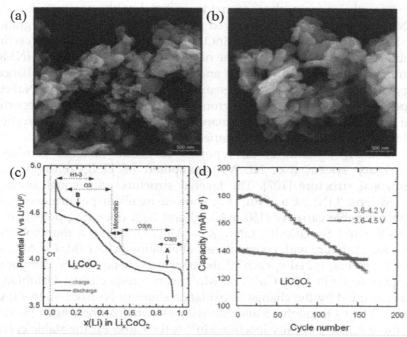

Fig. 10. (a-b) SEM Images of LiCoO₂, (c) electrochemical features of the LiCoO₂ vs Li⁺ cycled in the voltage range 3.0–4.8 V, (d) cycling stability of LiCoO₂ compared at different voltage intervals. (Copyright 2003 Elsevier)

LiMnO₂ layered compound with different synthesis methods. Thackeray et al. prepared LiMnO₂ from Li₂MnO₃ by leaching Li₂O from the rock-salt structure (Li₂O.MnO₂) with acid, subsequently relithiating the MnO₂ with lithium to attain more stability [103, 104]. These methods lead to increase in small amount of lithium (9%) within the Mn layers resulting in Li₁.₀₉Mn₀.₉₁O₂, thus closely resembling the structure of LCO and LNO. Monoclinic LiMnO₂ on cycling in between 3-4 V vs Li⁰/⁺ transforms to a more thermodynamically stable spinel phase. Cation doping (Ni²⁺) was used to stabilize the layered phase which exhibited promising behaviour for high energy applications.

Ohzuku et al. synthesized a binary LiNi₀.₅Mn₀.₅O₂, an intermediate compound between LiNiO₂ and LiMnO₂ with similar properties and practical capacity of ~140 mAh/g [93, 105]. Structure of LiNi₀.₅Mn₀.₅O₂ is believed to be in the form of layered LiMO₂ with the Li ions in between the metallic layers. Layered LiNi₀.₅Mn₀.₅O₂ adopts a hexagonal unit cell with O₃ phase identical to LCO. During lithiation and delithiation Mn retains its oxidation state of +4 and Ni in +2 state with electrochemically active ions. X-ray absorption spectroscopy confirmed that Mn⁺⁴ keeps the structure stable even at working potentials of around 3.6-4.3 V [102].

Recently, a new class of oxide material with a composition of $LiNi_{1-x-y}Co_xMn_yO_2$ (NCM), received great attention due to the integration of $LiNiO_2$ (high capacity), $LiMnO_2$ (low price) and $LiCoO_2$ (cycling stability). Liu et al. synthesized the new layered Li (Ni, Mn, Co)O_2 (NMC) compound having high thermal and structural stability with enhanced capacities. $LiNi_{1-x-y}Co_xMn_yO_2$ resembles the structure of alpha-$NaFeO_2$ type structure with $R\overline{3}m$ space group [93, 102, 106]. Dahn et al. reported the high electrochemical performance of this compound with excellent calendar life required for EV batteries.

In 2001, Ohzuku et al. reported an oxide material consisting of ternary system (Co, Ni, Mn) composed of $LiNi_xCo_xMn_xO_2$ with hexagonal structure [107]. The layered structure with equal amounts of TM ions, $LiNi_{1/3}Co_{1/3}Mn_{1/3}O_2$, showed excellent performance with high reversible capacity (160 mAh/g) and high operating potentials of (~4.6 V) [108]. Specifically, $LiNi_{1/3}Co_{1/3}Mn_{1/3}O_2$ adopts a rhombohedral rock-salt structure with consecutively alternating $[MO_2]^-$ (M=Co, Mn, Ni) and Li^+ layers. Initially, each of the transition metals of the compound happens to be in Ni^{2+}, Co^{3+} and Mn^{4+} states respectively. Delithiation was preceded by the change in oxidation states of Ni from +2 to +4, Co from +3 to +4 through +3 and Mn with +4 remains unchanged [8, 109]. So this electrochemically inactive Mn^{4+} is the cause for the stable crystal structure. Specific capacity of this cathode is mainly dependant on active redox species like Ni-content which can be improved to attain high energy densities. Considering these facts, a cathode material with Ni-rich phase was structured with $LiNi_{0.68}Co_{0.18}Mn_{0.18}O_2$, which is suitable for EVs and HEV batteries [110, 111].

6.2 Spinel Oxides

Early in 1980s, Michael Thackeray proposed spinel oxides with the chemical formula LiM_2O_4 as an effective cathode for LIB. $[M_2]O_4$ structure from the spinel LiM_2O_4 is an attractive host network for movement of Li during the electrochemical reactions. This spinel structure offers a 3-D framework of face-sharing tetrahedral and octahedra with a cubic symmetry $Fd\overline{3}m$ [105] for an effective diffusion process. In spinel (M_2) O_4, 75% of the metal cations occupy alternate layers between the ccp oxygen planes while the remaining 25% are located in the adjacent layers. Therefore, in each layer there exist sufficient M cations for maintaining an ideal ccp oxygen array during delithiation to deliver high specific energies. Mostly the crystal structures of the 3-D network expand and contract isotropically during the insertion and extraction of Li that leads to stable electrochemical cycling. Also the spinel compounds with $Li_4M_5O_{12}$ (or $Li(M_{1.67}Li_{0.33})O_4$) composition can be tolerant to cycling for the expansion and contraction of the cubic unit cell by less than 1% of controlled compositional limits.

Spinel $LiMn_2O_4$ is the most attractive compound of the lithium spinel's for its abundant availability, low cost, excellent rate capability and also environmental friendliness. $LiMn_2O_4$ is commonly synthesized at higher temperatures (above 700°C) to achieve a highly crystalline product. As reported, $LiMn_2O_4$ occupies the cubic spinel structure (Fig. 11) with $Fd\bar{3}m$ space group [8]. Anion lattice contains the closely-packed array of oxygen ions which is close to rock-salt layer structure, with different positions with the distribution of cations [112]. In $LiMn_2O_4$ oxide, ¼ of the octahedra sites in the lithium layer are occupied by the Mn cation, which remain vacant. These vacant octahedra sites are shared by the lithium

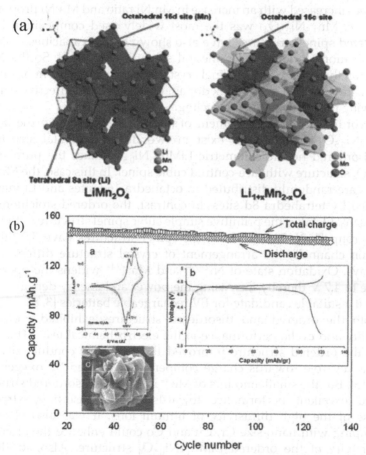

Fig. 11. (a) 3-D crystal structure of $LiMn_2O_4$ compounds, (b) cycling data reflecting the high stability of $LiMn_{1.5}Ni_{0.5}O_4$ at elevated temperature, where (a) typical SSCV curves of pristine and aged $LiMn_{1.5}Ni_{0.5}O_4$ spinel electrodes (70 C), (b) voltage profiles and (c) SEM images of $LiMn_{1.5}Ni_{0.5}O_4$ particles. (Copyright 2011, 2015 Royal Society of Chemistry)

ions within the TM layer since they occupy the tetrahedral sites. This cathode material is intrinsically highly stable and its 3D diffusion process for Li ions provides the unique MnO_2 spinel framework [8, 112]. This 3D diffusion path in $LiMn_2O_4$ provides superior rate performance and cycling stability, provides a stable high potential of 4 V versus lithium. Additionally, it offers a stable cubic $[Mn_2]O_4$ network over the whole 4 V compositional range $0 \leq x \leq 1$ in $Li_xMn_2O_4$.

Some of the transition metal oxides like iron, cobalt and nickel have been used as dopants with $LiMn_2O_4$. However, the addition of nickel into the crystal structure has resulted in promising cathode material for high power applications. Capacities were dependant upon the Mn content, and capacities increased with an increase in Mn:Ni ratio and Mn:Ni proportion of 3:1 (i.e. $Mn_{1.5}Ni_{0.5}O_4$) was the most widely used composition [113]. Disordered spinel structures have also shown higher capacities with Mn and Ni cations ordered in octahedral sublattice [113-115]. So the Ni as dopant on the surface of the spinel crystal structure in the form of a nano-sized coating instead of as a bulk dopant proved to be an effective way for improving the capacity during cycling.

According to the arrangement of Mn^{4+} and Ni^{2+} ions in the lattice, $LiMn_{1.5}Ni_{0.5}O_4$ (LMNO) can exist in two different crystal structures. The disordered nonstoichiometric $LiMn_{1.5}Ni_{0.5}O_4$ takes the pure spinel $LiMn_2O_4$ structure with face-centred cubic spinel. In this case, the Mn and Ni ions are randomly distributed in octahedra 16d sites and Li ions are occupied by tetrahedra 8d sites. In contrast, the ordered stoichiometric $LiMn_{1.5}Ni_{0.5}O_4$ takes the primitive simple cubic spinel structure with $P4_332$ space group [112]. Though both the spinel structures have 3D lithium diffusion channels, the arrangement of crystal structure differs in the pathways. Oxidation state of $Ni^{2+/4+}$ and $Mn^{3+/4+}$ widens the operating voltage to 4.7 V thereby increasing the power and energy density which makes it a suitable candidate for EV rechargeable batteries [3, 8].

Both the ordered and disordered structures exhibited good rate capability and cyclic performance in the electrochemical tests. The Mn^{3+} ion in the ordered structure improved the electronic conductivity and plays a key role towards charge compensation caused by oxygen loss [114-116]. But the small amounts of Mn^{3+} ions in the disordered structure showed excellent performance towards lithium insertion/extraction because of the high diffusivity of lithium ions. It was also observed that doping with nano-size Cr, Cu and Co could enhance the electronic conductivity of the ordered $LiMn_{1.5}Ni_{0.5}O_4$ structure. Also, Ru-doped $LiMn_{1.5}Ni_{0.5}O_4$ displayed excellent electrochemical performance even at higher rates with longer life [117]. However, stoichiometric composition of $LiNi_{0.45}Mn_{1.45}Cr_{0.1}O_4$ with Cr doping improved the disordering of the crystal structure [116].

6.3 Olivine Phosphates

Another promising class of cathode compounds are built with polyanionic groups such as $(SO_4)^{2-}$, $(PO_4)^{3-}$, $(MoO_4)^{2-}$ or $(WO_4)^{2-}$ which are considered potential electrodes for commercial LIBs [118]. Due to their stability towards overcharging and short circuits and an ability to resist high temperatures made phosphates unique. Polyanions $(XO_4)^{n-}$ arrangement drops the redox energy of the 3d-metals to suit the Fermi level of the Li anode. Olivine phosphates structure with composition $LiMPO_4$ have phosphorous and metal (M) occupying tetrahedral and octahedral sites respectively while lithium forms 1-D chains along the [010] direction. So far, in this family, olivine phosphate and Nasicon-like networks arepresently the subject of research. Among the olivine phosphates, widely used cathode is lithium iron phosphate ($LiFePO_4$) which was proposed by J.B. Goodenough and co-workers in 1997 [4].

In particular, $LiFePO_4$ (LFP) has received much more attention due to its high specific capacity (~170 mAh/g) at moderate current densities. Additionally, LFP offers low volume expansion, improved coulombic efficiency, flat charge-discharge plateau, low capacity fade and environmental compatibility [90, 118]. Figure 12 shows the structure of $LiFePO_4$ that belongs to the *Pnma* space group, with Li^+ and Fe^{2+}

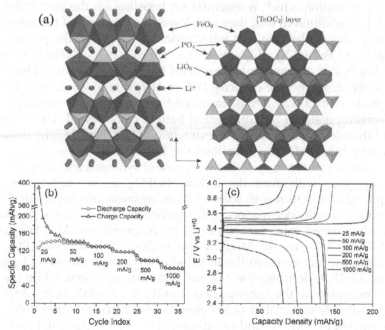

Fig. 12. (a) Typical olivine structure of $LiFePO_4$, (b) galvanostatic rate performance, (c) voltage profiles at different rates for $LiFePO_4$.

occupying octahedral 4a and 4c sites respectively. However, the P atom is located in tetrahedral 4c sites in a slightly distorted hexagonal close-packed (HCP) oxygen array. Herein, LiO_6 octahedra form a linear edge-shared chain while FeO_6 octahedra shares edges with two LiO_6 octahedra and one PO_4 tetrahedra. So the removal of oxygen is protected due to the strong covalent bonding between oxygen and phosphorous atoms in the phosphate [90]. Due to the inductive effect of the Fe-O-P, the olivine material gains the higher potential by lowering the Fermi level of $Fe^{2+/3+}$ redox couple. The unique structure of polyanion PO_4^{3-} unit stabilizes the LFP olivine compound whose practical capacity approaches close to its theoretical value of 170 mAh/g [4, 30]. Also, upon lithiation/delithiation the volume change is limited to 6.8% which gives superior cathode stability during charging and discharging of batteries. This extreme property of olivine LFP makes it a suitable candidate for EVs and HEVs [119].

In 1997, Goodenough et al. reported a new class of NASICON (sodium super ionic conductor) based cathode materials for the first time [120]. NASICON materials are a family compounds with $M_2(XO_4)_3$ structures made up of XO_4 tetrahedra sharing corners with MO_6 octahedra. This $M_2(XO_4)_3$ host network is chemically versatile that can be stabilized with a different transition metal cation M and polyanions. The $(XO_4)^{n-}$ polyanions not only form a stable network, but their structure can also support fast Li-ion conduction which is essential for boosting up the power density [3, 121]. In addition to this, they also lower the positions of the transition metal redox couples due to a polarization of oxygen in the polyanionic complex. $Li_2FeTi(PO_4)_3$, $Li_2NaV_2(PO_4)_3$, $LiFeNb(PO_4)_3$, $LiNa_2FeV(PO_4)_3$, based cathodes are main members with NASICON structure. However, due to the high specific capacity (197 mAh/g), monoclinic $Li_3V_2(PO_4)_3$, a derivative of $Li_2NaV_2(PO_4)_3$ reported by Nazar et al. is considered as a promising material for commercial battery cathodes [3]. This material with intrinsic properties, due to extraction of the three Li-ions from the structure, unique 3D network allows fast Li-ion migration leading to high power densityand long cycle-life with ease of synthesis and upscaling.

Triphylite with formula $Li(Mn, Fe)PO_4$, is an orthophosphate compound rich in iron with some of its manganese ions in M2 sites. As a function of the site occupation by magnetic ions, these olivine structures are distinguished into three classes. The magnetic ion (Mn, Fe) lies in the M1 and M2 sites for Mn_2SiS_4 and Fe_2SiS_4 while for $NaCoPO_4$ and $NaFePO_4$, the magnetic ion lies in M1 site only. The third class phospho-olivine $LiMPO_4$ occupies the M2 site with magnetic ions while the M1 site is occupied by the non-magnetic ion (Li^+). $LiMnPO_4$ is an attractive olivine cathode material in Li batteries for its low cost, higher voltage of 4.1 V caused by $Mn^{2+/3+}$ and its sustainability for the electrolyte solution [121]. $LiMnPO_4$ crystallizes into an ordered olivine structure indexed by orthorhombic $Pnmb$ space group. Doping of Fe in the matrix $LiMnPO_4$

to obtain $LiMn_yFe_{1-y}PO_4$ (LMFP) showed high performance (160-165 mAh/g) due to its wide potential of 3.4-4.1 V which is not high enough to decompose the organic electrolyte. Carbon-coated nano-$LiMnPO_4$ was also proved to be a very stable material with a flat redox potential of 4.1 V. Cobalt with Lithium cobalt phosphate (LCP) in the olivine structure also offered a promising behaviour of crystallizing with orthorhombic symmetry (*Pnma* space group). LCP with stability up to 4.8 V exhibited good theoretical capacity (167 mA/g) with minimum structural volume changes [122].

6.4 Next Generation Intercalation Cathodes

6.4.1 Lithium-Rich Layered Materials

In recent years, lithium-rich layered oxides are delivering high specific capacities at a high operating potential windows compared to stoichiometric layered $LiMO_2$ oxides which are much more suitable for high power applications like EVs. Structurally compatible Li-rich $xLi_2MnO_3.(1-x).LiMO_2$ (LR-NMC) compound was developed from the parent $LiMO_2$ lamellar structure by a simple substitution of excess Li^+ for M^{3+} in the [MO_2] layers (Fig. 13) [102, 123]. Composites of layered Li_2MnO_3 and $LiMO_2$ compounds are most suitable for designing electrodes in which Li is cycled in and out of the layered $LiMO_2$ component. Due to the symmetry of Li_2MnO_3 and $LiMO_2$ within the framework, the resulting composite holds the Li^+ ion layer alternate to layers of Li^+, Mn^{4+}, and M^{3+} ions in a cubic close-packed oxygen array. Existence of Li^+ and Mn^{4+} in the transition metallic layers moderates the structure of Li_2MnO_3 from the $R\bar{3}m$ group to C2/m.

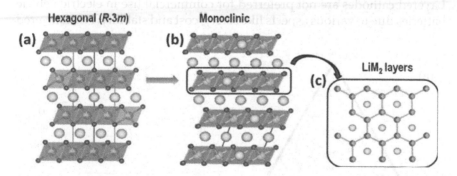

Fig. 13. Structural representation of (a) O_3-type layered oxides; (b) the overall cell of Li-rich layered oxides described as monoclinic; and (c) M / Li ordering within LiM_2 layer leading to a honeycomb pattern. (Copyright 2015 The Electrochemical Society)

Structural compatibility and stable capacity retention was provided by electrochemically inert Li_2MnO_3 maintaining stability of the electrode at a high voltage [124, 125]. Structure similar to the arrangement of Li_2MnO_3 was used as the partial substituents in the lithiated transition metal oxides further improve the performance. Moreover, nano Li_2MnO_3 stabilised with $Li[Mn_xNi_yCo_z]O_2$ electrode revealed excellent electrochemical behavior above 4.5 V. Introducing excess Li ions beyond the limitation of one Li ion per MO_2 formula in this material attained exceptionally high reversible capacities and energy densities (>300 mAh/g and 900 Wh/kg respectively) when cycled between 2–4.8 V.

Lithium-rich (LR-NMC) layered oxides with general formula $xLi_2MnO_3.(1-x).LiMO_2$ have been exploited for synthesis of several compositions and high capacities were obtained for x>0.3. Jin et al. designed a phase diagram comprising of $LiNi_{1/2}Mn_{1/2}O_2$, $LiCoO_2$ and Li_2MnO_3, as shown in Fig. 14. By using their positions in the phase diagram the layered oxide composition can be expressed as $Li[Co_x(Li_{1/3}Mn_{2/3})_y(Ni_{1/2}Mn_{1/2})1-x-y]O_2$ with +4 oxidation state of Mn.

From the X-ray diffraction (XRD) shown in Fig. 14(b), the patterns at different locations in the triangular phase could be indexed to a single phase of alpha-$NaFeO_2$ type with $R\bar{3}m$ space grouping. Thus by calculating the lattice parameters existing in these structures, the structural properties can be varied accordingly with composition [124-127]. Typical structural designs enabled thermal stability and cycling capacities even at elevated temperatures and higher potentials making this material a promising candidate for EVs.

6.4.2 *Nickel-Rich $Li[Ni_xCo_yMn_z]O_2$ Layered Materials (x>0.5, x+y+z=1)*

Layered cathodes are not preferred for commercial use in electric vehicle batteries due to various aspects like safety, cost and stability. NMC layered

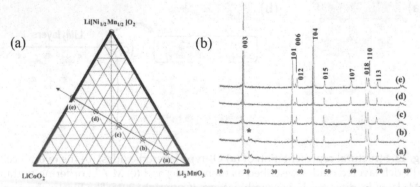

Fig. 14. (a) A new triangle phase diagram of $Li[Ni_{1/2}Mn_{1/2}]O_2$–$LiCoO_2$–Li_2MnO_3 system. (b) XRD patterns for samples (a–e). (Copyright 2005 Elsevier B.V.)

materials showed high specific capacity compared to other cathodes, depending on the Ni-content as the main active redox species. So nickel-rich layered lithium TMOs, have been under intense investigation as high-energy cathodes for electric vehicles for their high specific capacity and relatively low cost. Though the Ni-rich materials $Li[Ni_{0.8}Co_{0.1}Mn_{0.1}]O_2$ (NCM) have safety issues, their high practical capacity of 184 mAh/g with excellent capacity retention makes them attractive for use in high energy batteries [102, 123].

Stability and electrochemical performance of the NCM material are highly dependent on the synthesis method, degree of crystallinity, cation disorder and phase purity. Synthesis of low Ni-content NCM materials can be achieved through sol-gel, solution combustion, co-precipitation and spray pyrolysis methods but it is difficult to synthesize Ni-rich materials due to incomplete oxidation of Ni^{2+} to Ni^{3+}. Recently, a core shell nanostructured $Li[(Ni_{0.8}Co_{0.1}Mn_{0.1})_{0.8}(Ni_{0.5}Mn_{0.5})_{0.2}]O_2$ material was reported by Yang-Kook sun et al., via a co-precipitation method to improve the safety and cycle life of the battery. In these materials the core shell with $Li[Ni_{0.8}Co_{0.1}Mn_{0.1}]O_2$ delivers a high discharge capacity while the outer shell with $Li[(Ni_{0.5}Mn_{0.5})_{0.2}]O_2$ provides the structural and thermal stability in deep delithiated states. Herein, the level of Li^+/Ni^{2+} disorder decreases with an increase in Ni^{2+} content, thereby maintaining good structural stability and electrochemical performance. Ni-rich NCM material $Li[Ni_{0.8}Co_{0.1}Mn_{0.1}]O_2$ proved to be a suitable cathode for hybrid and electric vehicle batteries [126, 128].

7. Challenges

Although the above discussed electrode materials have gained more attention in the field of lithium-ion batteries, some of the major challenges associated with these materials restrict their application in electric vehicles. Achieving a good cell cycle life requires the electrode host compounds to demonstrate a fairly good structural stability and minimal volume change over the entire Li insertion/extraction operational voltage range [129]. There is a high likelihood of future utilisation of high capacities and attributes of the nanostructured materials with further alterations to make them suitable for EVs [3-7, 10-12]. Principal challenges facing for the development of batteries for electric vehicles are cost, safety, power density, rate of charge/discharge, stability and service life.

7.1 Structural Stability

Although the efficiency of energy conversion for LIBs depends on various factors, their overall performance strongly relies on the structural properties of the electrode material. Development of advanced LIBs to meet the emerging EV market demands was mainly dependent on the stability of the lattice at higher potentials [129]. In the past two decades, graphite and LTO as anodes with layered $LiCoO_2$ and olivine $LiFePO_4$ as cathodes have been widely used in the lithium battery market. However, in spite of all the improvements LIBs face challenges preventing their use as a market energy resource today.

Application of graphite as an anode in LIBs has become a technological bottleneck. Owing to the structural limitation, 6 carbon atoms are required to hold one lithium atom which eventually lowers the lithium storage capacities to less than 372 mAh/g. Presently nano structuring is also not beneficial for improving the capacity limits, since in the nano regime, Li intercalation voltage drops down and hence results in limited control over the lithiation. This eventually promotes unfavourable lithium plating compared to storing the guest atoms in the interlayers [10]. Being a well exploited material many attempts are going on in the field but have not yet come up with consistent outputs. The other carbon materials, that usually exhibit better lithium storage, for example, graphene, RGO, and mesoporous structures to mention a few, usually have the advantage of very large lithium storage locations in the form of surface defects or

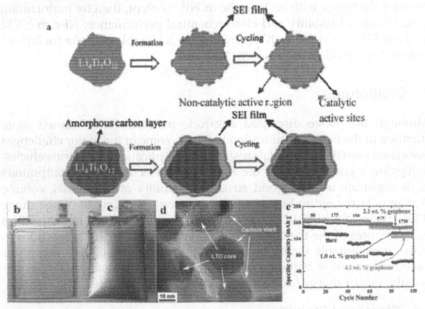

Fig. 15. (a) Schematic of surface activity of coated and uncoated LTO, (b-c) photographs of LTO cells as prepared and after 1000 cycles, (d) HRTEM carbon coated LTO, (e) rate performance of bare and graphene-wrapped LTO. (Copyright 2017 The Electrochemical Society, 2014 WILEY-VCH and 2014 The Royal Society of Chemistry)

vacancies. However, viably these defect locations also act as sources for electrochemical side reactions that could cause thicker SEI films, and in effect larger irreversible capacity lose [13, 130]. Recently, composites with other carbon materials or active materials like metal or metal nanoparticles were found to be very stable in an approach to reduce the irreversible capacity loss and to improve the overall performances [30, 131].

The second most promising anode LTO, in spite of all its advantageous characteristics faces a severe drawback of gas evolving reactions (Fig. 15). Detailed analysis of LTO anodes showed that, the bare LTO surface acts as sources for electrolyte reduction and it causes the gas release as byproducts mainly H_2 and considerable amounts of CO and CO_2 [132]. Usually, additional coatings of optimum thickness by carbon or ZnO nanoparticles could effectively shield the LTO surface from the gas evolution reactions [133]. Usage of suitable electrolyte additive was also adopted to hinder the gassing phenomena.

Alloying and conversion materials share some typical drawbacks as possible LIB anodes. Most noticeable effect is the drastic volume change associated with the lithiation reaction. For example, silicon experiences nearly 400% of volume change on complete lithiation. Similarly, Al -100%, Sn- 260% and all other members of the group also exhibit substantially higher volumetric expansion. High strain on volumetric change leads to pulverisation of the anodes and delamination led electrical isolation. This repeated pulverisation would disturb the SEI and hence results in unstable SEI layer formation [134, 135]. All these factors together cause very fast capacity degradation in a few cycles of operation (Fig. 16). The second noticeable challenge is the agglomeration effect. Metallic nanoparticles in the nanoscale usually have the tendency to move each other and to agglomerate during the electrochemical cycling. This is largely dependent on the ductility of the metal phase. Unfavourable agglomeration usually disturbs the anode matrix and causes gradual capacity fading [136]. In addition to these the conversion materials, in general owing to the stable electronic configurations exhibit poor conductivity that can impact on the rate performance of the lithium battery [137]. Usually, the active materials of these classes stimulate more than one of the above mentioned drawbacks hence, it would be better to treat the solutions in general. The most plausible approach would be a specifically designed active material – carbon hybrid. More than a simple mixture of both phases a combination with proper hierarchy could only provide the expected results. For example, in the case of Si anodes, once the silicon is made into a composite, the carbon would perform as a buffer to accommodate the volumetric expansion effects. All allotropes of carbon are being studied in this respect i.e., graphene, CNT, CNF, RGO, mesoporous structures etc. [17, 138] . If it is a metal oxide based conversion material, the carbon, with buffer action would enhance the conductivity forming a good anode combination.

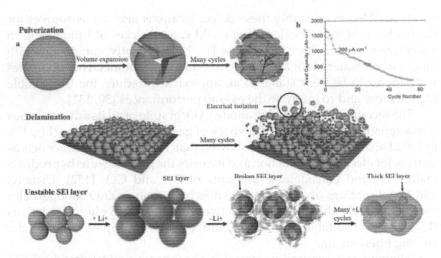

Fig. 16. (a) Challenges of Si anodes, (b) cycling performance of bulk Si anodes. (Copyright 2016 Springer Nature, 2015. American Chemical Society)

Other than the carbon, highly branched polymeric binders or core shell designs on other materials like TiO_2 are also attempted as possible solutions to deal with the volume strain assisted anode deformation [139, 140]. Embedding with carbon or highly stable polymer matrix is also found to provide satisfactory results in resisting the agglomerative effects too. Possible solutions to all of the new generation anode systems could only be achieved with the nano structuring technique. This is because, in the macro scales it is still difficult to solve the challenges effectively.

Similarly, several factors affect the matrix of cathode materials like cation disordering, Jahn–Teller distortion, slow manganese dissolution into the organic electrolyte, lattice instability and increased particle size of active material. $LiCoO_2$ was the first and one of the most promising cathode materials with fair electrical conductivity and Li^+ mobility. A strong bond of the low-spin $Co^{3+/4+}$ couple for the octahedral sites provides good structural stability and high electronic conductivity for Li_xCoO_2. However, only 0.5 Li+ ions per LCO formula ($x<0.5$) could be reversibly delithiated during cycling that leads to practical specific capacity of 140 mAh/g, which is only 50% of its theoretical capacity (280 mAh/g). Capacity fade and limitation of practical capacity was attributed to an ordering of lithium ions and consequent structural distortions from hexagonal to monoclinic transformation around $x<=0.5$ [141]. Li_xNiO_2 iso-structural with LCO has reversible capacity higher than that of latter, since the amount of lithium extracted during lithiation is 0.55 ($x>0.55$). Upon delithiation, Li_xNiO_2 undergoes several phase transformations accompanied by substantial volume changes that degrades the material integrity and compresses the cycling stability. Additionally, during the synthesis, the migration of Li ions to the TM layers was difficult to control due to which cation disordering occurred. In order to compensate the structural stability, Mn^{4+} was substituted for $LiNiO_2$ as $LiNi_{0.5}Mn_{0.5}O_2$. This structure was believed to be iso-structural to layered $LiMO_2$, but then the presence of 8-10% Li/Ni disorder during the initial synthesis alters the crystal structure to unstable pseudo-spinel. Since the major issue is with Li/Ni disordering in $LiNi_{0.5}Mn_{0.5}O_2$ compounds, Ohzuku in 2001 developed a new composition of $LiCo_xNi_xMn_{1-x-y}O_2$, and the presence of Co ions in the structure is believed to reduce the disordering to 6% for $x=y=1/3$.

However, the application of spinel $LiMn_2O_4$ as cathode also faces some issues with the dissolution of manganese from the spinel electrodes via a disproportionation reaction $(2Mn^{3+} \rightarrow Mn^{2+} + Mn^{4+})$ at higher potential, and Jahn-Teller distortion during deep discharge. The presence of Jahn-teller (J-T) active ions in the spinel-manganese oxides led to the structural distortions. Hence, irreversible crystallographic transformations from cubic spinel to distorted tetragonal with 6.5% increase in volume of unit cell was caused by the J-T Mn^{3+} ions [8, 142]. Numerous techniques have been put forward to improve the lattice stability by suppressing the Jahn-Teller distortion and Mn dissolution. For these, various divalent, trivalent and tetravalent dopants such as Ni, Al, Ti, Fe etc. played a crucial role to suppress these J-T active Mn^{3+} ions [143]. In the case of $LiMn_2O_4$, better stability can be achieved by coating the electrode materials with nanosized stabilizing surface layer that alleviates these issues.

Low electronic conductivity for the layered and olivine compounds is the other major limitation. Slow 1D migration of Li ions was limited by the close-packed hexagonal oxygen atoms that resulted in the low conductivity. Different techniques have been adopted to circumvent the low conductivity of these materials: reduce particle/grain size to nanoscale, applying a conductive layer like carbon and cation/anion doping. Synthesis of LFP through conventional solid state methods results in the larger particle size but by the novel techniques like hydrothermal and sol-gel, LFP nanoparticles can be synthesised. Huang et al. reported that the enhancement in the electrical conductivity can be achieved through coating carbon nanoparticles. Reducing the particle size to nanosize and carbon coating on the particle surface reduces the tap density of LFP that lowers the volumetric energy density. Additionally, the flat discharge plateau at ~3.4 V is relatively lower than the high voltage cathode material that limits its power density from the 800 W/kg which is ideally required for EVs.

7.2 Electrochemical Challenges and Nano-structuring

LIBs have been intensively studied for use as power supplies of electric vehicles which require high energy and power density. They have attractive power-storage capability owing to their high energy density. However, their power density is relatively low because of a large polarization at high current rates. This polarization is caused by slow lithium diffusion in the active material and increase in the resistance of the electrolyte when the charging-discharging rate is increased. However, the above discussed structural instability within the materials restricts these electrodes from meeting the high energy and power densities for EV rechargeable batteries. Except for a limited number of materials, almost all anodes show a severe drawback of irreversible capacity loss. It is generally considered because of electrolyte decomposition and subsequent SEI layer formation. Though it is a continued process which repeats in each cycle, it is very prominent in the initial cycles usually hence named as initial irreversible capacity loss [144]. The irreversible capacity loss would be reflected in the SEI layer formation in terms of its thickness. Several investigations show that it is less affected in the insertion materials. However, a 40 to 70% capacity loss is observed in alloying as well as conversion materials [145]. Moreover, once we come to the nano regime, owing to the increased surface area of anodes the electrolyte – electrode interaction would increase which eventually pushes the capacity loss to even higher ranges in which insertion materials also show a very drastic increase in the undesirable reactions. Some recent reports say that though the reversible capacities are large, the irreversible capacity loss was nearly 50% of the initial value for graphene and similar highly defective structures. To overcome the irreversible capacity loss many methods have been implemented off which one of the simplest methods is physical blending in which the anodes are covered with thin Li layer [146, 147]. This showed very drastic improvements in Si, hard carbon based anodes. Another approach is the use of sacrificial salts. An additional salt-like Li_3N would perform as a reservoir of Li ions to compensate the initial losses [148, 149]. Transition metals doped with Li_3N were found to be more effective in many commonly use anodes like MCMB, LTO, Co_3O_4 etc. [150-152]. A similar method is the artificial SEI layer formation. Here the SEI would be formed via some electrolyte additives or even with binder [153, 154]. This eventually drops the consumption of primary salt for the SEI layer formation. Besides these, a number of other approaches like Chemical Lithiation, Spontaneous Lithiation Mechanism, and Electrochemical Lithiation to mention a few are also found to be effective.

The most widely used layered $LiCoO_2$ cathode uses only 140 mAh/g from its full capacity of 280 mAh/g due to limitation of Li ions from the structure. This material demonstrates hundreds of cycles within this limit but deintercalation of lithium results in huge capacity decay. Its unstable nature promotes the physical crumbling of LCO particles through the shrinkage of inter CoO_2 layers from ~1.422 to 1.288 nm, oxygen evolution, cobalt dissolution, deposition on the anode and electrolyte decomposition

[128, 155]. Spinel $LiMn_{1.5}Ni_{0.5}O_4$, which possesses relatively high operating voltage (~4.7 V vs. Li/Li^+) with capacities similar to LCO and intrinsically 3-D Li^+ diffusion channels, is a promising high energy and high power cathode material for future vehicle applications. However, the commercialization of $LiMn_{1.5}Ni_{0.5}O_4$ still remains challenging due to its poor cycle stability, particularly at elevated operating temperatures, and inferior rate capability. 3D diffusion path in the spinel compound provides superior rate performance and cycling stability but suffers from severe capacity fading. Further lithium-rich layered oxides Li[Li, Mn, Ni, Co]O_2 offer a ~4 V working potential with much higher capacity values of ~250 mAh/g than those of LCO and LMNO cathodes. However, there is often a huge irreversible capacity loss initially during lithiation associated with the oxygen and lithium loss from the host structure. To overcome this issue, a new class of oxide material with $LiCo_xNi_xMn_{1-x-y}O_2$ showed a better electrochemical performance with high operating voltages of around 4.7 V. It demonstrated a high reversible capacity of 200 mAh/g in 2.8-4.6 V while 160 mAh/g in 2.5-4.4 V, which makes it a promising cathode for high energy and high power LIBs. Although the Li-rich oxides can deliver very high specific capacity, more than 250 mAh/g, they encounter the disadvantages of large irreversible capacity loss and poor rate capability preventing their application [110].

To improve the efficiency some metal ions were used for doping in the lattice and surfaces were modified by Al and Cr coating. An alternative approach is to focus on designing the nanostructured cathode materials with well-defined morphologies, such as porous structures and hollow spheres. Choice of fabricating nano-sized materials potentially enhances the mass transport of the cathode by increasing the surface area for providing higher intercalation/deintercalation rates [156]. Besides the advantage of having a high surface area, the nanoparticles still face challenges with some issues with the lower thermodynamic stability and side reactions at the electrode/electrolyte solution interface. One of the possible ways to mitigate from this concern is to build nanoparticles with a core shell structure and a suitable selection of compositions that provide better stability on account of some specific capacity. Major issues like irreversible capacity loss, cycling stability and low rate capability can be significantly reduced by coating with insulating materials. Since the high surface area associated with the nanostructured lithium-excess layered oxides could have a high surface reactivity to induce the side reaction between the electrode and electrolyte the insulating technique is favourable. Likewise, the cycle life and rate capability of doped $LiMn_{1.5}Ni_{0.5}O_4$ (spinel structure) could be enhanced significantly by cationic substitutions (Co, Cr, Fe, Ga or Zn) and surface modification and/ or strong cationic coating. Also for olivine compounds, it was a common practice to manipulate the active material by adding carbon additives and the surface coating of nanoparticles with thin layers of carbon or reducing the particle size [121]. A reversible capacity of 160 mAh/g is delivered by the nanostructured cathode particles coated with carbon.

Portable electric power sources have lower energy and power densities largely resulting from poor charge and mass transport properties. New materials that are chemically modified through molecular or atomic engineering and/or possess unique microstructures would offer significantly enhanced properties for more efficient energy conversion devices and high-density energy/power storage. The polarization at higher rates caused by slow lithium diffusion in the electrode material can be resolved through designing and fabricating nanostructured materials.

8. Conclusion

Current battery technology is not far from its theoretical energy density limits. However, for the realisation of industrial applications such as EVs, rapid progress is required in miniaturisation of batteries with enhanced power and energy capabilities. Further understanding the mechanisms of Li ion storage in nanosized materials and kinetic transport properties on the interface between electrode and electrolyte has to be explored. In order to develop green and viable energy storage devices, nanomaterials and nanotechnology have to be clubbed with novel synthesis methods with low cost and environmental friendliness. In 2016, 43% of the total LIBs manufactured were used in the electric vehicle sector and forecasted to be ~50% by 2025. They were estimated to have more than double the battery pack energy density from 100 Wh/kg to 250 Wh/kg with the use of high capacity electrode materials. Various intercalation electrodes have already been trending in the market, and conversion materials are slowly proving their potentials for a widespread commercialisation.

References

1. Armand, M. and Tarascon, J. 2008. Building better batteries. Nature 451, 652-657.
2. Kurzweil, P. 2015. Advances in battery technologies for electric vehicles: Post-lithium-ion battery chemistries for hybrid electric vehicles and battery electric vehicles. Elsevier BV, 127-172.
3. Etacheri, V., Marom, R., Elazari, R., Salitra, G. and Aurbach, D. 2011. Challenges in the development of advanced Li-ion batteries: A review. Energy Environ. Sci. 4, 3243-3262.
4. Goodenough, J.B. and Kim, Y. 2010. Challenges for rechargeable Li batteries. Chem. Mater. 22, 587-603.

5. Young, K., Wang, C., Wang, L.Y. and Strunz, K. 2012. Electric Vehicle Integration into Modern Power Networks: Electric Vehicle Battery Technologies, Springer, New York, NY, 15-56.

6. Nitta, N., Wu, F., Lee, J.T. and Yushin, G. 2015. Li-ion battery materials: Present and future. Mater. Today 18, 252-264.

7. Symons, P.C. and Butler, P.C. 2004. Handbook of batteries: Advanced batteries for electric vehicles and emerging applications, 3rd edition. McGraw-Hill, NY.

8. Daniel, C., Mohanty, D., Li, J. and Wood, D.L. 2014. Cathode materials review. AIP Conf. Proc. 1597, 26-43.

9. Drive, U.S. 2013. Electrochemical Energy Storage Technical Team Roadmap. https://www1.eere.energy.gov/vehiclesandfuels/pdfs/program/eestt_roadmap_june2013.pdf

10. Wang, K., Wei, M., Morris, M.A., Zhou, H. and Holmes, J.D. 2007. Mesoporous titania nanotubes: Their preparation and application as electrode materials for rechargeable lithium batteries. Adv. Mater. 19, 3016-3020.

11. Ji, L., Lin, Z., Alcoutlabi, M. and Zhang, X. 2011. Recent developments in nanostructured anode materials for rechargeable lithium-ion batteries. Energy Environ. Sci. 4, 2682.

12. Deng, D. 2015. Li-ion batteries: Basics, progress, and challenges. Energy Sci. Eng. 3, 385-418.

13. Chen, D., Tang, L. and Li, J. 2010. Graphene-based materials in electrochemistry. Chem. Soc. Rev. 39, 3157-3180.

14. Poinsignon, C. and Armand, M.B. 1993. Insertion materials for batteries. Europhys. News 24, 10-12.

15. Aurbach, D., Zaban, A., Ein-Eli, Y., Weissman, I., Chusid, O., Markovsky, B., Levi, M., Levi, E., Schechter, A. and Granot, E. 1997. Recent studies on the correlation between surface chemistry, morphology, three-dimensional structures and performance of Li and Li-C intercalation anodes in several important electrolyte systems. J. Power Sources 68, 91-98.

16. Deng, D. and Lee, J.Y. 2014. Meso-oblate spheroids of thermal-stabile linker free aggregates with size-tunable subunits for reversible lithium storage. ACS Appl. Mater. Interfaces 6, 1173-1179.

17. De Las Casas, C. and Li, W. 2012. A review of application of carbon nanotubes for lithium ion battery anode material. J. Power Sources 208, 74-85.

18. Wang, G., Shen, X., Yao, J. and Park, J. 2009. Graphene nanosheets for enhanced lithium storage in lithium ion batteries. Carbon N. Y. 47, 2049-2053.

19. Shukla, A.K. and Prem Kumar, T. 2008. Materials for next-generation lithium batteries. Curr. Sci. 94, 314-331.

20. Buiel, E. and Dahn, J.R. 1999. Li-insertion in hard carbon anode materials for Li-ion batteries. Electrochim. Acta 45, 121-130.

21. Gao, B., Bower, C., Lorentzen, J.D., Fleming, L., Kleinhammes, A., Tang, X.P., McNeil, L.E., Wu, Y. and Zhou, O. 2000. Enhanced saturation lithium composition in ball-milled single-walled carbon nanotubes. Chem. Phys. Lett. 327, 69-75.

22. Mi, C.H., Cao, G.S. and Zhao, X.B. 2004. A non-GIC mechanism of lithium storage in chemical etched MWNTs. J. Electroanal. Chem. 562, 217-221.

23. Yoo, E.J., Kim, J., Hosono, E., Zhou, H.S., Kudo, T. and Honma, I. 2008. Large reversible Li storage of graphene nanosheet families for use in rechargeable lithium ion batteries. Nano Lett. 8, 2277-2282.

24. Ji, F., Li, Y.L., Feng, J.M., Su, D., Wen, Y.Y., Feng, Y. and Hou, F. 2009. Electrochemical performance of graphene nanosheets and ceramic composites as anodes for lithium batteries. J. Mater. Chem. 19, 9063.

25. Liang, M. and Zhi, L. 2009. Graphene-based electrode materials for rechargeable lithium batteries. J. Mater. Chem. 19, 5871-5878.

26. Wu, Z.S., Ren, W., Xu, L., Li, F. and Cheng, H.M. 2011. Doped graphene sheets as anode materials with superhigh rate and large capacity for lithium ion batteries. ACS Nano 5, 5643-5471.

27. Bolotin, K.I., Sikes, K.J., Jiang, Z., Klima, M., Fudenberg, G., Hone, J., Kim, P. and Stormer, H.L. 2008. Ultrahigh electron mobility in suspended graphene. Solid State Commun. 146, 351-355.

28. Du, X., Skachko, I., Barker, A. and Andrei, E.Y. 2008. Approaching ballistic transport in suspended graphene. Nat. Nanotechnol. 3, 491-495.

29. Wang, H., Yang, Y., Liang, Y., Robinson, J.T., Li, Y., Jackson, A., Cui, Y. and Dai, H. 2011. Graphene-wrapped sulfur particles as a rechargeable lithium-sulfur battery cathode material with high capacity and cycling stability. Nano Lett. 11, 2644-2647.

30. Zhu, X., Zhu, Y., Murali, S., Stoller, M.D. and Ruoff, R.S. 2011. Nanostructured reduced graphene oxide/Fe_2O_3 composite as a high-performance anode material for lithium ion batteries. ACS Nano 5, 3333-3338.

31. Wang, H., Cui, L.F., Yang, Y., Casalongue, H.S., Robinson, J.T., Liang, Y., Cui, Y. and Dai, H. 2010. Mn_3O_4 - Graphene hybrid as a high-capacity anode material for lithium ion. J. Am. Chem. Soc. 132, 13978-13980.

32. Rowley-Neale, S.J., Randviir, E.P., Abo Dena, A.S. and Banks, C.E. 2018. An overview of recent applications of reduced graphene oxide as a basis of electroanalytical sensing platforms. Appl. Mater. Today 10, 218-226.

33. Ali, G., Mehmood, A., Ha, H.Y., Kim, J. and Chung, K.Y. 2017. Reduced graphene oxide as a stable and high-capacity cathode material for Na-ion batteries. Sci. Rep. 7, 1-8.

34. Fu, C., Zhao, G., Zhang, H. and Li, S. 2013. Evaluation and characterization of reduced graphene oxide nanosheets as anode materials for lithium-ion batteries. Int. J. Electrochem. Sci. 8, 6269-6280.

35. Li, Y., Hu, Y.S., Li, H., Chen, L. and Huang, X. 2015. A superior low-cost amorphous carbon anode made from pitch and lignin for sodium-ion batteries. J. Mater. Chem. A 4, 96-104.

36. Chang, K., Chen, W., Ma, L., Li, H., Li, H., Huang, F., Xu, Z., Zhang, Q. and Lee, J.Y. 2011. Graphene-like MoS_2/amorphous carbon composites with high capacity and excellent stability as anode materials for lithium ion batteries. J. Mater. Chem. 21, 6251-6257.

37. Ng, S.H., Wang, J., Wexler, D., Chew, S.Y. and Liu, H.K. 2007. Amorphous carbon-coated silicon nanocomposites: A low-temperature synthesis via spray pyrolysis and their application as high-capacity anodes for lithium-ion batteries. J. Phys. Chem. C 111, 11131-11138.

38. Xia, T., Zhang, W., Wang, Z., Zhang, Y., Song, X., Murowchick, J., Battaglia, V., Liu, G. and Chen, X. 2014. Amorphous carbon-coated TiO_2 nanocrystals for improved lithium-ion battery and photocatalytic performance. Nano Energy 6, 109-118.

39. Loeffler, B.N., Bresser, D., Passerini, S. and Copley, M. 2015. Secondary lithium-ion battery anodes: From first commercial batteries to recent research activities. Johnson Matthey Technol. Rev. 59, 34-44.

40. Wang, Q., Li, H., Chen, L. and Huang, X. 2001. Monodispersed hard carbon spherules with uniform nanopores. Carbon N. Y. 39, 2211-2214.

41. Zheng, H., Qu, Q., Zhang, L., Liu, G. and Battaglia, V.S. 2012. Hard carbon: A promising lithium-ion battery anode for high temperature applications with ionic electrolyte. RSC Adv. 2, 4904-4912.

42. Zheng, P., Liu, T. and Guo, S. 2016. Micro-nanostructure hard carbon as a high performance anode material for sodium-ion batteries. Sci. Rep. 6, 1-7.

43. Tang, Y., Yang, L., Qiu, Z. and Huang, J. 2009. Template-free synthesis of mesoporous spinel lithium titanate microspheres and their application in high-rate lithium ion batteries. J. Mater. Chem. 19, 5980-5984.

44. Lv, W., Gu, J., Niu, Y., Wen, K. and He, W. 2017. Review - Gassing mechanism and suppressing solutions in $Li_4Ti_5O_{12}$-based lithium-ion batteries. J. Electrochem. Soc. 164, A2213-A2224.

45. Jansen, A.N., Kahaian, A.J., Kepler, K.D., Nelson, P.A., Amine, K., Dees, D.W., Vissers, D.R. and Thackeray, M.M. 1999. Development of a high-power lithium-ion battery. J. Power Sources 81-82, 902-905.

46. Verde, M.G., Baggetto, L., Balke, N., Veith, G.M., Seo, J.K., Wang, Z. and Meng, Y.S. 2016. Elucidating the phase transformation of $Li_4Ti_5O_{12}$ lithiation at the nanoscale. ACS Nano 10, 4312-4321.

47. Charette, K., Zhu, J., Salley, S.O., Ng, K.Y.S. and Deng, D. 2014. Gram-scale synthesis of high-temperature (900 °C) stable anatase TiO_2 nanostructures assembled by tunable building subunits for safer lithium ion batteries. RSC Adv. 4, 2557-2562.

48. Aravindan, V., Lee, Y.S. and Madhavi, S. 2015. Research progress on negative electrodes for practical Li-ion batteries: Beyond carbonaceous anodes. Adv. Energy Mater. 5, 13.

49. Saravanan, K., Ananthanarayanan, K. and Balaya, P. 2010. Mesoporous TiO_2 with high packing density for superior lithium storage. Energy Environ. Sci. 3, 939.

50. Guo, Y.G., Hu, Y.S., Sigle, W. and Maier, J. 2007. Superior electrode performance of nanostructured mesoporous TiO_2 (Anatase) through efficient hierarchical mixed conducting networks. Adv. Mater. 19, 2087-2091.

51. Zhu, K., Wang, Q., Kim, J.-H., Pesaran, A.A. and Frank, A.J. 2012. Pseudocapacitive lithium-ion storage in oriented anatase TiO_2 nanotube arrays. J. Phys. Chem. C 116.

52. Sun, L.-Q., Li, M.-J., Sun, K., Yu, S.-H., Wang, R.-S. and Xie, H.-M. 2012. Electrochemical activity of black phosphorus as an anode material for lithium-ion batteries. J. Phys. Chem. C 116 14772-14779.

53. Park, C.-M. and Sohn, H.-J. 2007. Black phosphorus and its composite for lithium rechargeable batteries. Adv. Mater. 19, 2465-2468.

54. Li, G., Wang, X. and Ma, X. 2013. Nb_2O_5-carbon core-shell nanocomposite as anode material for lithium ion battery. J. Energy Chem. 22, 357-362.

55. Chen, Z., Li, H., Lu, X., Wu, L., Jiang, J., Jiang, S., Wang, J., Dou, H. and Zhang, X. 2018. Nitrogenated urchin-like Nb_2O_5 microspheres with extraordinary pseudocapacitive properties for lithium-ion capacitors. ChemElectroChem 5, 1516-1524.

56. Sasidharan, M., Gunawardhana, N., Yoshio, M. and Nakashima, K. 2012. Nb_2O_5 hollow nanospheres as anode material for enhanced performance in lithium ion batteries. Mater. Res. Bull. 47, 2161-2164.

57. Chan, C.K., Peng, H., Liu, G., McIllwrath, K., Zhang, X.F., Huggins, R.A. and Cui, Y. 2008. High-performance lithium battery anodes using silicon nanowires. Nat. Nanotechnol. 3, 31-35.

58. Wu, H., Chan, G., Choi, J.W., Ryu, I., Yao, Y., McDowell, M.T., Lee, S.W., Jackson, A., Yang, Y., Hu, L. and Cui, Y. 2012. Stable cycling of double-walled silicon nanotube battery anodes through solid-electrolyte interphase control. Nat. Nanotechnol. 7, 310-315.

59. Yao, Y., McDowel, M.T., Ryu, I., Wu, H., Liu, N., Hu, L., Nix, W.D. and Cui, Y. 2011. Supporting information interconnected silicon hollow nanospheres for lithium-ion battery anodes with long cycle life. Nano Lett. 11, 2949-2954.

60. Lu, Z., Liu, N., Lee, H.W., Zhao, J., Li, W., Li, Y. and Cui, Y. 2015. Nonfilling carbon coating of porous silicon micrometer-sized particles for high-performance lithium battery anodes. ACS Nano 9, 2540-2547.

61. Chou, S.L., Wang, J.Z., Choucair, M., Liu, H.K., Stride, J.A. and Dou, S.X. 2010. Enhanced reversible lithium storage in a nanosize silicon/graphene composite. Electrochem. Commun. 12, 303-306.

62. Liu, N., Lu, Z., Zhao, J., McDowell, M.T., Lee, H.W., Zhao, W. and Cui, Y. 2014. A pomegranate-inspired nanoscale design for large-volume-change lithium battery anodes. Nat. Nanotechnol. 9, 187-192.

63. Park, M.H., Cho, Y., Kim, K., Kim, J., Liu, M. and Cho, J. 2011. Germanium nanotubes prepared by using the Kirkendall effect as anodes for high-rate lithium batteries. Angew. Chemie - Int. Ed. 50, 9647-9650.

64. Kennedy, T., Mullane, E., Geaney, H., Osiak, M., O'Dwyer, C. and Ryan, K.M. 2014. High-performance germanium nanowire-based lithium-ion battery anodes extending over 1000 cycles through in situ formation of a continuous porous network. Nano Lett. 14, 716-723.

65. Yu, Y., Gu, L., Wang, C., Dhanabalan, A., Van Aken, P.A. and Maier, J. 2009. Encapsulation of Sn@carbon nanoparticles in bamboo-like hollow carbon nanofibers as an anode material in lithium-based batteries. Angew. Chemie - Int. Ed. 48, 6485-6489.

66. Wang, Z., Tian, W. and Li, X. 2007. Synthesis and electrochemistry properties of Sn-Sb ultrafine particles as anode of lithium-ion batteries. J. Alloys Compd. 439, 350-354.

67. Au, M., McWhorter, S., Ajo, H., Adams, T., Zhao, Y. and Gibbs, J. 2010. Free standing aluminum nanostructures as anodes for Li-ion rechargeable batteries. J. Power Sources 195, 3333-3337.

68. Park, J.H., Hudaya, C., Kim, A.Y., Rhee, D.K., Yeo, S.J., Choi, W., Yoo, P.J. and Lee, J.K. 2014. Al-C hybrid nanoclustered anodes for lithium ion batteries with high electrical capacity and cyclic stability. Chem. Commun. 50, 2837-2840.

69. Li, S., Niu, J., Zhao Y.C., So, K.P., Wang, C., Wang, C.A. and Li, J. 2015. High-rate aluminium yolk-shell nanoparticle anode for Li-ion battery with long cycle life and ultrahigh capacity. Nat. Commun. 6, 1-7.
70. Huang, G., Xu, S., Lu, S., Li, L. and Sun, H. 2014. Micro-/Nanostructured Co_3O_4 anode with enhanced rate capability for lithium-ion batteries. ACS Appl. Mater. Interfaces 6, 7236-7243.
71. Wang, J., Wang, J., Yang, N., Tang, H., Dong, Z., Jin, Q., Yang, M., Kisailus, D., Zhao, H., Tang, Z. and Wang, D. 2013. Accurate control of multishelled Co_3O_4 hollow microspheres as high-performance anode materials in lithium-ion batteries. Angew. Chemie 125, 6545-6548.
72. Ge, X., Gu, C.D., Wang, X.L. and Tu, J.P. 2014. Correlation between microstructure and electrochemical behavior of the mesoporous Co_3O_4 Sheet and its ionothermal synthesized hydrotalcite-like α-Co(OH)$_2$ precursor. J. Phys. Chem. C 118, 911-923.
73. Yao, X., Guo, G., Zhao, Y., Zhang, Y., Tan, S.Y., Zeng, Y., Zou, R., Yan, Q. and Zhao, Y. 2016. Synergistic effect of mesoporous Co_3O_4 nanowires confined by N-doped graphene aerogel for enhanced lithium storage. Small 12, 3849-3860.
74. Wang, B., Chen, J.S., Wu, H.B., Wang, Z. and Lou, X.W. 2011. Quasiemulsion-templated formation of alpha-Fe_2O_3 hollow spheres with enhanced lithium storage properties. J. Am. Chem. Soc. 133, 17146-17148.
75. Xu, X., Cao, R., Jeong, S. and Cho, J. 2012. Spindle-like mesoporous α-Fe_2O_3 anode material prepared from MOF template for high-rate lithium batteries. Nano Lett. 12, 4988-4991.
76. Liu, X., Si, W., Zhang, J., Sun, X., Deng, J., Baunack, S., Oswald, S., Liu, L., Yan, C. and Schmidt, O.G. 2015. Free-standing Fe_2O_3 nanomembranes enabling ultra-long cycling life and high rate capability for Li-ion batteries. Sci. Rep. 4, 7452.
77. Wang, Z., Luan, D., Madhavi, S., Hu, Y. and Lou, X.W. 2012. Assembling carbon-coated-Fe_2O_3 hollow nanohorns on the CNT backbone for superior lithium storage capability. Energy Environ. Sci. 5, 5252-5256.
78. Zhang, W.-M., Wu, X.-L., Hu, J.-S., Guo, Y.-G. and Wan, L.-J. 2008. Carbon coated Fe_3O_4 nanospindles as a superior anode material for lithium-ion batteries. Adv. Funct. Mater. 18, 3941-3946.
79. Jiang, H., Hu, Y., Guo, S., Yan, C., Lee, P.S. and Li, C. 2014. Rational design of MnO/carbon nanopeapods with internal void space for high-rate and long-life li-ion batteries. ACS Nano 8, 6038-6046.
80. Yuan, T., Jiang, Y., Sun, W., Xiang, B., Li, Y., Yan, M., Xu, B. and Dou, S. 2016. Ever-increasing pseudocapacitance in RGO-MnO-RGO sandwich nanostructures for ultrahigh-rate lithium storage. Adv. Funct. Mater. 26, 2198-2206.
81. Cao, K.Z., Jiao, L., Xu, H., Liu, H., Kang, H., Zhao, Y., Liu, Y., Wang, Y. and Yuan, H. 2015. Reconstruction of mini-hollow polyhedron Mn_2O_3-derived from MOFs as a high-performance lithium anode material. Adv. Sci. 3, 1-7.
82. Huang, S.Z., Cai, Y., Jin, J., Liu, J., Li, Y., Yu, Y., Wang, H.E., Chen, L.H., Su, B.L. 2015. Hierarchical mesoporous urchin-like Mn_3O_4/carbon microspheres with highly enhanced lithium battery performance by in-situ carbonization of new lamellar manganese alkoxide (Mn-DEG). Nano Energy 12, 833-844.

83. Liu, B., Hu, X., Xu, H., Luo, W., Sun, Y. and Huang, Y. 2014. Encapsulation of MnO nanocrystals in electrospun carbon nanofibers as high-performance anode materials for lithium-ion batteries. Sci. Rep. 4, 1-6.

84. Liu, J., Chen, N. and Pan, Q. 2015. Embedding MnO nanoparticles in robust carbon microsheets for excellent lithium storage properties. J. Power Sources 299, 265-272.

85. Whittingham, M.S. 1976. Showed that lithium self-diffusion is. Science (80) 192, 1126-1127.

86. Whittingham, M.S. 1976. The role of ternary phases in cathode reactions. J. Electrochem. Soc. 123, 315.

87. Liu, C., Neale, Z.G. and Cao, G. 2016. Understanding electrochemical potentials of cathode materials in rechargeable batteries. Biochem. Pharmacol. 19, 109-123.

88. Mekonnen, Y., Sundararajan, A. and Sarwat, A.I. 2016. A review of cathode and anode materials for lithium-ion batteries. SoutheastCon 2016, 1-6.

89. Yuan, L.X., Wang, Z.H., Zhang, W.X., Hu, X.L., Chen, J.T., Huang, Y.H. and Goodenough, J.B. 2011. Development and challenges of $LiFePO_4$ cathode material for lithium-ion batteries. Energy Environ. Sci. 4, 269-284.

90. Whittingham, M.S. 2004. Lithium batteries and cathode materials. Chem. Rev. 104, 4271-4301.

91. Pitchai, R., Thavasi, V., Mhaisalkar, S.G. and Ramakrishna, S. 2011. Nanostructured cathode materials: A key for better performance in Li-ion batteries. J. Mater. Chem. 21, 11040-11051.

92. Mizushima, K. 1980. $LixCoO_2$ ($0<x<-1$): A new cathode material for batteries of high energy density. Mater. Res. Bull. 15.

93. Ohzuku, T. and Makimura, Y. 2001. Layered lithium insertion material of $LiNi_{1/2}Mn_{1/2}O_2$: A possible alternative to $LiCoO_2$ for advanced lithium-ion batteries. Chem. Lett. 2, 744-745.

94. Yoon, W.-S., Grey, C.P., Balasubramanian, M., Yang, X.-Q. and McBreen, J. 2003. In situ X-ray absorption spectroscopic study on $LiNi_{0.5}Mn_{0.5}O_2$ cathode material during electrochemical cycling. Chem. Mater. 15, 3161-3169.

95. Ammundsen, B., Desilvestro, J., Groutso, T., Hassell, D., Metson, J.B., Regan, E., Steiner, R. and Pickering, P.J. 2000. Formation and structural properties of layered $LiMnO_2$ cathode materials. J. Electrochem. Soc. 147, 4078.

96. Shao-Horn, Y. 1999. Structural characterization of layered $LiMnO_2$ electrodes by electron diffraction and lattice imaging. J. Electrochem. Soc. 146, 2404.

97. Armstrong, A., Robertson, A. and Bruce, P. 1999. Structural transformation on cycling layered $Li(Mn_{1-y}Co_y)O_2$ cathode materials. Electrochim. Acta 45, 285-294.

98. Bruce, P.G., Armstrong, A.R. and Gitzendanner, R.L. 1999. New intercalation compounds for lithium batteries: Layered $LiMnO_2$. J. Mater. Chem. 9, 193-198.

99. Obrovac, M. 1998. Structure and electrochemistry of $LiMO_2$ (M=Ti, Mn, Fe, Co, Ni) prepared by mechanochemical synthesis. Solid State Ionics 112, 9-19.

100. Kim, J.K. and Manthiram, A. 1997. A manganese oxyiodide cathode for rechargeable lithium batteries. Nature 390, 265-267.
101. Gummow, R.J. 1994. An investigation of spinel-related and orthorhombic $LiMnO_2$ cathodes for rechargeable lithium batteries. J. Electrochem. Soc. 141, 1178.
102. He, P., Yu, H., Li, D. and Zhou, H. 2012. Layered lithium transition metal oxide cathodes towards high energy. J. Mater. Chem. 22, 3680-3695.
103. Rossouw, M.H. and Thackeray, M.M. 1991. Lithium manganese oxides from Li_2MnO_3 for rechargeable lithium battery applications. Mater. Res. Bull. 26, 463-473.
104. Thackeray, M.M., Rossouw, M.H., Kock, A.D., Harpe, A.P., Gummow, R.J., Pearce, K. and Liles, D.C. 1993. The versatility of MnO_2 for lithium battery applications. J. Power Sources 43, 289-300.
105. Koyama, Y., Makimura, Y., Tanaka, I., Adachi, H. and Ohzuku, T. 2004. Systematic research on insertion materials based on superlattice models in a phase triangle of $LiCoO_2$-$LiNiO_2$-$LiMnO_2$. J. Electrochem. Soc. 151, A1499.
106. Liu, Z., Yu, A. and Lee, J.Y. 1999. Synthesis and characterization of $LiNi_{1-x-y}Co_xMn_yO_2$ as the cathode materials of secondary lithium batteries. J. Power Sources 81–82, 416-419.
107. Ohzuku, T. and Makimura, Y. 2001. Layered lithium insertion material of $LiCo_{1/3}Ni_{1/3}Mn_{1/3}O_2$ for lithium-ion batteries. Chem. Lett. 30, 642-643.
108. Belharouak, I., Sun, Y.K., Liu, J. and Amine, K. 2003. $Li(Ni_{1/3}Co_{1/3}Mn_{1/3})O_2$ as a suitable cathode for high power applications. J. Power Sources 123, 247-252.
109. Whitfield, P.S., Davidson, I.J., Cranswick, L.M.D., Swainson, I.P. and Stephens, P.W. 2005. Investigation of possible superstructure and cation disorder in the lithium battery cathode material $LiMn_{1/3}Ni_{1/3}Co_{1/3}O_2$ using neutron and anomalous dispersion powder diffraction. Solid State Ionics 176, 463-471.
110. Yoon, W.-S., Grey, C.P., Balasubramanian, M., Yang, X.-Q., Fischer, D.A. and McBreen, J. 2004. Combined NMR and XAS study on local environments and electronic structures of electrochemically Li-ion deintercalated $Li_{1-x}Co_{1/3}Ni_{1/3}Mn_{1/3}O_2$ electrode system. Electrochem. Solid-State Lett. 7, A53.
111. Koyama, Y., Tanaka, I., Adachi, H., Makimura, Y. and Ohzuku, T. 2003. Crystal and electronic structures of superstructural $Li_{1-x}[Co_{1/3}Ni_{1/3}Mn_{1/3}]O_2$ ($0 \leq x \leq 1$). J. Power Sources 119-121, 644-648.
112. Lu, J. and Lee, K.S. 2016. Spinel cathodes for advanced lithium ion batteries: A review of challenges and recent progress. Materials Technology 31, 628-641.
113. Ohzuku, T., Takeda, S. and Iwanaga, M. 1999. Solid-state redox potentials for $Li[Me_{1/2}Mn_{3/2}]O_4$ (Me: 3d-transition metal) having spinel-framework structures: A series of 5 volt materials for advanced lithium-ion batteries. J. Power Sources 81-82, 90-94.
114. Takahashi, K., Saitoh, M., Sano, M., Fujita, M. and Kifune, K. 2004. Electrochemical and structural properties of a 4.7 V-Class $LiNi_{0.5}Mn_{1.5}O_4$ positive electrode material prepared with a self-reaction method. J. Electrochem. Soc. 151, A173.

115. Myung, S.-T., Komaba, S., Kumagai, N., Yashiro, H., Chung, H.-T. and Cho, T.-H. 2002. Nano-crystalline $LiNi_{0.5}Mn_{1.5}O_4$ synthesized by emulsion drying method. Electrochim. Acta 47, 2543-2549.

116. Kiziltas-Yavuz, N., Bhaskar, A., Dixon, D., Yavuz, M., Nikolowski, K., Lu, L., Eichel, R. and Ehrenberg, H. 2014. Improving the rate capability of high voltage lithium-ion battery cathode material $LiNi_{0.5}Mn_{1.5}O_4$ by ruthenium doping. J. Power Sources 267, 533-541.

117. Zhu, W., Liu, D., Trottier, J., Gagnon, C., Mauger, A., Julien, C.M. and Zaghib, K. 2013. In-situ X-ray diffraction study of the phase evolution in undoped and Cr-doped $Li_xMn_{1.5}Ni_{0.5}O_4$ (0.1 ≤ x ≤ 1.0) 5-V cathode materials. J. Power Sources 242, 236-243.

118. Padhi, A.K., Nanjundaswamy, K.S., Goodenough, J.B. 1997. Phospho-olivines as positive-electrode materials for rechargeable lithium batteries. Journal of the Electrochemical Society 144, 1188-1194.

119. Chung, S.Y., Bloking, J.T. and Chiang, Y.M. 2002. Electronically conductive phospho-olivines as lithium storage electrodes. Nat. Mater. 1, 123-128.

120. Padhi, A.K., Nanjundaswamy, K.S., Masquelier, C. and Goodenough, J.B. 1997. Mapping of transition metal redox energies in phosphates with NASICON structure by lithium intercalation. J. Electrochem. Soc. 144, 2581-2586.

121. Zaghib, K., Mauger, A. and Julien, C.M. 2015. Olivine-Based Cathode Materials. Rechargeable Batteries. Green Energy and Technology. Springer, Cham.

122. Ravet, N., Chouinard, Y., Magnan, J.F., Besner, S., Gauthier, M. and Armand, M. 2001. Electroactivity of natural and synthetic triphylite. J. Power Sources 97-98, 503-507.

123. Rozier, P. and Tarascon, J.M. 2015. Review-Li-rich layered oxide cathodes for next-generation Li-ion batteries: Chances and challenges. J. Electrochem. Soc. 162, A2490-A2499.

124. Kim, J.S., Johnson, C.S., Vaughey, J.T., Thackeray, M.M., Hackney, S.A., Yoon, W. and Grey, C.P. 2004. Electrochemical and structural properties of $xLi_2MO_3(1-x)LiMn_{0.5}Ni_{0.5}O_2$ electrodes for lithium batteries (M = Ti, Mn, Zr; 0 ≤ x ≤ 0.3). Chem. Mater. 16, 1996-2006.

125. Yu, S.H., Yu, S.H., Yoon, T., Mun, J., Park, S., Kang, Y.S., Park, J.H., Oh, S.M. and Sung, Y.E. 2013. Continuous activation of Li_2MnO_3 component upon cycling in $Li_{1.167}Ni_{0.233}Co_{0.100}Mn_{0.467}Mo_{0.033}O_2$ cathode material for lithium ion batteries. J. Mater. Chem. A 1, 2833-2839.

126. Nazri, G.A. and Pistoia, G. 2003. Science and Technology: Lithium Batteries. Springer US.

127. Zhonghua Lu, J.R.D., Zhaohui Chen. 2003. Lack of cation clustering in $Li[Ni_xLi_{1/3-2x/3}Mn_{2/3-x/3}]O_2$ (0 < x ≤ 1/2) and $Li[Cr_xLi_{(1-x)/3}Mn_{(2-2x)/3}]O_2$ (0 < x < 1). Chem. Mater. 15, 3214-3220.

128. Li, M., Lu, J., Chen, Z. and Amine, K. 2018. 30 Years of lithium-ion batteries. Adv. Mater. 30, 1-24.

129. Xu, J., Dou, S., Liu, H. and Dai, L. 2013. Cathode materials for next generation lithium ion batteries. Nano Energy 1-4.

130. Raccichini, R., Varzi, A., Passerini, S. and Scrosati, B. 2015. The role of graphene for electrochemical energy storage. Nat. Mater. 14, 271-279.

131. Wu, Z., Ren, W., Wen, L., Gao, L., Zhao, J., Chen, Z., Zhou, G., Li, F. and Cheng, H.M. 2010. Graphene anchored with Co_3O_4 nanoparticles as anode of lithium ion capacity and cyclic performance. ACS Nano 4, 3187-3194.

132. He, Y.B., Li, B., Liu, M., Zhang, C., Lv, W., Yang, C., Li, J., Du, H., Zhang, B., Yang, Q.H., Kim, J.K. and Kang, F. 2012. Gassing in $Li_4Ti_5O_{12}$-based batteries and its remedy. Sci. Rep. 2, 33-35.

133. Han, C., He, Y.B., Li, H., Li, B., Du, H., Qin, X. and Kang, F. 2015. Suppression of interfacial reactions between $Li_4Ti_5O_{12}$ electrode and electrolyte solution via zinc oxide coating. Electrochim. Acta 157, 266-273.

134. Choi, J.W. and Aurbach, D. 2016. Promise and reality of post-lithium-ion batteries with high energy densities. Nat. Rev. Mater. 1, 16013.

135. Han, Y., Qi, P., Zhou, J., Feng, X., Li, S., Fu, X., Zhao, J., Yu, D., Wang, B. 2015. Metal-organic frameworks (MOFs) as sandwich coating cushion for silicon anode in lithium ion batteries. ACS Appl. Mater. Interfaces 7, 26608-26613.

136. Tarascon, J., Poizot, P., Laruelle, S., Grugeon, S. and Dupont, L. 2000. Nano-sized transition-metal oxides as negative-electrode materials for lithium-ion batteries. Nature 407, 496-499.

137. Wu, H. Bin, Chen, J.S., Hng, H.H. and Lou, X.W. 2012. Nanostructured metal oxide-based materials as advanced anodes for lithium-ion batteries. Nanoscale 4, 2526-2542.

138. Cui, L., Yang, Y., Hsu, C. and Cui, Y. 2009. Carbon-silicon core-shell nanowires as high capacity electrode for lithium ion batteries 2009. Nano Lett. 9, 1-5.

139. Koo, B., Kim, H., Cho, Y., Lee, K.T., Choi, N.S. and Cho, J. 2012. A highly cross-linked polymeric binder for high-performance silicon negative electrodes in lithium ion batteries. Angew. Chemie - Int. Ed. 51, 8762-8767.

140. Jeong, G., Kim, J.G., Park, M.S., Seo, M., Hwang, S.M., Kim, Y.U., Kim, Y.J., Kim, J.H. and Dou, S.X. 2014. Core-shell structured silicon nanoparticles@ TiO_{2-x}/Carbon mesoporous microfiber composite as a safe and high-performance lithium-ion battery anode. ACS Nano 8, 2977-2985.

141. Reimers, J.N. 1992. Electrochemical and in situ X-Ray diffraction studies of lithium intercalation in Li_xCoO_2. J. Electrochem. Soc. 139, 2091.

142. Wang, Y. and Cao, G. 2008. Developments in nanostructured cathode materials for high-performance lithium-ion batteries. Adv. Mater. 20, 2251-2269.

143. Etacheri, V. 2017. Sol-Gel Materials for Energy, Environment and Electronic Applications: Sol-gel processed cathode materials for lithium-ion batteries. Adv. Sol-Gel Deriv. Materials Technol. Springer, Cham, 155-195.

144. Matsumura, Y. 1995. Mechanism leading to irreversible capacity loss in Li ion rechargeable batteries. J. Electrochem. Soc. 142, 2914.

145. Aravindan, V., Lee, Y.S. and Madhavi, S. 2017. Best practices for mitigating irreversible capacity loss of negative electrodes in Li-ion batteries. Adv. Energy Mater. 7, 1-17.

146. Kulova, T.L., Skundin, A.M., Pleskov, Y.V., Terukov, E.I. and Kon'kov, O.I. 2007. Lithium insertion into amorphous silicon thin-film electrodes. J. Electroanal. Chem. 600, 217-225.

147. Kulova, T.L. 2011. Minimal irreversible capacity caused by the lithium insertion into materials of negative electrodes in lithium-ion batteries. Russ. J. Electrochem. 47, 965-967.

148. Wen, Z., Wang, K., Chen, L. and Xie, J. 2006. A new ternary composite lithium silicon nitride as anode material for lithium ion batteries. Electrochem. Commun. 8, 1349-1352.

149. Yersak, T.A., Trevey, J.E. and Lee, S.H. 2011. In situ lithiation of TiS_2 enabled by spontaneous decomposition of Li3N. J. Power Sources 196, 9830-9834.

150. Liu, Y., Horikawa, K., Fujiyosi, M., Imanishi, N., Hirano, A. and Takeda, Y. 2004. Layered lithium transition metal nitrides as novel anodes for lithium secondary batteries. Electrochim. Acta 49, 3487-3496.

151. Liu, Y., Horikawa, K., Fujiyoshi, M., Matsumura, T., Imanishi, N. and Takeda, Y. 2004. Novel composite anodes based on layered lithium transition metal nitrides for lithium secondary batteries. Solid State Ionics 172, 69-72.

152. Liu, Y., Takeda, Y., Matsumura, T., Yang, J., Imanishi, N., Hirano, A. and Yamamoto, O. 2006. Novel composite anodes consisting of lithium transition-metal nitrides and transition metal oxides for rechargeable Li-ion batteries. J. Electrochem. Soc. 153, 437.

153. Zhang, S.S. 2006. A review on electrolyte additives for lithium-ion batteries. J. Power Sources 162, 1379-1394.

154. Menkin, S., Golodnitsky, D. and Peled, E. 2009. Artificial solid-electrolyte interphase (SEI) for improved cycleability and safety of lithium-ion cells for EV applications. Electrochem. Commun. 11, 1789-1791.

155. Kraytsberg, A. and Ein-Eli, Y. 2012. Higher, stronger, better – A review of 5 volt cathode materials for advanced lithium-ion batteries. Adv. Energy Mater. 2, 922-939.

156. Ding, Y.L., Xie, J., Cao, G.S., Zhu, T.J., Yu, H.M. and Zhao, X.B. 2011. Single-crystalline $LiMn_2O_4$ nanotubes synthesized via template-engaged reaction as cathodes for high-power lithium ion batteries. Adv. Funct. Mater. 21, 348-355.

Advanced High Voltage Cathode Materials for Rechargeable Lithium Ion Batteries

J. Richards Joshua[1], T. Maiyalagan[2]*, N. Sivakumar[1]* and Y.S. Lee[3]

[1] PG & Research Department of Physics, Chikkaiah Naicker College, Erode, Tamilnadu, India

[2] Department of Chemistry, SRM Institute of Science and Technology, Chennai, India

[3] Department of Chemical Engineering, Chonnam National University, Gwangju 500-757, Republic of Korea

1. Introduction

Depletion of natural resources, climate changes and increase in global population have led to significant enhancement of worldwide energy demand which is a cause of concern for researchers into innovating new technologies for energy production. In 2015, a conference COP21 was held in Paris. It is believed that, in 21st century, energy demand is the main problem for the human race. To combat this challenge, we have to increase the energy production through various renewable energy resources. The researchers, predict that the energy demand will double by 2040. To take care of world electricity demand 11.4 trillion dollars will be invested in power production by 2040 off which up to 60% of the investment will be for renewable energy [1]. Currently, the renewable energy resources satisfy the resource requirement of large-scale facilities that cater to electricity demand Unfortunately, we are able to store only 1% of the energy produced, so it shows the need for improving large energy storage devices [2]. The solution is to develop better energy storage devices by converting chemical energy into electrical energy through electrochemical

*Corresponding authors: maiyalagan@gmail.com; nskdnp@gmail.com

devices, such as batteries and super capacitors [3]. From the various battery technologies, Lithium ion batteries hold a dominant position in comparison to other batteries.

In the early 1990s with the commercialization and introduction of LIBs in the battery world, a revolution took place in the production of portable electronic devices. With the natural combination of high energy and power densities, LIBs become a major choice for portable electronics, power tools and electrical vehicles (EVs) [4]. By replacing gasoline engines with LIBs the emission of greenhouse gases has been greatly reduced [5]. As mentioned above, currently we have capacity to store only 1% of the energy produced. To resolve this storage issue, using high efficiency of LIBs in various electrical grids and also by improving the quality of energy produced from various natural renewable resources (wind, solar and geothermal resources) can possibly reduce the energy demand. For the past few years, excessive interest has been shown in LIBs by industry, government funding agencies and researchers in this field. Presently, the cost of LIBs is high and due to a shortage of Li, it is doubtful that LIBs are capable of satisfying the power needs of several electronic devices. Current use of various transition metals in cathodes and is going to become an issue one day [6]. With these disadvantages focusing on other aspects LIBs have convincing fundamental advantages over other chemistries. They are (i) Off all elements Li has a low reduction potential, which allows LIBs to have a high cell potential; (ii) From the various elements in the periodic table, Li is the third lightest element and its charged ions have a small ionic radius. These are the reasons that allow LIBs to deliver high energy and power density. Additionally, the high charge per ion in multivalent cations, produces an additional charge that greatly reduces their mobility.

The major disadvantage of Li is the limited resources available in the earth's crust, and it seems there will be a significant shortage of Li in the future [7, 8]. This is to be met by developing new exploration mining technologies of Li sources. Based on quantities, the amount of Li resources available in the earth's crust are sufficient to meet the global power requirement [9]. But on the other hand, the increase in cost has been a major factor in limiting Li applications in large scale grid level energy storage applications. In battery fabrication Li is used as cathode and electrolyte which consumes only 30-40% off the cost of battery fabrication including the cost of Co. thus forming a major component of the price of LIBs [10, 11]. Based on availability and charge capacity, LIBs will dominate the electrochemical energy storage applications for few more years to come.

Among the various rechargeable batteries, Li has been the first choice for electrochemical energy storage. The extensive research on LIBs has greatly improved the performance and increased the usage of LIBs in various energy storage applications. Developing electrode materials with high specific capacity, high charge capacity and high operating voltages greatly improves the power and energy densities of the LIBs which makes them smaller and cheaper. Figure 1a shows the availability of various elements in the earths crust and their cost. Relatively Mn is cheaper than Co, causing a major cost difference between Mn and Co cathodes. The availability and abundance determines the cost of the material, the following carts show advantages of some elements. P and S are the most abundant materials and they are more conductive than any other element. The conversion reactions with Li are shown in the Fig. 1b.

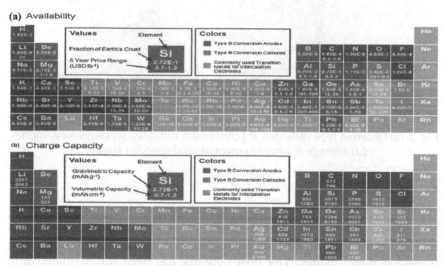

Fig. 1. (a) Availability, (b) capacities of elements that host Li as electrodes.

The properties of various elements in the cart are not suitable for Li reactions. Figure 2(a-d) shows the various elements suitable for Li reactions. The cart gives information about the cathode materials, in terms of their operating voltages and theoretical capacities. The acronyms are LCO – lithium cobalt oxide, LMO – "lithium manganese oxide", NCM – "nickel cobalt manganese", NCA – "nickel cobalt aluminium oxide", LFP – "lithium iron phosphate", LFSF – "lithium iron fluorosulphate" and LTS – "lithium titanium sulphide".

In this review we are going to discuss various cathode materials for the LIBs like transition metal oxides, spinel oxides, fluorophosphates and sulphur based materials.

Accordingly, lots of review articles on transition and spinel metal oxides have been utilized for this report. In this report, we have focused on advanced cathode materials for LIBs like conversion materials, fluorine and chlorine based compounds, sulphur and lithium sulphide, selenium and tellurium based materials, fluorophosphates and fluorosulphates.

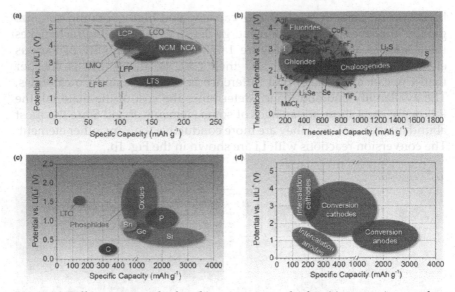

Fig. 2. (a) Different type cathodes, (b) conversion cathodes, (c) conversion anodes, (d) average discharge potentials and specific capacity.

2. Cathodes

The reaction mechanism in the cathodes involves the storage of guest ions in the solid host network. In LIBs Li metal forms the guest ions and the host network consists of chalcogens, polyanionic compounds, and transition metal oxides. The host network has been differentiated into various structures like olivine, layered, spinel etc. Table 1 summarize the different cathodes and carbon coated materials with a specific capacity of 100-200 mAh/g at 3-5 V vs Li/Li$^+$.

3. Conversion Cathode Materials

To improve the electrochemical properties of the cathodes, the electrochemical reactions are used to capitalize all the oxidation states of the transition metals used in the cathodes [125, 126].

The conversion cathodes undergo solid state redox reactions during the intercalation of Li ions. The reversible electrochemical reactions of the conversion cathodes are given as follows:

$$\text{Type A } MX_s + {}_yLi \longleftrightarrow M + zLi_{(y/s)}X \qquad (1)$$

$$\text{Type B } _yLi + X \longleftrightarrow Li_y X \qquad (2)$$

Metal fluorides are mainly used for research because of high stability and ability to produce ions suitable for efficient charge transfer between

Table 1. Summary of cathode and carbon coated cathode materials for LIBs

Material	Research group (year)	Synthesis method	Initial capacity ($mAhg^{-1}$)	Number of cycles capacity loss
Layered and Carbon/Metal Coated Metal Oxides				
Carbon-coated $LiCoO_2$	Cao et al. [12]	Commercial $LiCoO_2$ powder with milling for 24 h at 300 rpm	130 at 0.1 °C	–
Carbon-coated $LiCoO_2$ thin film with PVdF-HFP gel electrolyte	Park et al. [13]	Sol-gel process and screen printing	110	–
ZnO-coated $LiCoO_2$	Chang et al. [14]	Plasma-enhanced chemical vapour deposition	~178 at 1 °C	30:~33%
Al_2O_3-coated $LiCoO_2$	Lu et al. [15]	In situ sol-gel process followed by calcination at 1123K for 12 h	195 (1.0 wt% Al_2O_3)	30:15%
$Li_4Ti_5O_{12}$-coated $LiCoO_2$	Yi et al. [16]	Commercial $LiCoO_2$ powder coated with sol-gel process	142.7 (3 wt% Al_2O_3)	40:2.7%
$FePO_4$-coated $LiCoO_2$	Li et al. [17]	Co-precipitation method followed by high-temperature treatment	146 (3 wt% $FePO_4$)	400:11.3%
ZnO-coated $LiCoO_2$	Fang et al. [18]	Commercial $LiCoO_2$ powders coated with wet chemical method	~188 at 0.2 wt% ZnO)	30:10.4%
ZnO-coated $LiCoO_2$	Fang and Duh [19]	Wet chemical method with a calcination process	~185(calcinations at 650 °C)	30:~13.5%

(Contd.)

Table 1. (Contd.)

Material	Research group (year)	Synthesis method	Initial capacity (mAhg^{-1})	Number of cycles : capacity loss
SrO/Li$_2$O$_3$/Ta$_2$O$_5$/TiO$_2$-coated LiCoO$_2$	Wang et al. [20]	Commercial LiCoO$_2$ powders coated with Sol-gel method	~140 at 1 °C (1.0 wt% SrO/Li$_2$O/La$_2$O$_3$/ Ta$_2$O$_5$/ TiO$_2$)	900:~13.5%
M-coated LiCoO$_2$ (M=TiO$_2$ZrO$_2$ or ZrTiO$_4$)	Fey et al. [21]	Sol-gel and mechano-thermal processes	167(ZrO$_2$-coated) 165(TiO$_2$-coated) 168(ZrTiO$_4$-coated)	110:20% 24:20% 37:20%
Li$_2$PO$_2$N-coated LiCoO$_2$	Chol et al. [22]	RF magnetron sputtering method	~185 at 0.2 °C	50:~16%
LiCoO$_2$ thin film	Park et al. [23]	Screen-printing method using ethyl-cellulose-based paste	133	–
Ag-doped LiCoO$_2$	Huang et al. [24]	Commercial LiCoO$_2$ powder with milling for 4 h in alcohol	172.3 at 1 °C 133.1 at 10 °C	50:22.29% at 1 °C 50:4.88% at10 °C
LiNiO$_2$	Yamada et al. [25]	Solid-State reaction at 500-900 °C for 5 h	197 at 1 °C (at 700 °C in O$_2$) 173 at 1°C (at 700 °C in air)	12:~0%
LiNiO$_2$	Kalyani et al. [26]	Microwave synthesis with O$_2$ heat pre-treatment	163	<5%
Co-coated LiNiO$_2$	Sheng-wen et al. [27]	Co-precipitation method	180	20:0.07% / cycle

Material	Reference	Method	Capacity	Capacity fade
Ga-doped $LiNiO_2$	Nishida et al. [28]	Sol-gel method	190	100:5%
$LiCo_{1-x}Ni_xO_2$ $(0 \leq X \leq 0.2)$	Gummow and Thackery [29]	Solid-state reaction at 400 °C for 48-120 h	-	-
$LiNi_{0.5}Mn_{0.5}O_2$	Abdel-Ghany et al. [30]	Wet-chemical method	166	30:5% at 900 °C
$LiNi_{0.9}Co_{0.1}O_2$	Shi et al. [31]	Rheological phase reaction method	199 (synthesized at 800 °C)	15:27.1%
$LiNi_{0.8}Co_{0.2}O_2$ powder	Sivaprakash et al. [32]	Solid-state reaction at 800 °C for 15-48 h with intermittent grinding	~157	20:~11%
$Li_{1+x}Ni_{0.75}Co_{0.25}Mg_xO_{2(1+x)}$	Chang et al. [33]	Sol-gel method	~170(x=0.1)	30:~12%
$LiAl_xCo_{1-x}O_2$ $(0 \leq x \leq 0.3)$	Myung et al. [34]	Emulsion-drying method	110 (x=0.1)	-
Macroporous $LiNi_{0.5}Mn_{0.5}O_2$	Wang et al. [35]	Solid-state reaction method at 900 °C for 10 h	174 at 1 °C	10:<5%
Carbon-coated $LiNi_{1/3}Mn_{1/3}Co_{1/3}O_2$	Kim et al. [36]	Co-precipitation at 350 °C for 3 h	150 at 1 wt% coating	50:2% at 1 wt% coating
Nanocrystalline $LiCoO_2$-coated $Li_{1.05}Ni_{0.35}Co_{0.25}Mn_{0.4}O_2$	Son and Caims [37]	Sol-gel method	164.1 at 0.1 °C / 151.2 at 0.2 °C	20:~0% at 0.1 °C / 20:7.8% at 0.2 °C
La_2O_3-coated $LiNi_{0.8}Co_{0.2}O_2$	Fey et al. [38]			

(Contd.)

Table 1. (*Contd.*)

Spinel Lithium Metal Oxides

Material	Research group (year)	Synthesis method	Initial capacity (mAhg^{-1})	Number of cycles capacity loss
$LiMn_2O_4$	Pistoia and Rosati [39]	Solid-state reaction at 730°C for 6 h	~120	80:~16%
$Carbon/Li_{1+x}Mn_2O_4$ $(0.9 \leq x \leq 1.2)$	Guyomard and Tarascon [40]	Solid-State reaction at 800°C for 24 h	~125	200:~12%
$LiCr_xMn_{2x}O_4 (x=0), 0.04, 0.06$ or 0.1	Wang et al. [41]	Pechini method	122(x=0.04)	50:7%
$LiM_{0.05}Mn_{1.95}O_4$ (M=Li,Al,Co,Ni,or B)	Lee et al. [42]	Citrate gel method	125(Ni-doped) 122(Co-doped)	110:~2% 110:~4%
$LiAl_{0.24}Mn_{1.76}O_{3.98}S_{0.02}$	Sun et al. [43]	Sol-gel method	104 at 50 °C 98.5 at 80 °C	70:2.4% at 50 °C 70:8.4% at 80 °C
$LiAl_{0.18}Mn_{1.82}O_{3.97}S_{0.03}$	Sun et al. [44]	Sol-gel method	107 at 25 and 50 °C 100 at 80 °C	50:3% at 25 and 50 °C
$LiAl_{0.05}Mn_{1.95}O_4$	Yi et al. [45]	Adipic acid-assisted sol-gel method at 800 °C	128(D50 = 17.3 pm)	50:8.30%
$LiNi_{0.4}Mn_{1.6}O_4$	Patoux et al.	Solid- state reaction at 600-900 °C for 10 h	129	1000:20%
$LiMn_{1.5}Ni_{0.42}Ga_{0.08}O_4$	Shin and Manthiram [46]	Hydroxide precursor method	124 at 25 °C 121 at 55 °C	100:~1%

Material	Reference	Synthesis method	Capacity	Ratio/%
LiCr$_x$Mn$_{2-x}$O$_4$	Song et al. [47]	Solid-state reaction at 450-750°C for 2h	120 (synthesized at 650°C)	100:13%
LiFePO$_4$-coated LiMn$_{1.5}$Ni$_{0.5}$O$_4$	Liu et al. [48]	Sol-gel method using citric acid	~110 at 1°C	140:25%at1°C
Gold-coated LiMn$_2$O$_4$	Tu et al. [49]	Solid-state reaction at 750°C for 20 h and coating by ion sputtering method	~126	50:~6.3%
Li$_3$PO$_4$-coated LiMn$_2$O$_4$	Li et al. [50]	Sol-gel method and coating dry ball-milling method	112.4 at 55°C	100:15% at 55°C
ZnO-coated LiNi$_{0.5}$Mn$_{1.5}$O$_4$ powder	Sun et al. [51]	Sol-gel method and coating by in situ mixing for 4 h	137 at 55°C	50:~15% at 55°C
Co-doped LiCo$_y$Mn$_{2-y}$O$_4$ (y=0.05-0.33)	Arora et al. [52]	Solid-state reaction at 600°C for 6 h and 750°C for 72 h	~105(y=0.16)	25:~3%
LiMn$_2$O$_4$/C	Huang and Bruce [53]	Modified sol-gel method	150 at 0.5°C (synthesized at 200°C)	300:40%
LiMn$_2$O$_4$	Santiago et al. [54]	Combustion of manganese nitrate tetrahydrate and urea	107	–
LiMn$_2$O$_4$	Yi et al. [55]	Adipic acid-assisted sol-gel method	90.7 (synthesized at 350°C) 130.1 (synthesized at 800°C)	50:6.4% 50:14.8%

(Contd.)

Table 1. (*Contd.*)

Material	Research group (year)	Synthesis method	Initial capacity (mAhg^{-1})	Number of cycles Capacity loss
Olivine Transition Metal Phosphates and Silicates				
Mesoporous LiFePO$_4$	Ren and Bruce [56]	Solid-state reaction at 550 °C for 5 h	155 at 0.1 °C 127 at 1 °C	30:~0%
LiFePO$_4$ nanorods	Changa et al. [57]	Hydrothermal synthesis	~144 at 0.5 °C	20:~8%
LiFePO$_4$/graphene/C composite	Su et al. [58]	In situ solvo thermal method	163.7 at 0.1 °C 114 at 5 °C	30:3% at 0.1 °C (LiFePO$_4$/graphene 2%/C6%)
Nanoporous LiFePO$_4$/C composite	Yang and Gao [59]	Impregnation from ethanol solution	68 at 50 °C	50:12%
Nanoporous LiFePO$_4$/C composite	Su et al. [60]	Hydrothermal synthesis	155.3	50:3%
LiFePO$_4$/C	Yang et al. [61]		150 at 1°C and 50°C (15 wt% C)	30:14.5%
LiFePO$_4$-C/solid polymer electrolyte	Jin et al. [62]	Hydrothermal synthesis at 150 °C	128 (5 wt%)	30:0.78%
C-LiFePO$_4$/polypyrrole composite	Boyano et al. [63]	In situ electro deposition method	154 at 0.1 °C (20% PPy)	120:~0%

Material	Reference	Method	Capacity	Cyclability/Rate
LiFePO$_4$/C composite fiber	Toprakci et al. [64]	Combination of electro spinning and sol-gel methods	166 at 0.05 °C (synthesized at 700 °C)	50:~0% at 0.1 °C
LiFePO$_4$/C composite	Huang et al. [65]	Stearic acid-assisted rheological phase reaction at 600 °C for 4 h	160 at 0.5 °C 155 at 1 °C	200:<3% at 2 °C
Carbon-coated LiFePO$_4$	Dong et al. [66]	Solid-state reaction at 650-800 °C for 10 h followed by slow cooling	156.7 at 0.1 °C (synthesized at 650 °C)	50:3.5%
Carbon-coated LiFePO$_4$	Shin et al. [67]	Mechanochemical activation method	150 at 0.054 °C 135 at 1 °C	0.1-0.3 mAhg^{-1}/cycle
Carbon-coated nanocrystalline LiFePO$_4$	Zheheva et al. [68]	Freeze-drying method	122 at 0.1 °C and 500 °C	85:11.5%
SiO$_2$ coated LiFePO$_4$	Li et al. [69]	Sol-gel method	158 at 0.1 °C and 55 °C 145 at 1 °C and 55 °C	100.6% at 0.5 °C and 55 °C
ZrO$_2$-nanocoated LiFePO$_4$	Liu et al. [70]	Precipitation method	146 at 0.1 °C 97 at 1 °C	100:1.8% at 0.1 °C 100:3.1% at 1 °C
CeO$_2$-coated LiFePO$_4$	Yao et al. [71]	Commercially LiFePO$_4$/C powder	153.8 0.1 °C and 20 °C 99.7 at 0.1 °C and-20 °C	30:1.4% at -20 °C
Ru-doped LiFePO$_4$/C	Wang et al. [72]	Rheological phase reaction at 350 °C for 10 h and at 750 °C for a few hours	156 at 0.1 °C	30:~0%

(Contd.)

Table 1. (*Contd.*)

Material	Research group (year)	Synthesis method	Initial capacity ($mAhg^{-1}$)	Number of cycles capacity loss
Co-doped $LiFe_{1-x}Co_xPO_4$	Zhao et al. [73]	Hydrothermal synthesis	170 at 0.1 °C (x=1/4)	-
$LiFePO_4/C$ nanocomposites	Fey et al. [74]	Combination of carbothermal reduction and molten salt Sol-gel method	141 at 0.2 °C	-
$LiCoPO_4/C$	Xing et al. [75]	Sol-gel method	136.2	30:32%
$LiCoPO_4/C$ nanocomposites	Doan and Tangiuschi [76]	Combination of spray pyrolysis and wet ball-milling techniques followed by heat treatment	109 at 0.05 °C 142 at 20 °C	40:13% at 0.1 °C
Vanadium based compounds				
LiV_3O_8	West et al. [77]	Freeze-drying and spray-drying	-	50:0.5%/cycle
$Li_6V_{10}O_{28}$	Xie et al. [78]	Hydrothermal synthesis and annealing dehydration treatment	132	15:~24%
LiV_3O_8	Xiong et al. [79]	Spray-drying method	340.2 at 25 mAg^{-1}	100:15.2% at 125mAg^{-1}

Material	Reference	Method	Capacity	Retention
$Li_3V_2(PO_4)_3/C$	Ge et al. [80]	Sol-gel method	179.8 at 700 °C (d=30-50 nm and L=800 nm)	50:~21%
$Li_3V_2(PO_4)_3/C$	Wang et al. [81]	Electrostatic spray deposition		
$Li_3V_2(PO_4)_3/C$	Yuan et al. [82]	Sol-gel method	127.9 at 1 °C (synthesized at 750 °C) 124 at 5 °C (synthesized at 750 °C)	100:~0% at 1 °C and 5 °C
$Li_3V_2(PO_4)_3/C$	Chen et al. [83]	Carbo thermal reduction process	122 at 0.5 °C at −25 °C	100:5% at 20 °C
Cr-doped LiV_3O_8	Feng et al. [84]	Sol-gel method	269.9 at 150 mAg^{-1}	100:5.6%
Polypyrrole-LiV_3O_8 composite	Feng et al. [85]	Solution dispersion in ethanol followed by co-heating process	292 at 40 mAg^{-1} (20% PPy)	40:~14
LiV_3O_8/C nanosheet	Idris et al. [86]	Hydrothermal synthesis followed by carbon coating	227 at 0.2 °C	100:14.5%
Nanoporous LiV_3O_8	Ma et al. [87]	Tartaric acid-assisted sol-gel process	301 at 40 mAg^{-1}	50:3.7%
Polypyrrole-LiV_3O_8 composite	Tian et al. [88]	Oxidative polymerization of pyrrole using ferric chloride	169 at 0.5 °C	50:~0%
Amorphous $Fe_2V_4O_{13}$	Si et al. [89]	Liquid precipitation method	235	40:14.5%

(*Contd.*)

Table 1. (*Contd.*)

Material	Research group (year)	Synthesis method	Initial capacity (mAhg^{-1})	Number of cycles capacity loss
LiV$_3$O$_8$ nanorods	Xu et al. [90]	Hydrothermal synthesis	302 at 300 °C	30:~8%
Nanostructured composites				
LiFePO$_4$ powder	Ogihara et al. [91]	Spray pyrolysis	150	500:16%
LiF /Fe nanocomposites	Li et al. [92]	Mechanical ball-milling method	~568	20.46%
LiMn$_2$O$_4$/CNT nanocomposites	Ding et al. [93]	In-situ hydrothermal method	116 at 1 °C 84 at 10 °C	100:0.009% at 1 °C 100:5.8% at 10 °C
LiMn$_2$O$_4$/CNT nanocomposites	Xia et al. [94]	In-situ hydrothermal method	124 at 1 °C 106 at 10 °C	500:8% at 1 °C 1000:23% at 10 °C
LiFePO$_4$/multiwalled CNT nanocomposites	Li et al. [95]	Hydrogen arc discharge method followed by mixing	155 at 0.1 °C 147 at 1 °C	50:~5% at 0.1 °C 50:~6% at 1 °C
V$_2$O$_5$/polypyrrole composites	Ren et al. [96]	Sol-gel method Using	271.8 at 0.1 °C (2.5 wt% PPy)	50:17.1%
TEOS/[Bmim] [NTf$_2$]/ Carbon-Titania composites	Wang and Dai [97]	Sol-gel method Using	-	-

Advanced Cathode Materials

PTMA	Nakahara et al. [98]	Radically polymerizing	-	1000:11%
Ellagic acid	Goriparti et al. [99]	Lithiation, delithiation; Intercalation, deintercalation	50	10:25%
PAN/sulphur	Wang et al. [100]	Heating process	850 at	50:5% month
PEO	Marmorstein et al. [101]	Mixing powder	1600 at 90-100 °C	10:30%
Li_2S	Yang et al. [102]	Ball milled	800	10:0.25%
Conversion Cathodes-Metal Fluoride				
FeF_3/C	Li et al. [103]	Ball milling	712	10:16%
$Li_{1-x}Ni_{1-x}O_{2-y}F_y$	Kubo et al. [104]	Solid-state reaction	240	30:-
$Li_4Mn_5O_{12-n}Fn$	Choi and Manthiram [105]	Solid-state reaction	150 at 500 and 600 °C	50:-
$LiMn_{1.8}Li_{0.1}Ni_{0.1}O_{4-x}F_x$	Mastsumoto et al. [106]	Solid-state reaction	Around 88 and 93	-
Li_2O/VF_3 amd TiF_3	Poizot et al. [107]	Li-alloying, solid-state	700 500-600	100:25%

(Contd.)

Table 1. (*Contd.*)

Material	Research group (year)	Synthesis method	Initial capacity ($mAhg^{-1}$)	Number of cycles capacity loss
Carbon Fluoride				
FeF_3, FeF_2, and BiF_3	Amatucci et al. [108]	Solid-state prelithiation process resulted	243	0.18 at 24 °C
$(CF)_n/(C_2F)_n$	Malini [109]	Mild reaction conditions	-	-
IF_5/HF	Kim [110, 111]	RT and HT	-	-
Graphite	.Nansé et al. [112]	Graphene paper prepared chemically with graphite	298 680 528	50:19% 2:87%
$LiBF_4$	Cui X et al. [113]	Solid state	865	-
CFx(RT)/LiC104 Fluorophosphates	Broussely et al. [114]	-	667	60%
$LiFePO_4/LiFe_{0.9}Mg_{0.1}PO_4$	Delabarre et al. [115]	Novel carbothermal reduction (CTR)	150	-
Nano-carbon coated $LiVPO_4F$	Julien et al. [116]	Chemical lithiation; ball milling	143	50:-
$LiVPO_4F$	Gover et al. [117]	Sol-gel, calcinations	130	30:4.6%

$LiVPO_4F/C$	Hamwi et al. [118]	Annealing	136 at 0.1 °C rate / 123 at 1 °C rate	100:4.7%
$LiVPO_4F/LiCoO_2$	Liu et al. [119]	Carbothermal reduction	140	300 at 45 °C / 280 at 60 °C
$LiVPO_4F/C$	Zhang et al.	Hydrothermal method	143	50:2.8%
$LiFePO_4F$	Huang et al.	Sol-gel; Ceramic	145	30:4.7% and 3.9%
$LiFePO_4F$	Ramesh et al. [120]	Solid state	145	40 at room temperature and 55 °C
$LiFePO_4F$	Wang et al. [121]	Novel sol-gel	145	30:5%
$LiCoPO_4F$	Trad K et al. [122]	Two-step method combining a sol-gel route; solid state reaction	91	20:25.3%
$LiFeSO_4F$	Ramzan et al. [123]	Ceramic preparation;ball milled	85	25:5%
$LiFeSO_4F_2$	Khasanova et al. [124]	Electronically conductive coatings, high-energy ball milling	88	-

electrodes producing high operating voltages and reversible capacities [127]. Iron Fluoride FeF_3, is cost effective and is characterized by high theoretical capacity of about 712 mAh/g. In 1990, Aria et al. first reported the electrochemical properties of trifluorides with specific capacity of about 80 mAh/g [128]. Figure 6a shows the reactions take place in FeF_2 particles.

S, Se, Te and I follow the Type B reactions of the conversion cathode materials. Off these materials S has been focused on due to its high theoretical capacity of about (1675 mAh/g) low cost and abundance in nature. Oxygen also comes under type B reactions which is used for lithium air batteries because of its gaseous form. Figure 6b shows the steps of S conversion involved in intermediate polysulphides soluble in organic electrolytes. Figure 6s shows the discharge profiles of the various conversion cathode materials.

3.1 Metal Fluoride and Metal Chloride Based Compounds

Recently, metal fluoride (MF) and metal chlorides (MCl) attracted more attention because of their intermediate operating voltages and high theoretical capacitance. Because of their larger band gap, MF including FeF_3 and FeF_2 suffer from poor electronic conductivity induced by high ionic character of the halogen bond. Chlorine compounds also suffer from poor conductivity. All the precious reports on MF and MCL show them exhibiting high operating voltages but suffering from poor electronic conductivity. Additionally, Type A conversion cathode materials form metal nanoparticles in the full lithiated state, and BiF_3 and FeF_2 have been reported to have catalysed the decomposition of ions at higher voltages and reducing the cycle life [129]. On the other hand, Cu nanoparticles have been diffused into the electrolyte after being converted to Cu^{2+}. Even though such unwanted reactions have been reduced, they may affect the operating voltages and cyclic performances [130, 131].

To overcome the disadvantages of poor electronic conductivity the synthesis of nano particles for conversion cathode materials is essential to improve the transport of electrons and ions. To improve the conductivity of the MF and MCl they are often wrapped in some conductive materials to modify their surfaces which improves the conductivity of the conversion cathode materials. The materials forming the composite of MF and MCl such as FeF_3/CNT [132], FeF_3/grapheme [133], AgCl/actelylene black [134], and BiF_3/MoS_2/CNT [135] reduce the unwanted side reactions between the electrolyte and active materials during various stages of charge and discharge.

3.2 Sulphur and Lithium Sulphide

Sulphur is considered as a promising cathode material for LIBs because of its high theoretical capacity (1675 mAh/g), cost effectiveness and abundance in the earth's crust. The disadvantage is that S suffers from low operating potentials compared to Li/Li^+, low electronic conductivity and electrolyte decomposition by formation of polysulphides in side reactions and very low vaporization temperature which greatly affects the structure of the active material while drying under vacuum It also undergoes an 80% volumetric change which may destroy electrical conduction with the carbon-based anode composites. [136]. To overcome the volume expansion and structural changes it can be encapsulated in a hollow structure with high void space. The highly conductive polyvinyl pyrrolidone polymer [137], carbon [138], and TiO_2 [139] have been impregnated by using chemical precipitation. While checking with half cells these composites deliver a cyclic life of almost 1000 cycles.

The electrolyte modification is also a remedy for reducing the polysulfide formation. $LiNO_3$ [140], P_2S_2 [141] additives are used to form a good SEI on the surface of the Li and reduces the formation of polysulphides. To decrease the decomposition of electrolyte the lithium polysulphide can be used [142]. Numerous reports have been published using high molarity electrolytes [143, 144] and solid-state electrolytes have also been used to enhance the performance of the S and increasing the safety by avoiding Li short circuiting [145, 146].

3.3 Selenium and Tellurium

Higher theoretical capacities of 1630 mAh/g and 1280 mAh/g of Se and Te compared to that of S has attracted a lot of attention. Due to the high electronic conductivity of Se and Te they make great potenntial cathode materials for utilization in LIBs. No active materials have been reported in Se and Te based cathode materials because like sulphur, Se and Te has low melting point and suffer from the formation of polyselenides [147] which results in capacity fade, poor conductivity and columbic efficiency. The volume expansion is also a issue for Se and Te based cathode materials. These materials also should be wrapped with highly conductive polymers and carbon to improve the performance of the Se and Te cathodes. By reducing the volume expansion and enhancing the electrochemical performance of Se and Te cathodes, they should prove to be the most promising cathodes for LIBs in the years to come.

3.4 Fluorophosphates and Oxyphosphates

Baker et al., 2008 [148] performed the first research on fluorophosphate compounds and reported the electrochemical performance of $LiVPO_4F$ and later on the crystal structure was explained by Croguennec et al., 2009 [149] in 2012. Recently, fluorine-based phosphates have attracted all the interest due to their high electronegativity and also the electron withdrawing ability of F-ion into phosphate which greatly increases the stability of the material. The crystal structure of $LiVPO_4F$ shows that the octahedral shape which shares a common fluorine atom with vanadium so as to form .....V.....F......V...F with infinite chains. These chains show a spacious 3D frame work, connected by sharing tetrahedral phosphates along the [1 0 0], [0 1 0] and [0 0 1] directions, causing Li^+ isotopic diffusion which increases the stability of the materials significantly [150].

At operating voltages of 4.2 V and 1.8 V vs Li^+ /Li^0 the lithiation/ delithiation of $LiV_{(3+)}PO_4F$ shows high reversibility with corresponding redox couples. Using $LiV_{(3+)}PO_4F$ as cathode and graphite as anode the cell delivers the capacity of about 140 mAh/g at C/10 with the capacity retention of about 15% after 400 cycles. Similarly, like $LiV_{(3+)}PO_4F$, the α-$LiV_{(4+)}OPO_4$ was synthesized by solid-state method which shows similar cell structure with $LiVPO_4F$. The different cell performances of $LiVPO_4F$ and α-$LiV_{(4+)}OPO_4$ are shown in Fig. 8.

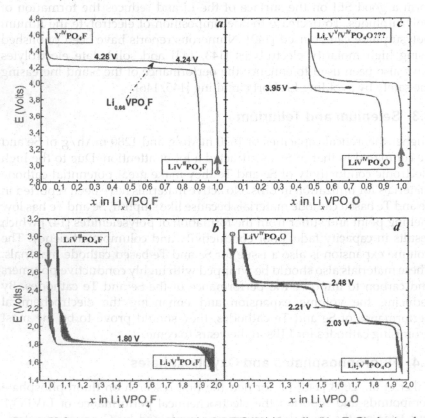

Fig. 3. Voltage–composition plots for LiVPO4X | | Li cells (X = F, O) obtained during first cycles at C/50 upon oxidation (a and c) and in GITT mode upon reduction (b and d).

Though LiVPO$_4$F performs well, its availability and toxicity have made it difficult to commercialize it for use in LIBs as a reliable electrode material. So focus has been shifted to Fe based polyanionic compound such as LiFePO$_4$F, whose synthesis and electrochemical performances were first reported by Tarascon's group [151, 152] and also by Nazar's group. The LiFePO$_4$F delivers a reversible capacity of about 145 mA·h/g at 2.8 V vs Li$^+$/Li^0 and was described by two-phase mechanism leading to Li$_2$FePO$_4$F phase with tsavorite framework [153, 154].

The redox potentials of LiFe$_{(3+)}$PO$_4$F (2.8 V) and Li$_2$Fe$_{(2+)}$PO$_4$F polymorphs (3.4 V) are different. The operating potential of olivine LiFePO$_4$ is much higher than that of LiFePO$_4$F. From this it is evident that the addition of a more electronegative element into the composition does not enhance the potential of the redox couples at all the times. There are some other facts like, the nature of the neighbouring atoms, and structural factors which affect the redox potential. Arroyo et al. was the first to perform the principle calculations on F-based compounds and their electrochemical performances depending on the particular crystallographic site occupied by F. Recently, Rousse also conducted studies on polymorphs of Li$_2$Fe$_{(2+)}$PO$_4$F with octahedral face sharing which are usually more ionic and gain higher potential.

3.5 Fluorosulphate

Sebastian et al. reported that new cathode materials such as LiMgSO$_4$F, demonstrate the benefits of LiMSO$_4$F (M=Mn, Fe, Co) involving MII/MIII oxidation states [155-157]. Then Savage et al. discovered tsavorite type LiFePO$_4$OH. The high voltage cathode materials such as LiFeSO$_4$F showed a capacity of about 140 mAh/g with a potential of about 3.6 V [158-165]. Lithium fluorosulphate compounds show great electrochemical performances and chemistry depending on the type of metal ions and on the synthesis method [166-175].

3.6 Fluoro Sulphate Polyanionic Compounds

The electrode materials based on fluorosulphate polyanionic LiMSO$_4$F compounds possess various properties of both electrochemical and eco-friendly nature. Fluorosulphate possess a wide range of operating voltages of about 4.25, 4.95 and 5.25 [176-178]. By comparison various fluorosulphate polyanionic compounds and Li$_2$CoP$_2$O$_7$ posses an operating voltage of about 4.9 V and Li$_2$NiP$_2$O$_7$ has an operating voltage of 5.4 V for electrode materials [179]. In contrast within the operating galvanostatic potential the redox reactions of Fe^{2+}/Fe^{3+}, Ni^{2+}/Ni^{3+}, Mn^{2+}/Mn^{3+} and Co^{2+}/Co^{3+} do not take place [180]. Similar to LiFeSO$_4$F, the LiMSO$_4$F (M=Co, Ni, Mn, Zn) was synthesized at low temperatures by solid-state method [181-183], polymer assisted methods [184]. In addition various fluorosulphate materials like Li(Fe1-xMx) SO_4F (M=Mn, Zn), $NaMSO_4F$ (M=Fe, Co, Ni, Mn, Mg, Zn, Cu), $NaMSO_4F·2H_2O$ (M=Fe, Co, Ni, Zn) [185, 186] have also has been studied.

4. Summary

High voltage cathode materials for LIBs are discussed in this chapter. It is believed that high voltage cathode materials are an ideal choice for the EVs and HEVs in the future. But still there is lots of scope for improvement to enhance the performance of the cathode materials. To increase the performance of the batteries highly efficient cathode materials need to be developed. The carbon based cathode materials can greatly reduce the cost of LIBs which are most suitable for EVs and HEVs. Though lots of research can be done in the future for augmenting the performance of LIBs at present they dominate the world for energy storage applications.

References

1. New Energy Outlook. 2016. Bloomberg New Energy Finance.
2. Larcher, D. and Tarascon, J.M. 2015. Towards Greener and More Sustainable Batteries for Electrical Energy Storage. Nature. Chem. 7, 19-29.
3. Winter, M. and Brodd, R.J. 2004. What Are Batteries, Fuel Cells, and Supercapacitors? Chem. Rev. 104, 4245-4270.
4. Abraham, K.M., Amine, K. and Arbizzani, C. 2013. Lithium Batteries: Advanced Technologies and Applications. John Wiley & Sons, Inc. Hoboken, New Jersey.
5. Palacin, M.R. 2009. Recent Advances in Rechargeable Battery Materials: A Chemist's Perspective. Chem. Soc. Rev. 38, 2565-2575.
6. Reynaud, M. 2013. Design of New Sulfate-based Positive Electrode Materials for Li- and Na-ion Batteries. University of Picardie Jules Verne, Amiens, France.
7. Armand, M. and Tarascon, J.M. 2008. Building Better Batteries. Nature 652-657.
8. Tarascon, J.M. and Armand, M. 2001. Issues and Challenges Facing Rechargeable Lithium Batteries. Nature 414, 359-367.
9. Planté, G. 1860. Nouvelle Pile Secondaire D'une Grande Puissance. C. r. hebd. Séances Acad. Sci. 50, 640-642.
10. Winter, M., Besenhard, J.O., Spahr, M.E. and Novák, P. 1998. Insertion Electrode Materials for Rechargeable Lithium Ion Batteries. Adv. Mater. 10, 725-763.
11. Bergstrom, S. 1952. Nickel–Cadmium Batteries – Pocket Type. J. Electrochem. Soc. 99, 248C-250C.
12. Jungner, E.W. 1903. Method of Producing Electrodes for Electric Accumulators. US Patent 731308 A.

13. Holleck, G.L., Driscoll, J.R. and Paul, B.E. 1980. The Use of LaNi5Hx-Type Hydrides in Ni-H2 batteries: Benefits and Problems. J. Less-Common Met. 74, 379-384.
14. Bullis, K. 2015. Old Battery Type Gets an Energy Boost. Technology Review.
15. Ikeda, H., Saito, T. and Tamura, H. 1975. Proc. Manganese Dioxide Symp. IC Sample Office, Cleveland: OH, Vol. 1.
16. Whittingham, M.S. 1977. Chalcogenide Battery. US Patent 4009052A.
17. Broadhead, J., Disalvo, F.J. and Trumbore, F.A. 1975. Non-Aqueous Battery Using Chalcogenide Electrode. US Patent 3864167.
18. Broahead, J. and Butherus, A.D. 1974. Rechargeable Non-aqueous Battery. US Patent 3791867, 134.
19. Murphy, D.W. and Christian, P.A. 1979. Solid State Electrodes for High Energy Batteries. Science 205, 651-656.
20. Mizushima, K., Jones, P.C., Wiseman, P.J. and Goodenough, J.B. 1980. LixCoO2 (0 < x ≤1): A New Cathode Material for Batteries of High Energy Density. Mater. Res. Bull. 15, 783-789.
21. Thackeray, M.M., David, W.I.F., Bruce, P.G. and Goodenough, J.B. 1983. Lithium Insertion into Manganese Spinels. Mater. Res. Bull. 18, 461-472.
22. Murphy, D.W., Di Salvo, F.J., Carides, J.N. and Waszczak, J.V. 1978. Topochemical Reactions of Rutile Related Structures with Lithium. Mater. Res. Bull. 13, 1395-1402.
23. Lazzari, M. and Scrosati, B. 1980. A Cyclable Lithium Organic Electrolyte Cell Based on Two Intercalation Electrodes. J. Electrochem. Soc. 127, 773-774.
24. Padhi, A.K., Nanjundaswamy, K.S. and Goodenough, J.B. 1997. J. Electrochem. Soc. 144, 1188.
25. Mohri, M. Yanagisawa, N., Tajima,Y., Tanaka, H., Mitate, T., Nakajima, S., Yoshida, M., Yoshimoto, Y., Suzuki, T. and Wada, H. 1989. 4th International Meetings on Lithium Batteries. Rechargeable lithium battery based on pyrolytic carbon as a negative electrode. J. Power Sources 26, 545-551.
26. Basu, S. 1983. Ambient Temperature Rechargeable Battery. US Patent 4423125A.
27. Guérard, D. and Hérold, A. 1972. New Method for the Preparation of Lithium Insertion Compounds in Graphite. C.R. Acad. Sci. C 275, 571-572.
28. Nishi, Y., Azuma, H. and Omaru, A. Non Aqueous Electrolyte Cell. US Patent 4959281.
29. Nagaura, T. and Tozawa, K. 1990. Lithium Ion Rechargeable Battery. Prog. Batteries Solar Cells 9, 209-217.
30. Ozawa, K. 1994. Lithium-ion rechargeable batteries with LiCoO2 and carbon electrodes: The LiCoO2/C system. Solid State Ionics 69, 212-221.
31. Armand, M., Chabagno, J.M. and Duclot, M.J. 1979. Fast Ion Transport in Solids: Electrodes and Electrolytes. pp. 131-136. *In*: Proceedings of the International Conference on Fast Ion Transport in Solids. Vashishta, P., Mundy, J.N., Shenoy, G.K. (Eds.). Lake Geneva: Wisconsin, U.S.A.
32. Armand, M. and Duclot, M. 1980. Nouveaux Matériaux Eastomères à Conduction Ionique. FR Patent 2442512.
33. Armand, M. and Duclot, M. 1981. Electrochemical Generators for Producing Current and New Materials for Their Manufacture. US Patent 4303748A. & CA Patent 1165814A1.

34. Feuillade, G. and Perche, P. 1975. Ion-conductive Macromolecular Gels and Membranes for Solid Lithium Cells. J. Appl. Electrochem. 5, 63-69.
35. Kelly, I., Owen, J.R. and Steele, B.C.H. 1984. Mixed Polyether Lithium-ion Conductors. J. Electroanal. Chem. Interfacial Electrochem. 168, 467-478.
36. Kelly, I.E., Owen, J.R. and Steele, B.C.H. 1985. Poly(ethylene oxide) Electrolytes for Operation at near Room Temperature. J. Power Sources 14, 13-21.
37. Gozdz, A.S., Schmutz, C., Tarascon, J.-M. and Warren, P.C. 1995. Method of making an electrolyte activatable lithium-ion rechargeable battery cell. US Patent 5456000.
38. Gozdz, A.S., Schmutz, C. and Tarascon, J.-M. 1994. Rechargeable Lithium Intercalation Battery with Hybrid Polymeric Electrolyte. US Patent 5296318.
39. Gozdz, A.S., Schmutz, C., Tarascon, J.-M. and Warren, P.C. 1995. Polymeric electrolytic cell separator membrane. US Patent 5418091.
40. Tarascon, J.M., Gozdz, A.S., Schmutz, C., Shokoohi, F. and Warren, P.C. 1996. Performance of Bellcore's Plastic Rechargeable Li-ion Batteries. Solid State Ionics 86, 49-54.
41. Verma, P., Maire, P. and Novák, P. 2010. A review of the features and analyses of the solid electrolyte. Electrochim. Acta 55, 6332-6341. HYPERLINK "https://inis.iaea.org/search/search.aspx?orig_q=RN:42020597"
42. Goodenough, J.B. and Park, K.-S. 2013. The Li-ion Rechargeable Batteries: A Perspective. J. Am. Chem. Soc. 135, 1167-1176.
43. Goodenough, J.B. and Kim, Y. 2010. Rising to the Challenge and "Challenges for Rechargeable Li Batteries". Chem. Mater. 22, 587-603.
44. Padhi, A.K., Nanjundaswamy, K.S. and Goodenough, J.B. 1997. Phospho-Olivines as Positive Electrode Materials for Rechargeable Lithium Batteries. J. Electrochem. Soc. 144, 1188-1194.
45. Recham, N., Chotard, J.N., Dupont, L., Delacourt, C., Walker, W., Armand, M. and Tarascon, J.M. 2010. A 3.6 V Lithium-based Fluorosulphate Insertion Positive Electrode for Lithium-ion Batteries. Nat Mater 9, 68-74.
46. Goodenough, J.B. 1994. Design Considerations. Solid State Ionics 69, 184-198.
47. Whittingham, M.S. 2004. Lithium Batteries and Cathode Materials. Chem. Rev. 104, 4271-4302.
48. Hackney, S.A., Armstrong, A.R., Bruce, P.G., Gitzendanner, R. and Shao-Horn, Y. 1999. J. Electrochem. Soc. 146, 2404.
49. Masquelier, C. and Croguennec, L. 2013. Polyanionic (Phosphates, Silicates, Sulfates) Frameworks as Electrode Materials for Rechargeable Li (or Na) Batteries. Chem. Rev. 113, 6552-6591.
50. Rousse, G. and Tarascon, J.M. 2014. Sulfate-based Polyanionic Compounds for Li-Ion Batteries: Synthesis, Crystal Chemistry, and Electrochemistry Aspects. Chem. Mater. 26, 394-406.
51. Delmas, C., Maccario, M., Croguennec, L., Cras, F. Le and Weill, F. 2008. Nat. Mater. 7, 665-671.
52. Thackeray, M.M., Johnson, P.J., de Picciotto, L.A., Bruce, P.G. and Goodenough, J.B. 1984. Electrochemical Extraction of Lithium from $LiMn_2O_4$. Mater. Res. Bull. 19, 179-187.
53. Padhi, A.K., Nanjundaswamy, K.S. and Goodenough, J.B. 1997. Phospho-

Olivines as Positive Electrode Materials for Rechargeable Lithium Batteries. J. Electrochem. Soc. 144, 1188-1194.

54. Recham, N., Chotard, J.N., Dupont, L., Delacourt, C., Walker, W., Armand, M. and Tarascon, J.M. 2010. A 3.6 V Lithium-based Fluorosulphate Insertion Positive Electrode for Lithium-ion Batteries. Nat. Mater. 9, 68-74.

55. Li, W., Reimers, J.N. and Dahn, J.R. 1993. Solid State Ionics 67, 123-130.

56. Tarascon, J.M. and Armand, M. 2001. Issues and Challenges Facing Rechargeable Lithium Batteries. Nature 414, 359-367.

57. Goodenough, J.B. 1994. Design Considerations. Solid State Ionics 69, 184-198.

58. Whittingham, M.S. 2004. Lithium Batteries and Cathode Materials. Chem. Rev. 104, 4271-4302.

59. Pellow, M.A., Emmott, C.J.M., Barnhart, C.J. and Benson, S.M. 2015. Energy Environ. Sci. 8, 1938-1952.

60. Gu, M., Belharouak, I., Zheng, J., Wu, H., Xiao, J., Genc, A., Amine, K., Thevuthasan, S., Baer, D.R., Zhang, J.G., Browning, N.D., Liu, J. and Wang, C. 2013. ACS Nano 7, 760-767.

61. Rousse, G. and Tarascon, J.M. 2014. Sulfate-based Polyanionic Compounds for Li-ion Batteries: Synthesis, Crystal Chemistry, and Electrochemistry Aspects. Chem. Mater. 26, 394-406.

62. Nitta, N., Wu, F., Lee, J.T. and Yushin, G. 2015. Li-ion Battery Materials: Present and Future. Mater. Today 18, 252-264.

63. Mohanty, D., Sefat, A.S., Kalnaus, S., Li, J., Meisner, R.A., Payzant, E.A., Abraham, D.P., Wood, D.L. and Daniel, C. 2013. J. Mater. Chem. A 1, 6249.

64. Ohzuku, T. and Ueda, A. 1994. Solid-State Redox Reactions of $LiCoO_2$ (R$\bar{3}$m) for 4 Volt Secondary Lithium Cells. J. Electrochem. Soc. 141, 2972-2977.

65. Yang, X.Q., Sun, X. and McBreen, J. 2000. New Phases and Phase Transitions Observed in $Li_{1-x}CoO_2$ during Charge: In Situ Synchrotron X-ray Diffraction Studies. Electrochem. Commun. 2, 100-103.

66. Van der Ven, A., Aydinol, M.K. and Ceder, G. 1998. First-Principles Evidence for Stage Ordering in $LixCoO_2$. J. Electrochem. Soc. 145, 2149-2155.

67. Chen, Z., Lu, Z. and Dahn, J.R. 2002. Staging Phase Transitions in $LixCoO_2$. J. Electrochem. Soc. 149, A1604-A1609.

68. Amatucci, G.G., Tarascon, J.M. and Klein, L.C. 1996. CoO_2: The End Member of the $LixCoO_2$ Solid Solution. J. Electrochem. Soc. 143, 1114-1123.

69. Whittingham, M.S. 2004. Lithium Batteries and Cathode Materials. Chem. Rev. 104, 4271-4302.

70. Croy, J.R., Balasubramanian, M., Kim, D., Kang, S.-H. and Thackeray, M. M. 2011. Chem. Mater. 23, 5415-5424.

71. Wang, H., Jang, Y.I., Huang, B., Sadoway, D.R. and Chiang, Y.M. 1999. TEM Study of Electrochemical Cycling-induced Damage and Disorder in $LiCoO_2$ Cathodes for Rechargeable Lithium Batteries. J. Electrochem. Soc. 146, 473-480.

72. Endo, E., Yasuda, T., Kita, A., Yamaura, K. and Sekai, K. 2000. A $LiCoO_2$ Cathode Modified by Plasma Chemical Vapor Deposition for Higher Voltage Performance. J. Electrochem. Soc. 147, 1291-1294.

73. Chung, K.Y., Yoon, W.-S., McBreen, J., Yang, X.-Q., Oh, S.H., Shin, H.C., Cho, W.I. and Cho, B.W. 2006. Structural Studies on the Effects of ZrO_2 Coating on $LiCoO_2$ during Cycling Using In Situ X-Ray Diffraction Technique. J. Electrochem. Soc. 153, A2152-A2157.

74. Amatucci, G.G., Tarascon, J.M. and Klein, L.C. 1996. Cobalt Dissolution in $LiCoO_2$-based Non-aqueous Rechargeable Batteries. Solid State Ionics 83, 167-173.

75. Thomas, M.G.S.R., David, W.I.F., Goodenough, J.B. and Groves, P. 1985. Synthesis and Structural Characterization of the Normal Spinel $Li[Ni_2]O_4$. Mater. Res. Bull. 20, 1137-1146.

76. Dahn, J.R., von Sacken, U. and Michal, C.A. 1990. Structure and Electrochemistry of $Li_{1\pm y}NiO_2$ and a New Li_2NiO_2 Phase with the $Ni(OH)_2$ Structure. Solid State Ionics 44, 8797.

77. Dahn, J.R., von Sacken, U., Juzkow, M.W. and Al-Janaby, H. 1991, Rechargeable $LiNiO_2$/Carbon Cells. J. Electrochem. Soc. 138, 2207-2211.

78. Morales, J., Pérez-Vicente, C. and Tirado, J.L. 1990. Cation Distribution and Chemical Deintercalation of $Li_{1-x}Ni_{1+x}O_2$. Mater. Res. Bull. 25, 623-630.

79. Gummow, R.J. and Thackeray, M.M. 1992. Lithium-Cobalt-Nickel-Oxide Cathode Materials Prepared at 400 °C for Rechargeable Lithium Batteries. Solid State Ionics 53, 681-687.

80. I.J. Pickering, Lewandowski, J.T., Jacobson, A.J. and Goldstone, J.A. 1992. A Neutron Powder Diffraction Study of the Ordering in $Li_xNi_{1-x}O$. Solid State Ionics 53, 405-412.

81. Kanno, R., Kubo, H., Kawamoto, Y., Kamiyama, T., Izumi, F., Takeda, Y. and Takano, M. 1994. Phase Relationship and Lithium Deintercalation in Lithium Nickel Oxides. J. Solid State Chem. 110, 216-225.

82. Rougier, A., Gravereau, P. and Delmas, C. 1996. Optimization of the Composition of the $Li_{1-z}Ni_{1+z}O_2$ Electrode Materials: Structural, Magnetic, and Electrochemical Studies. J. Electrochem. Soc. 143, 1168-1175.

83. Arai, H., Okada, S., Sakurai, Y. and Yamaki, J.-i. 1998. Thermal Behavior of $Li_{1-y}NiO_2$ and the Decomposition Mechanism. Solid State Ionics 109, 295-302.

84. Armstrong, A.R. and Bruce, P.G. 1996. Synthesis of Layered $LiMnO_2$ as an Electrode for Rechargeable Lithium Batteries. Nature 381, 499-500.

85. Capitaine, F., Gravereau, P. and Delmas, C. 1996. A new Variety of $LiMnO_2$ with a Layered Structure. Solid State Ionics 89, 197-202.

86. Gu, M., Belharouak, I., Zheng, J., Wu, H., Xiao, J., Genc, A., Amine, K., Thevuthasan, S., Baer, D.R., Zhang, J.-G., Browning, N.D., Liu, J. and Wang, C. 2013. Formation of the Spinel Phase in the Layered Composite Cathode Used in Li-Ion Batteries. ACS Nano 7, 760-767.

87. Tu, J., Zhao, X.B., Cao, G.S., Zhuang, D.G., Zhu, T.J. and Tu, J.P. 2006. Enhanced Cycling Stability of $LiMn_2O_4$ by Surface Modification with Melting Impregnation Method. Electrochim. Acta 51, 6456-6462.

88. Kim, J.S., Johnson, C.S. and Thackeray, M.M. 2002. Layered $xLiMO_2 \cdot (1-x)$ $Li_2M'O_3$ Electrodes for Lithium Batteries: A Study of $0.95LiMn_{0.5}Ni_{0.5}O_2 \cdot 0.05Li_2TiO_3$. Electrochem. Commun. 4, 205-209.

89. Johnson, C.S., Kim, J.S., Lefief, C., Li, N., Vaughey, J.T. and Thackeray, M.M. 2004. The Significance of the Li_2MnO_3 Component in 'Composite'

$xLi_2MnO_3 \cdot (1-x)LiMn_{0.5}Ni_{0.5}O_2$ electrodes. Electrochem. Commun. 6, 1085-1091.

90. Kim, J.-S., Johnson, C.S., Vaughey, J.T., Thackeray, M.M., Hackney, S.A., Yoon, W. and Grey, C.P. 2004. Electrochemical and Structural Properties of $xLi_2M'O_3 \cdot (1-x)LiMn_{0.5}Ni_{0.5}O_2$ Electrodes for Lithium Batteries (M' = Ti, Mn, Zr; $0 \leq x \leq 0.3$). Chem. Mater. 16, 1996-2006.

91. Thackeray, M.M., Johnson, C.S., Vaughey, J.T., Li, N. and Hackney, S.A. 2005. Advances in Manganese-Oxide 'Composite' Electrodes for Lithium-ion Batteries. J. Mater. Chem. 15, 2257-2267.

92. Gu, R.-M., Yan, S.-Y., Sun, S., Wang, C.-Y. and Li, M.-W. 2015. Electrochemical Behavior of Lithium-rich Layered Oxide $Li[Li_{0.23}Ni_{0.15}Mn_{0.62}]O_2$ Cathode Material for Lithium-ion Battery. J. Solid State Electrochem. 19, 1659-1669.

93. Thackeray, M.M., Kang, S.H., Johnson, C.S., Vaughey, J.T. and Hackney, S.A. 2006. Comments on the Structural Complexity of Lithium-rich $Li_{1+x}M_{1-x}O_2$ Electrodes (M = Mn, Ni, Co) for Lithium Batteries. Electrochem. Commun. 8, 1531-1538.

94. Johnson, C.S., Li, N., Lefief, C., Vaughey, J.T. and Thackeray, M.M. 2008. Synthesis, Characterization and Electrochemistry of Lithium Battery Electrodes: $xLi_2MnO_3 \cdot (1-x)LiMn_{0.333}Ni_{0.333}Co_{0.333}O_2$ ($0 \leq x \leq 0.7$). Chem. Mater. 20, 6095-6106.

96. Rozier, P. and Tarascon, J.M. 2015. Review—Li-rich Layered Oxide Cathodes for Next Generation Li-ion Batteries: Chances and Challenges. J. Electrochem. Soc. 162, A2490-A2499.

97. Johnson, C.S., Kim, J.S., Lefief, C., Li, N., Vaughey, J.T. and Thackeray, M.M. 2004. The Significance of the Li_2MnO_3 Component in 'Composite' $xLi_2MnO_3 \cdot (1-x)LiMn_{0.5}Ni_{0.5}O_2$ electrodes. Electrochem. Commun. 6, 1085-1091.

98. Sathiya, M., Abakumov, A.M., Foix, D., Rousse, G., Ramesha, K., Saubanère, M., Doublet, M.L., Vezin, H., Laisa, C.P., Prakash, A.S., Gonbeau, D., Van Tendeloo, G. and Tarascon, J.M. 2015. Origin of Voltage Decay in High-Capacity Layered Oxide Electrodes. Nat Mater. 14, 230-238.

99. Tarascon, J.M., Wang, E., Shokoohi, F.K., McKinnon, W.R. and Colson, S. 1991. The Spinel Phase of $LiMn_2O_4$ as a Cathode in Secondary Lithium Cells. J. Electrochem. Soc. 138, 2859-2864.

100. Tarascon, J.M. and Guyomard, D. 1991. Li Metal-free Rechargeable Batteries Based on $Li_{1+x}Mn_2O_4$ Cathodes ($0 \leq x \leq 1$) and Carbon Anodes. J. Electrochem. Soc. 138, 2864-2868.

101. Jang, D.H., Shin, Y.J. and Oh, S.M. 1996. Dissolution of Spinel Oxides and Capacity Losses in $4V Li/LixMn_2O_4$ Cells. J. Electrochem. Soc. 143, 2204-2211.

102. Huang, H., Vincent, C.A. and Bruce, P.G. 1999. Correlating Capacity Loss of Stoichiometric and Nonstoichiometric Lithium Manganese Oxide Spinel Electrodes with Their Structural Integrity. J. Electrochem. Soc. 146, 3649-3654.

103. Shin, Y. and Manthiram, A. 2004. Factors Influencing the Capacity Fade of Spinel Lithium Manganese Oxides. J. Electrochem. Soc. 151, A204-A208.

104. Thackeray, M.M., Shao-Horn, Y., Kahaian, A.J., Kepler, K.D., Skinner, E., Vaughey, J.T. and Hackney, S.A. 1998. Structural Fatigue in Spinel

Electrodes in High Voltage (4 V) Li/LixMn$_2$O$_4$ Cells. Electrochem. Solid-State Lett. 1, 7-9.

105. Kim, D.K., Muralidharan, P., Lee, H.-W., Ruffo, R., Yang, Y., Chan, C.K., Peng, H., Huggins, R.A. and Cui, Y. 2008. Spinel LiMn$_2$O$_4$ Nanorods as Lithium Ion Battery Cathodes. Nano Lett. 8, 3948-3952.

106. Hosono, E., Kudo, T., Honma, I., Matsuda, H. and Zhou, H. 2009. Synthesis of Single Crystalline Spinel LiMn$_2$O$_4$ Nanowires for a Lithium Ion Battery with High Power Density. Nano Lett. 9, 1045-1051.

107. Lee, H.-W., Muralidharan, P., Ruffo, R., Mari, C.M., Cui, Y. and Kim, D.K. 2010. Ultrathin Spinel LiMn$_2$O$_4$ Nanowires as High Power Cathode Materials for Li-ion Batteries. Nano Lett. 10, 3852-3856.

108. Ding, Y.-L., Xie, J., Cao, G.-S., Zhu, T.-J., Yu, H.-M. and Zhao, X.-B. 2011. Single-crystalline LiMn$_2$O$_4$ Nanotubes Synthesized via Template-engaged Reaction as Cathodes for High-Power Lithium Ion Batteries. Adv. Funct. Mater. 21, 348-355.

109. Ellis, B.L., Lee, K.T. and Nazar, L.F. 2010. Positive Electrode Materials for Li-ion and Li Batteries. Chem. Mater. 22, 691-714.

110. Gong, Z. and Yang, Y. 2011. Recent Advances in the Research of Polyanion-type Cathode Materials for Li-ion Batteries. Energy Environ. Sci. 4, 3223-3242.

111. Ni, J., Zhang, L., Fu, S., Savilov, S.V., Aldoshin, S.M. and Lu, L. 2015. A Review on Integrating Nano-Carbons into Polyanion Phosphates and Silicates for Rechargeable Lithium Batteries. Carbon 92, 15-25.

112. Hong, H.Y.P. 1976. Crystal Structures and Crystal Chemistry in the System Na$_{1+x}$Zr$_2$SixP$_{3-x}$O$_{12}$. Mater. Res. Bull. 11, 173-182.

113. Anantharamulu, N., Koteswara Rao, K., Rambabu, G., Vijaya Kumar, B., Radha, V. and Vithal, M. 2011. A Wide-ranging Review on Nasicon Type Materials. J. Mater. Sci. 46, 2821-2837.

114. Weber, N. and Kummer, J.T. 1967. Proc. Annu. Power Sources Conf. 1967. 21, 37-39.

115. Whittingham, M.S. and Huggins, R.A. 1971. Measurement of Sodium Ion Transport in Beta Alumina Using Reversible Solid Electrodes. J. Chem. Phys. 54, 414-416.

116. Nadiri, A., Delmas, C., Salmon, R. and Hagenmuller, P. 1984. Chemical and electrochemical Alkali Metal Intercalation in the 3D-Framework of Fe$_2$(MoO$_4$)$_3$. Rev. Chim. Miner. 21.

117. Reiff, W.M., Zhang, J.H. and Torardi, C.C. 1986. Topochemical Lithium Insertion into Fe$_2$(MoO$_4$)$_3$: Structure and Magnetism of Li$_2$Fe$_2$(MoO$_4$)$_3$. J. Solid State Chem. 62, 231-240.

118. Delmas, C., Cherkaoui, F., Nadiri, A. and Hagenmuller, P. 1987. A Nasicon-type Phase as Intercalation Electrode: NaTi$_2$(PO$_4$)$_3$. Mater. Res. Bull. 22, 631-639.

119. Weppner, W., Schulz, H., Delmas, C., Nadiri, A. and Soubeyroux, J.L. 1988. Proceedings of the 6th International Conference on Solid State Ionics: The Nasicon-type Titanium Phosphates Ati$_2$(PO$_4$)$_3$ (A = Li, Na) as Electrode Materials. Solid State Ionics 28, 419-423.

120. Padhi, A.K., Manivannan, V. and Goodenough, J.B. 1998. Tuning the Position of the Redox Couples in Materials with NASICON Structure by Anionic Substitution. J. Electrochem. Soc. 145, 1518-1520.

121. Manthiram, A. and Goodenough, J.B. 1989. Lithium insertion into $Fe_2(SO_4)_3$ frameworks. J. Power Sources 26, 403-408.
122. Nanjundaswamy, K.S., Padhi, A.K., Goodenough, J.B., Okada, S., Ohtsuka, H., Arai, H. and Yamaki, J. 1996. Synthesis, Redox Potential Evaluation and Electrochemical Characteristics of NASICON-related-3D Framework Compounds. Solid State Ion 92, 1-10.
123. Padhi,A.K., Nanjundaswamy, K.S., Masquelier, C. and Goodenough, J.B. 1997. Mapping of Transition Metal Redox Energies in Phosphates with NASICON Structure by Lithium Intercalation. J. Electrochem. Soc. 144, 2581-2586.
124. Huang, H., Yin, S.C., Kerr, T., Taylor, N. and Nazar, L.F. 2002. Nanostructured Composites: A High Capacity, Fast Rate $Li_3V_2(PO_4)_3$/Carbon Cathode for Rechargeable Lithium Batteries. Adv. Mater. 14, 1525-1528.
125. Patoux, S., Wurm, C., Morcrette, M., Rousse, G. and Masquelier, C. 2003. A Comparative Structural and Electrochemical Study of Monoclinic $Li_3Fe_2(PO_4)_3$ and $Li_3V_2(PO_4)_3$. J. Power Sources 119-121, 278-284.
126. Castets, A., Carlier, D., Trad, K., Delmas, C. and Ménétrier, M. 2010. Analysis of the 7Li NMR signals in the Monoclinic $Li_3Fe_2(PO_4)_3$ and $Li_3V_2(PO_4)_3$ Phases. J. Phys. Chem. C 114, 19141-19150.
127. Davis, L.J.M., Heinmaa, I. and Goward, G.R. 2010. Study of Lithium Dynamics in Monoclinic $Li_3Fe_2(PO_4)_3$ using 6Li VT and 2D Exchange MAS NMR Spectroscopy. Chem. Mater. 22, 769-775.
128. Sato, M., Ohkawa, H., Yoshida, K., Saito, M., Uematsu, K. and Toda, K. 2000. Enhancement of Discharge Capacity of $Li_3V_2(PO_4)_3$ by stabilizing the orthorhombic phase at room temperature. Solid State Ionics 135, 137-142.
129. Huang, H., Faulkner, T., Barker, J. and Saidi, M.Y. 2009. Lithium Metal Phosphates, Power and Automotive Applications. J. Power Sources 189, 748-751.
130. Morcrette, M., Leriche, J.-B., Patoux, S., Wurm, C. and Masquelier, C. 2003. In Situ X-Ray Diffraction during Lithium Extraction from Rhombohedral and Monoclinic $Li_3V_2(PO_4)_3$. Electrochem. Solid-State Lett. 6, A80-A84.
131. Saidi, M.Y., Barker, J., Huang, H., Swoyer, J.L. and Adamson, G. 2003. Performance Characteristics of Lithium Vanadium Phosphate as a Cathode Material for Lithium-ion Batteries. J. Power Sources 119-121, 266-272.
132. Masquelier, C., Padhi, A.K., Nanjundaswamy, K.S. and Goodenough, J.B. 1998. New Cathode Materials for Rechargeable Lithium Batteries: The 3-D Framework Structures $Li_3Fe_2(XO_4)_3(X = P, As)$. J. Solid State Chem. 135, 228-234.
133. Morcrette, M., Wurm, C. and Masquelier, C. 2002. On the Way to the Optimization of $Li_3Fe_2(PO_4)_3$ Positive Electrode Materials. Solid State Sci. 4, 239-246.
134. Padhi, A.K., Nanjundaswamy, K.S., Masquelier, C., Okada, S. and Goodenough, J.B. 1997. Effect of Structure on the Fe_{3+}/Fe_{2+} Redox Couple in Iron Phosphates. J. Electrochem. Soc. 144, 1609-1613.
135. Ravet, N., Goodenough, J.B., Besner, S., Simoneau, M., Hovington, P. and Armand, M. 1999. 196th Meeting of the Electrochemical Society.
136. Ravet, N., Chouinard, Y., Magnan, J.F., Besner, S. Gauthier, M. and Armand, M. 2001. Electroactivity of Natural and Synthetic Triphylite. J. Power Sources 97-98, 503-507.

137. Huang, H., Yin, S.-C. and Nazar, L.F. 2001. Approaching Theoretical Capacity of LiFePO$_4$ at Room Temperature at High Rates. Electrochem. Solid-State Lett. 4, A170-A172.
138. Rousse, G., Rodriguez-Carvajal, J., Patoux, S. and Masquelier, C. 2003. Magnetic Structures of the Triphylite LiFePO$_4$ and of Its Delithiated Form FePO$_4$. Chem. Mater. 2003, 15, 4082-4090.
139. Fey, G.T.-K. and Lu, T.-L. 2008. Morphological Characterization of LiFePO$_4$/C Composite Cathode Materials Synthesized via a Carboxylic Acid Route. J. Power Sources 2008, 178, 807-814.
140. Iltchev, N., Chen, Y.K., Okada, S. and Yamaki, J. 2003. LiFePO$_4$ Storage at Room and Elevated Temperatures. J. Power Sources 119-121, 749-754.
141. Hsu, K.-F., Tsay, S.-Y. and Hwang, B.-J. 2004. Synthesis and Characterization of Nano-sized LiFePO4 Cathode Materials Prepared by a Citric Acid-Based Sol-gel Route. J. Mater. Chem. 14, 2690-2695.
142. Arnold, G., Garche, J., Hemmer, R., Ströbele, S., Vogler, C. and Wohlfahrt-Mehrens, M. 2003. Fine-particle Lithium Iron Phosphate LiFePO$_4$ Synthesized by a New Low-cost Aqueous Precipitation Technique. J. Power Sources 119-121, 247-251.
143. Park, K.S., Son, J.T., Chung, H.T., Kim, S.J., Lee, C.H. and Kim, H.G. 2003. Synthesis of LiFePO$_4$ by Co-precipitation and Microwave Heating. Electrochem. Commun. 5, 839-842.
144. Yang, S., Zavalij, P.Y. and Whittingham, M. Stanley. 2001. Electrochem. Commun. 3, 505-508.
145. Yang, S., Song, Y., Zavalij, P.Y. and Whittingham, M. Stanley. 2002. Reactivity, Stability and Electrochemical Behavior of Lithium Iron Phosphates. Electrochem. Commun. 4, 239-244.
146. Ferrari, S., Lavall, R.L., Capsoni, D., Quartarone, E., Magistris, A., Mustarelli, P. and Canton, P. 2010. Influence of Particle Size and Crystal Orientation on the Electrochemical Behavior of Carbon-Coated LiFePO$_4$. J. Phys. Chem. C 114, 12598-12603.
147. Murugan, A.V., Muraliganth, T. and Manthiram, A. 2008. Comparison of Microwave Assisted Solvothermal and Hydrothermal Syntheses of LiFePO$_4$/C Nanocomposite Cathodes for Lithium Ion Batteries. J. Phys. Chem. C 112, 14665-14671.
148. Vadivel Murugan, A., Muraliganth, T. and Manthiram, A. 2008. Rapid Microwave-Solvothermal Synthesis of Phospho-Olivine Nanorods and Their Coating with a Mixed Conducting Polymer for Lithium Ion Batteries. Electrochem. Commun. 10, 903-906.
149. Murugan, A.V., Muraliganth, T., Ferreira, P.J. and Manthiram, A. 2009. Dimensionally Modulated, Single-Crystalline LiMPO$_4$ (M = Mn, Fe, Co, and Ni) with Nano Thumblike Shapes for High-Power Energy Storage. Inorg. Chem. 48, 946-952.
150. Yang, S., Zhou, X., Zhang, J. and Liu, Z. 2010. Morphology-Controlled Solvothermal Synthesis of LiFePO$_4$ as a Cathode Material for Lithium-ion Batteries. J. Mater. Chem. 20, 8086-8091.
151. Doherty, C.M., Caruso, R.A., Smarsly, B.M. and Drummond, C. 2009. Colloidal Crystal Templating to Produce Hierarchically Porous LiFePO4 Electrode Materials for High Power Lithium Ion Batteries. Chem. Mater. 21, 2895-2903.

152. Hasegawa, G., Ishihara, Y., Kanamori, K., Miyazaki, K., Yamada, Y., Nakanishi, K. and Abe, T. 2011. Facile Preparation of Monolithic $LiFePO_4$/Carbon Composites with WellDefined Macropores for a Lithium-Ion Battery. Chem. Mater. 23, 5208-5216.

153. Vu, A. and Stein, A. 2011. Multiconstituent Synthesis of LiFePO4/C Composites with Hierarchical Porosity as Cathode Materials for Lithium Ion Batteries. Chem. Mater. 23, 3237-3245.

154. Recham, N., Armand, M., Laffont, L. and Tarascon, J.-M. 2009. Eco-Efficient Synthesis of $LiFePO_4$ with Different Morphologies for Li-Ion Batteries. Electrochem. Solid-State Lett. 12, A39-A44.

155. Recham, N., Dupont, L., Courty, M., Djellab, K., Larcher, D., Armand, M. and Tarascon, J.M. 2009. Ionothermal Synthesis of Tailor-Made LiFePO4 Powders for Li-Ion Battery Applications. Chem. Mater. 21, 1096-1107.

156. Tarascon, J.-M., Recham, N., Armand, M., Chotard, J.-N., Barpanda, P., Walker, W. and Dupont, L. 2010. Hunting for Better Li-Based Electrode Materials via Low Temperature Inorganic Synthesis. Chem. Mater. 22, 724-739.

157. Yang, S., Song, Y., Ngala, K., Zavalij, P.Y. and Whittingham, M. Stanley. 2003. Performance of $LiFePO_4$ as Lithium Battery Cathode and Comparison with Manganese and Vanadium Oxides. J. Power Sources 119-121, 239-246.

158. Islam, M.S., Driscoll, D.J., Fisher, C.A.J. and Slater, P.R. 2005. Atomic-scale Investigation of Defects, Dopants, and Lithium Transport in the $LiFePO_4$ Olivine-type Battery Material. Chem. Mater. 17, 5085-5092.

159. Gibot, P., Casas-Cabanas, M., Laffont, L., Levasseur, S., Carlach, P., Hamelet, S., Tarascon, J.-M. and Masquelier, C. 2008. Room-Temperature Single-Phase Li Insertion/Extraction in Nanoscale $LixFePO_4$. Nat. Mater. 7, 741-747.

160. Hamelet, S., Gibot, P., Casas-Cabanas, M., Bonnin, D., Grey, C.P., Cabana, J., Leriche, J.-B., Rodriguez-Carvajal, J., Courty, M., Levasseur, S., Carlach, P., Van Thournout, M., Tarascon, J.-M. and Masquelier, C. 2009. The Effects of Moderate Thermal Treatments Under Air on LiFePO4-based Nano Powders. J. Mater. Chem. 19, 3979-3991.

161. Malik, R., Burch, D., Bazant, M. and Ceder, G. 2010. Particle Size Dependence of the Ionic Diffusivity. Nano Lett. 10, 4123-4127.

162. Badi, S.-P., Wagemaker, M., Ellis, B.L., Singh, D.P., Borghols, W.J.H., Kan, W.H., Ryan, D.H., Mulder, F.M. and Nazar, L.F. 2011. Direct Synthesis of Nanocrystalline $Li_{0.90}FePO_4$: Observation of Phase Segregation of Anti-site Defects on Delithiation. J. Mater. Chem. 21, 10085-10093.

163. Chen, Z. and Dahn, J.R. 2002. Reducing Carbon in $LiFePO_4$/C Composite Electrodes to Maximize Specific Energy, Volumetric Energy, and Tap Density. J. Electrochem. Soc. 149, A1184-A1189.

164. Amine, K., Yasuda, H. and Yamachi, M. 2000. Olivine $LiCoPO_4$ as 4.8 V Electrode Material for Lithium Batteries. Electrochem. Solid-State Lett. 3, 178-179.

165. Okada, S., Sawa, S., Egashira, M., Yamaki, J.-i., Tabuchi, M., Kageyama, H., Konishi, T. and Yoshino, A. 2001. Cathode Properties of Phospho-olivine $LiMPO_4$ for Lithium Secondary Batteries. J. Power Sources 97-98, 430-432.

166. Yamada, A. and Chung, S.-C. 2001. Crystal Chemistry of the Olivine-type

Li(Mn_yFe_{1-y})PO_4 and (Mn_yFe_{1-y})PO_4 as Possible 4 V Cathode Materials for Lithium Batteries. J. Electrochem. Soc. 148, A960-A967.

167. Zhou, F., Cococcioni, M., Kang, K. and Ceder, G. 2004. The Li Intercalation Potential of $LiMPO_4$ and $LiMSiO_4$ Olivines with M = Fe, Mn, Co, Ni. Electrochem. Commun. 6, 1144-1148.

168. Delacourt, C., Poizot, P., Morcrette, M., Tarascon, J.M. and Masquelier, C. 2004. One-Step Low Temperature Route for the Preparation of Electrochemically Active $LiMnPO_4$ Powders. Chem. Mater. 16, 93-99.

169. Dominko, R., Bele, M., Gaberscek, M., Remskar, M., Hanzel, D., Goupil, J.M., Pejovnik, S. and Jamnik, J. 2006. Porous Olivine Composites Synthesized by Sol–gel Technique. J. Power Sources 153, 274-280.

170. Yang, J. and Xu, J.J. 2006. Synthesis and Characterization of Carbon-coated Lithium Transition Metal Phosphates $LiMPO_4$ (M = Fe, Mn, Co, Ni) Prepared via a Nonaqueous Sol-Gel Route. J. Electrochem. Soc. 153, A716-A723.

171. Barker, J., Saidi, M.Y. and Swoyer, J.L. 2001. Lithium Metal Fluorophosphate Materials and Preparation Thereof. Patent WO 2001/084655.

172. Barker, J., Saidi, M.Y. and Swoyer, J.L. 2003. Electrochemical Insertion Properties of the Novel Lithium Vanadium Fluorophosphate, $LiVPO_4F$. J. Electrochem. Soc. 150, A1394-A1398.

173. Barker, J., Saidi, M.Y. and Swoyer, J.L. 2004. A Comparative Investigation of the Li Insertion Properties of the Novel Fluorophosphate Phases, $NaVPO_4F$ and $LiVPO_4F$. J. Electrochem. Soc. 151, A1670-A1677.

174. Barker, J., Gover, R.K.B., Burns, P., Bryan, A., Saidi, M.Y. and Swoyer, J.L. 2005. Structural and Electrochemical Properties of Lithium Vanadium Fluorophosphate, $LiVPO_4F$. J. Power Sources 146, 516-520.

175. Barker, J., Gover, R.K.B., Burns, P., Bryan, A., Saidi, M.Y. and Swoyer, J.L. 2005. Performance Evaluation of Lithium Vanadium Fluorophosphate in Lithium Metal and Lithium-Ion Cells. J. Electrochem. Soc. 152, A1776-A1779.

176. Ateba Mba, J.-M., Masquelier, C., Suard, E. and Croguennec, L. 2012. Synthesis and Crystallographic Study of Homeotypic $LiVPO_4F$ and $LiVPO_4O$ Chem. Mater. 24, 1223-1234.

177. Pizarro-Sanz, J.L., Dance, J.M., Villeneuve, G. and Arriortua-Marcaida, M.I. 1994. The Natural and Synthetic Tavorite Minerals: Crystal Chemistry and Magnetic Properties. Mater. Lett. 18, 327-330.

178. Groat, L.A., Raudsepp, M., Hawthorne, F.C., Ercit, T.S., Sherriff, B.L. and Hartman, J.S. 1990. The Amblygonite-Montebrasite series: Characterization by Single-Crystal Structure Refinemenf Infrared Spectroscopy, and Multinuclear MAS-NMR Spectroscopy. Am. Mineral. 75, 992-1008.

179. Recham, N., Chotard, J.N., Jumas, J.C., Laffont, L., Armand, M. and Tarascon, J.M. 2010. Ionothermal Synthesis of Li-Based Fluorophosphates Electrodes. Chem. Mater. 22, 1142-1148.

180. Tarascon, J.-M., Recham, N. and Armand, M. 2010. Method for Producing Inorganic Compounds. Patent WO 2010/046608.

181. Ramesh, T.N., Lee, K.T., Ellis, B.L. and Nazar, L.F. 2010. Tavorite Lithium Iron Fluorophosphate Cathode Materials: Phase Transition and Electrochemistry of $LiFePO_4F-Li_2FePO_4F$. Electrochem. Solid-State Lett. 13, A43-A47.

182. Ellis, B.L., Ramesh, T.N., Rowan-Weetaluktuk, W.N., Ryan, D.H. and Nazar, L.F. 2012. Solvothermal Synthesis of Electroactive Lithium Iron Tavorites and Structure of Li2FePO4F. J. Mater. Chem. 22, 4759-4766.

183. de Dompablo, M.E.A.Y., Amador, U. and Tarascon, J.M. 2007. A Computational Investigation on Fluorinated-Polyanionic Compounds as Positive Electrode for Lithium Batteries. J. Power Sources 174, 1251-1257.

184. Wu, H.M., Tu, J.P., Yuan, Y.F., Xiang, J.Y. and Chen, X.T. 2007. Effects of abundant Co doping on the structure and electrochemical characteristics of $LiMn_{1.5}Ni_{0.5-}xCoxO_4$. Journal of Electroanalytical Chemistry 8-14.

185. Shin, D.W. and Manthiram, A. 2011. Surface-segregated, high-voltage spinel $LiMn_{1.5}Ni_{0.42}Ga_{0.08}O_4$ cathodes with superior high-temperature cyclability for lithium-ion batteries. Electrochem. Commun. 1213-1216.

186. Sun, X., Hu, X., Shi, Y., Li, S. and Zhou, Y. 2009. The study of novel multi-doped spinel $Li_{1.15}Mn_{1.96}Co_{0.03}Gd_{0.01}O_{4+\delta}$ as cathode material for Li-ion rechargeable batteries. Solid State Ionics 377-380.

Nickel-based Cathode for Li-ion Batteries

Matthew Li[1,2], Zhongwei Chen[2], Khalil Amine[1] and Jun Lu[1]*

[1] Chemical Sciences and Engineering Division, Argonne National Laboratory, 9700 Cass Ave, Lemont, IL 60439, USA
[2] Department of Chemical Engineering, Waterloo Institute of Nanotechnology, University of Waterloo, 200 University Ave West, Waterloo, ON N2L 3G1, Canada

1. Introduction

Of all the electrochemical energy systems, a lithium-ion battery (LIB) is arguably the single most impactful battery contributing to a quality human life. Powered by LIBs, advanced portable electronics with sophisticated functions now play a critical role in modern society. Electric vehicles and grid storage applications of LIBs will also most likely play a huge part in humanity's movement away from fossil fuel based energy sources. At present, Ni-based cathode materials possess electrochemistries which make them the dominant choice for LIBs. In this chapter, a discussion on the evolution of the cathode material for LIBs, namely the Ni-Mn-Co (NMC) oxide and Ni-Co-Al oxide (NCA) cathode systems will be undertaken. Key design parameters will be discussed with emphasis on element selection and composition in addition to a brief overview of the market of the electric vehicles and some of the targets for the material design.

2. Nickel Based Cathodes (Reproduced with permission [1] Copyright 2018, Wiley-VCH)

2.1 From LiNiO$_2$ to Nickel-Manganese-Cobalt Oxides

Perhaps the most impactful and fruitful cathode materials for EV

---

*Corresponding author: junlu@anl.gov

applications were the layered Ni-Mn-Co oxide (NMC) and Ni-Co-Al (NCA) oxides. The origin of these two materials can be traced back to the layered $LiNiO_2$ (LNO). LNO was first synthesized and isolated by L.D. Dyer in 1954 [2] but the first electrochemical testing of $LiNiO_2$ was conducted in 1985 by M. Thomas and Goodenough where a single voltage profile was shown [3].

LNO was a material that possessed a similar layered structure material compared to LCO [4] but with a higher capacity (220 mAh g^{-1}) and energy density (800 Wh kg^{-1}) [5]. LNO was originally investigated as a cheaper alternative to LCO due to the lower cost of Ni and slightly lower voltage vs Li/Li^+ compared to LCO which mitigated anodic electrolyte decomposition [6]. Unfortunately, LNO suffered from many problems such as: low 1st charge/discharge coulombic efficiency and inability to sustain deep charge cycles due to severe changes in crystal structure where Ni^{2+} tended to relocate irreversibly to Li^+ sites due to its similar atomic size (cation mixing) [7]. The cation mixing effect severely deteriorated the rate performance of LNO. Moreover, the thermal stability of LNO was extremely poor and worser than LCO [8]. These problems effectively rendered it unusable for the industry [9-11]. Based on these problems, researchers abandoned the single metal LNO system and moved onto dual metal systems. Researchers discovered that if Co [12-14] or Mn [13, 16-18] were partially substituted for Ni in LNO, the cation mixing between Ni^{2+} and Li^+ was reduced and the irreversible crystal structure transitions experienced at deep charges was prevented, resulting in enhanced cyclability and capacity. In terms of safety, dynamic scanning calorimetry became a standard safety test for next-gen metal oxide based cathodes, where the thermal stability was quantified by its decomposition temperatures. Researchers found that the onset temperature of thermal decomposition of LNO mixed with any of the three metals was delayed compared to LNO and LCO [19]. It was recognized that the incorporations of any singular selection of these three metals with Ni clearly offered substantial but distinct benefits. For example, the incorporation of Mn indeed enhanced thermal stability and cycle stability [20], but the practical reversible capacity was low [21] and the synthesis of $LiNi_{1-x}Mn_xO_2$ (0<x<1) was rather difficult to achieve [22]. Similarly, though improvements on thermal stability from adding Al was apparent, it was not as beneficial as the other metals in preventing the migration of Ni^{2+} to vacated Li^+ sites resulting in a large initial irreversible capacity and poor rate performances [23].

Finally, the mixing of Co with Ni undeniably prevented the migration of Ni^{2+} but still possessed a strong exothermic reaction peak upon heating to higher temperatures and did not satisfy thermal stability concerns [19, 24].

Metal was insufficient, thus researchers began to simultaneously substitute two different metals for Ni forming a ternary layered metal oxide and found synergistic effects [25]. One of the most common mix of metals of commercial relevance was the ternary NMC system. First invented and patent filed at 2001 (2004 granted) by researchers at Argonne National Laboratories, [26] the NMC system is the most widely used lithium-ion battery in EVs. The main benefits of NMC systems were their increased capacity, high decomposition temperature, use of lower toxicity material and the reduced cost due to the substitution of both Co and Ni with cheap Mn. Initially, the cycle stability and electronic conductivity was lower than LCO and posed concerns for the commercial application of ternary metal oxides. However, drastic changes in performance were found by tuning the ratio between the three metals and that led to intensive research into finding the best mix between the metals. Each metal had a different role in NMC, Ni was responsible for providing capacity as it was oxidized from $Ni^{2+} \rightarrow Ni^{3+} \rightarrow Ni^{4+}$, [27, 28] while Mn^{4+} was not electrochemically active but maintained the structural stability and lowered the cost [29]. Co was oxidized from $Co^{2+} \rightarrow Co^{3+}$ after the oxidation of Ni^{2+} and prevented the Nifrom migrating in the Li sites. Research on NMC cathode began with low Ni ratios. In 2001, T. Ohzuku's and Y. Makimura synthesized and tested $LiNi_{0.33}Mn_{0.33}Co_{0.33}O_2$ [30] while in the same year, Dahn [31] investigated $LiNi_xCo_{1-2x}Mn_xO_2$ (x=0.25 and 0.375), the work from both of these groups introduced the NMC class of material to the battery field. All ratios demonstrated reasonably high capacities (>150 mAh g^{-1}), good cycle stability (depending on the charge cut off voltage [5]) and enhanced thermal stability [32] NMC with x=1/3 has exhibited inferior rate performances due to the lower Li-ion diffusion compared to LCO and was often blended with the safer, higher rate performance and cheaper LMO. This blend of NMC-LMO has become a very popular cathode material among researchers and industry as the LMO provided extra safety to the system, in addition to lowering the cost and increasing the power density [33-35]. In 2011, Z. Li and Whittingham [36] sought to find an optimal NMC ratio, testing NMC 333, 442 and 992. Compared to NMC 333, NMC 442 possessed a higher capacity at low currents and comparable capacities at higher currents while reducing cost (lowered Co content from 33% to 20%). NMC 992 was found to possess significantly poorer capacity at high discharge rates and was less stable. In general, it was realized that the higher the Ni content in the cell, the higher the intrinsic specific capacity but at a cost of poorer stability and rate performance [37].

Specifically, NMC materials with Ni content >50% (known as Ni-rich compounds) were particularly problematic Ni-rich cathodes can almost be considered another class of layered cathode material. While Ni-rich compounds were theoretically denser in energy and cheaper (lower Co content), they were usually problematic to implement due to safety and cycle stability concerns. The relationship of Ni content to thermal stability and oxygen gas generation is shown in Fig. 1a. Following the convention: NMC 433 (4-Ni: 3-Mn: 3-Co molar ratio), the ratio with the least amount of Ni exhibited the least amount of gas generation and at a higher decomposition initiation temperature. Every subsequent increase in Ni content noticeably decreased the onset temperature for phase change from the original layered to spinel structure and rock salt which sped up the overall degradation phase transformation process [37, 38].

Additionally, the higher Ni content and reduced Co content promoted the irreversible migration of Ni^{2+} to Li^+ sites [39] which severely hindered Li^+ transfer throughout the structure of the NMC particle and increased the overall cell impedance [40].

Though the exact mechanism that caused these detrimental effects were not fully understood, it was clear that the combining effect of phase transition and impedance increase resulted in poorer cycle performance at higher Ni contents. A comprehensive relationship between Ni content, specific capacity, cycle stability and thermal stability is shown in Fig. 1b. Very recently, based on density functional theory, researchers concluded that more Ni^{4+} are present in Ni-rich materials which readily oxidize the electrolyte and oxygen ions due to their relatively low LUMO which promotes oxygen generation and electrolyte decomposition [41].

Strategies to achieve viable Ni-rich NMC materials typically revolved around preventing the transformation of the NMC layered phase. J. Yang has demonstrated that doping with Li_2MnO_3 can suppress the phase transition [45]. Adopting a more morphological themed approach, researchers at ANL developed the now popular strategy coating of less reactive (lower Ni content) NMC or other cathode materials over a Ni-rich particle (core-shell structure) which required specific expertise for precisely controlling the coverage [46-48]. This strategy mitigates the surface exposure of high Ni-content NMC to the electrolyte which is the main interface where undesired reactions occur. However, the key disadvantage of this strategy is the mismatch in volume change between the core and shell material cutting off charge transfer pathways [47]. This phenomenon is the main reason for the cycle degradation of core shell Ni-rich structures. Building upon this work, in 2009 ANL has created what is called a concentration gradient shelll (CGS) [49]. CGS uses a shell that consisted of a Ni concentration gradient with the Ni-content highest near the center and lowest at the surface covering a Ni-rich core. This material was synthesized by simply precipitating NMC with progressively lower Ni content onto a Ni-rich core (substrate). Such a strategy has been shown to completely eliminate the structural mismatch generated in the core-shell strategy but tended to degrade due to migration of Ni content towards the surface [50]. Further advancement in this technology created what is known as the full concentration gradient (FCG) [51]where the Ni and

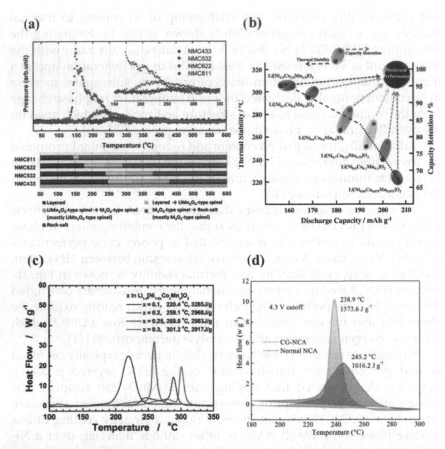

Fig. 1. (a) The dependence of thermal stability in gas evolution on Ni content in NMC. (Reproduced with permission [38]. Copyright 2014 American Chemical Society) (b) the dependence of Ni content on discharge capacity and capacity retention and thermal stability in NMC. (Reproduced with permission [42]. Copyright 2013 Elsevier. (c) Thermal stability in W g^{-1} of various Ni composition in NMC. (Reproduced with permission [43]. Copyright 2007 Electrochemical Society, Inc.) (d) Thermal stability of regular NCA and modified NCA (Ni: Co: Al=0.8:0.15:0.05). (Reproduced with permission [44]. Copyright 2017 American Chemical Society, All thermal stabilities measurements were conducted after charging to 4.3 V)

Mn concentrations varied all the way from the core to the surface which further led to the two-sloped full concentration gradient (TSFCG) [52] where a smooth concentration profile of Ni, Co and Mn was created from the core to near the surface with an abrupt concentration change near the surface. Both FCG and TSFCG offered substantially higher performance benefits over the CGS materials, with the TSFCG material demonstrating excellent stability over 500 cycles in a full cell configuration (Si-based anode) delivering 350 Wh kg^{-1} on the first cycle [52].

Initially, the lower Ni content NMCs have found significant application in the industry, representing ~26% by mass of all battery cathodes [53] sold in 2016 and has become by far the most popular cathode material for EVs. Currently, NMC 333 and 532 represents a large portion of the NMC market with NMC 622 and 811 still a minority, [53] but higher Ni content cathodes are expected to be more and more prevalent in the near future [54].

2.2 Nickel-Cobalt-Aluminum Oxide

In addition to Ni-rich NMC, another Ni-rich ternary metal cathode material was the Ni-Co-Al (NCA) system. NCA was slightly better than Ni-rich NMC in terms of energy density but had a slightly lower discharge voltage due to slightly different active redox pairs [55].

The first appearance of the NCA system was in 2001 by K.K. Lee's group [56] and has become a very popular cathode material with Panasonic implementing it in the Tesla vehicles [57, 58]. The most popular mix was: $LiNi_{0.8}Co_{0.15}Al_{0.05}$ and has been known colloquially among industry and researchers as NCA [59]. One should note that NCA incorporates relatively low level of its third metal (Al at 5-10%), much lower than the Mn in NMC (10-40%). This was because the use of higher levels of Al (>10%) resulted in severe capacity decay and poor Li^+ ion diffusion throughout its structure [19, 25]. Because NCA can only operate acceptably at low Al content, all metal ratios of NCA are usually considered Ni-rich and also suffer from the corresponding disadvantages. A common problem between all Ni-rich systems is the chemical sensitivity of Ni^{3+} which along with residual excess Li (from its lithiation manufacturing steps) promotes adsorption of moisture and CO_2 forming LiOH and Li_2CO_3 which have been shown to limit cycle life. During storage in ambient air, the surface Ni-ions react with CO_2 and moisture forming insulating layers on its surface [60]. Furthermore, similar to the Ni-rich NMC cathode, the Ni cation in NCA is known to dissolve into the electrolyte due to the HF created by the reaction between $LiPF_6$ (electrolyte salt) with trace amounts of water [5].

This phenomenon decayed the cathode and had very recently led material scientists to apply more resistive coatings composed of ZnO [59], $FePO_4$ [61], Li_3PO_4 [62], $AlPO_4$ [63], $LiMnPO_4$ [64]. Among many other similar concepts from research done on surface coatings for NMC. Al offered good thermal stability compared to LCO and LNO but still underwent severe exothermic reactions at higher temperatures (200-250°C) when NCA was in its delithiated state [44, 65]. The thermal stability of NCA was inferior to NMC 333 [55] due to the higher Ni content which rendered it problematic for commercial use. However, when compared to NMC 811, the thermal stability of NCA was far superior. As shown in Fig. 1c, thermal decomposition of NMC 811 began to occur at ~190 °C and peaked at ~220 °C releasing 3285 J g^{-1}. For NCA, the thermal decomposition began to increase at ~200 °C peaking at ~239 °C and only released 1573 J g^{-1} (Fig. 1d). Besides Panasonic/Tesla Inc., no other EV company uses a pure NCA cathode, but Automotive Energy Supply Corporation (AESC) does supply NCA mixed with LMO cathodes for the Nissan Leaf EV [33]. In 2016, NCA only occupied ~9% (16, 200 tons) of all battery cathode materials sold (180,000 tons) and is forecasted to reach about 40, 000 tons (10%) in 2025 if Tesla Inc. continues to incorporate NCA into their cathodes [53].

3. The Current Status of Electric Vehicles and Lithium-ion Batteries (Reproduced with permission [1] Copyright 2018,Wiley-VCH)

It is typically recognized that materials with <60% Ni content have been fully commercialized while >80% are still under development [66]. Overall, the commercialization of Ni-based layered oxide cathode materials (NMC and NCA) have been quite successful, representing a total of 35% of the battery market in 2016 (Fig. 2) and instrumental in improving the range and cost of EVs. From its early stage at the Electric Vehicle Company (2000 units sold), to the current increase in interest by almost all automotive companies, the incorporation of these new cathodes had undoubtedly made an impact on the commercialization of xEVs. The market penetration of xEVs increased drastically over the recent years. Just in the year of 2014, the total sales of xEVs has reached 320, 000 vehicles with over half being pure EVs (182,400 vehicles). Although this only represents 0.3% of all passenger vehicles sold during this time span, it was orders of magnitude higher than the EVs sold in the past decades [67]. Companies such as Toyota, Honda, GM, Tesla Inc. have heavily invested in a xEV based future. Volvo Cars have even recently announced that they will be looking to completely electrify (either EV or hybrid) its passenger vehicles line by 2019 [68]. Several European countries have set goals to eliminate the internal combustion engine by 2040 [69]. With a global market size of >$2.4 billion in 2011, >$8.9 billion in 2015 and a forecasted >$14 billion by 2020, [70] it is now hard to imagine a future society without EVs. However, there is still much room for improvement. The energy density requirements of EVs have always been a topic of much discussion. While it played a crucial role in the development of LIBs it appears that the current driving range for even pure EVs are approaching satisfactory levels for the consumer. The energy density targets (volume and mass basis) have nearly been reached by current LIBs. Cell-level

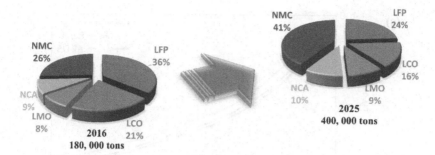

Fig. 2. Mass percent of all LIBs market shares of each leading materials in 2016 and forecasted to 2025. (Reproduced with permission [53]. Copyright 2018, Wiley-VCH)

volumetric energy density was targeted at 750 Wh L^{-1} [71] and the state-of-the-art high tension cylindrical 18650 cells for Tesla Inc. are already at 600-650 Wh L^{-1} (20% less for pouch and prismatic cell configurations) [72]. Cell-level gravimetric energy densities of current LIBs (248 Wh kg^{-1} in Tesla Model S 2014 [33]) are reasonably close to the targeted 350 Wh kg^{-1} [71].

A study was conducted by Z. Needell and J. Trancik on the impact of higher range EVs where they proposed the metric: daily vehicle adsorption potential (DAP) as shown in Fig. 3a [73]. The DAP is defined as the % of days that a pure-EV would not be able to make the driver's daily trips on a single charge. Inversely, 1-DAP represents the % of days that drivers of EVs will be required to recharge within the day. Shown in Fig, 3b, in 2013, the 1-DAP of the Nissan Leaf (88 Wh kg^{-1}) was about 10%. This means that in 2013, around 10% of the time the drivers will not be able to return home without requiring a recharge. The 2016 Tesla Model S-90 has a 1-DAP of only about 1-2%. It was argued that such a low percentage of 1-DAP is already sufficient for the widespread adoption of EVs into the markets for that specific geographical location of study. While the exact implications on the EV market penetration of a 1-DAP value of 1% are subjective without any concrete input from the consumers, it is still quite surprising that future solutions to the very popular problem of energy density will most likely experience diminishing returns. This suggests that the most important challenge for the electrification of transportation is probably not its energy density (albeit still important for other LIB applications), but instead the cost of LIBs. However, it should be noted that the cost and energy density are intrinsically related. This is because higher energy density electrode requires less active material and therefore, lowering the cost of manufacturing. Nevertheless, the cost targets of <150 $US/kWh for battery packs [71, 74, 75] have yet to be met by any of the battery manufacturing companies.

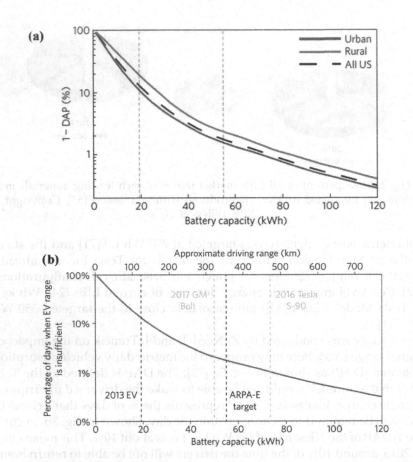

Fig. 3. (a) The impact of type of driving (urban, rural) in the US on the 1-DAP value. Reproduced with permission [73]. (Copyright 2016 Springer Nature). (b) The relationship between the % of days when EV range is insufficient (1-DAP) versus the battery capacity and range with the 2013 EV (Nissan Leaf @ 88 Wh kg^{-1}) and 2016 Tesla Model S-90 labeled on graph. (Reproduced with permission [76]. Copyright 2016 Springer Nature)

The cumulative cost benefits thus far of these new cathode and anode materials can perhaps be summarized in the modeled cost of productions by the US Department of Energy (DOE), Vehicle Technologies Office. Over a 7-year span, the modelled cost has decreased ~70% to ~268 US$/kWh (2015) [77].

While there are still a large volume of estimates from literature that ranges from 1700-1000 US$ (YR:2014) as summarized by B. Nykvist in Fig. 3a, there were a number of corroborating optimistic estimates claiming ~220-400 US$ (YR:2014)/kWh (Fig. 3b) [74] and 250 US$/kWh in 2015 [53].

Table 1. List of battery company, their vehicle of application, and
battery material [57]

Battery company	Vehicle of application [Company/Model]	Battery material [Anode | | Cathode]
AESC	Nissan/Leaf	Graphite | | LMO-NCA
LG Chem	Renault/Zoe	Graphite | | NMC-LMO
Li-Tec	Daimler/Smart	Graphite | | NMC
Li Energy Japan	Mitsubishi/i-MiEV	Graphite | | LMO-NMC
Samsung	Fiat/500	Graphite | | NMC-LMO
Lishen Tianjin	Coda/EV	Graphite | | LFP
Toshiba	Honda/Fit	LTO | | NMC
Panasonic	Tesla/Model S	Graphite | | NCA

Reported in 2014 by Advanced Automotive Batteries (data presented in Table 1), the main battery technologies used in EVs were NCA, NMC and LMO usually blended with NCA or NMC due to LMO's low specific capacity [57].

LFP is currently a very popular cathode material for BYD Auto, a major EV producer in China [54]. In contrast to the wide selection of cathode technologies, the anode of choice for passenger xEVs remains to be carbon based with only Toshiba among the shown companies utilizing LTO as the anode. LTO has found more application in the electric bus market with companies such as Microvast© demonstrating significant interest in the safety aspect of this technology [78].

Interestingly, there is no reported use of LCO in any of the EVs, possibly suggesting that the cost benefits of adding the allegedly cheaper low-Co cathodes should have already been reflected in the price of EVs. This implies that most recent cost reduction was achieved solely through process optimization such as the current movement away from the 18,650 cylindrical cells to the 21,700 format [79]. Estimates from industry indicate, while maintaining the use of NMC cathode materials, there was still a substantial drop in the cost of a cathode from ~64 US\$/kWh to ~40 US\$/kWh in the time frame of 2010 to 2015 which was due to the increased NMC production efficiency [53]. Cost reductions were also experienced in the separator, electrolyte, anode technologies and from economies of scale (recycling, overhead, utilities) even though the same base materials were used. However, without another major change in materials, these cost reductions were ultimately limited as reflected in the asymptotic approach to 150 US\$/kWh of forecasted cost curves (Fig. 4). Furthermore, since the cost of nickel and cobalt are quite volatile, (Co increased from ~15 US\$/lb in 2005 to 52 US\$/lb in 2007 and recent increases from ~10 US\$/lb in 2015 to ~27 US\$/lb in 2017) [75] the marginally acceptable forecasted prices could be easily invalidated. In addition to the simple cost of raw material, the safety concerns could also have severe monetary penalties for battery manufacturers. In 2006, Sony's recall of its LIB in laptops has resulted in a 40% loss that year in their battery business [53].

After lacking a sufficient recovery, Sony has recently sold its battery business to Murata in 2016 which could be an indication of the financial severity of a recall [80, 81].

4. Next-Generation Li-ion Battery Research (Reproduced with permission [1] Copyright 2018, Wiley-VCH)

Next generation Li-ion battery technologies with potentially huge cost and performance benefits could be the answer to reaching the cost and energy density targets. Systems such as Li- and Mn-rich cathodes, 5 V cathodes, and silicon anode have been extensively researched. 5 V cathode materials were considered as a potential candidate for "next-generation" LIBs. This class of material includes the spinel $LiMn_{1.5}Ni_{0.5}O_4$ (published by Dahn in 1997 [82] and patent filed in 1997 by K. Amine [83], inverse spinel $LiNiVO_4$ (by Dahn in 1994) [84] and the olivine $LiCoPO_4$ (by Amine in 2000) [85] among others which all possessed an extremely high discharge voltage profile of >4.5 V vs Li/Li$^+$ as shown in the cyclic voltammetry (Fig. 5a) and charge/discharge voltage profiles (Fig. 5b). As the energy density is a product of capacity and discharge voltage, this class of material is envisioned to achieve energy densities well above 400 Wh kg^{-1}. One of the main problems associated such a high voltage cathode is the oxidation of the electrolyte. Traditional carbonate-based electrolytes cannot withstand potentials higher than 4.3 V vs Li/Li$^+$ and will be oxidized if used with 5V cathodes. Accordingly, researchers have employed more anodically stable electrolytes such as ionic liquids, sulfones, nitrile, carbonates derivatives and carbonates with additives [86].

Additional problems with 5 V cathodes were unique to the type of the material but similar to its parental archetype. For example, similar to the spinel $LiMn_2O_4$, Mn^{2+} dissolution is a concern for the spinel $LiMn_{1.5}Ni_{0.5}O_4$ cathode especially at high temperatures [87] while the $LiCoPO_4$ olivine cathode also suffers from poor electronic conductivity much like the olivine LFP [88]. Although the nature of the problems and solutions were similar to its parental archetype [89] the 5 V cathode materials have not made it to the market as yet.

Another class of material that also employs high voltages is the Li- and Mn-rich (LMR) cathode or sometimes known as layer-layered/

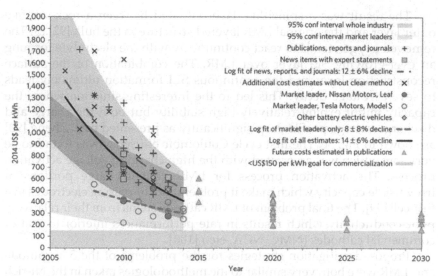

Fig. 4. Estimated cost of battery packs in 214 US$/kWh from 2005 to 2014 and forecasts to 2030. (Reproduced with permission [74]. Copyright 2015 Springer Nature.)

layered-spinel material. Pioneered at ANL, the concept was to substitute entire crystal units rather than just cations [91]. Versions of this material are composed of a layered R-3m structured $LiMO_2$ (where typically M=Ni) stabilized by a C2/m structured monoclinic Li_2MnO_3. Together, this combination was often written as $xLi_2MnO_3 \cdot (1-x)LiMO_2$ where $0<x<1$ [92, 93].

The first patent on LMR was filed by Thackeray and Amine in 2001 [26] and later published by Lu and Dahn in 2001 [16], which demonstrated a reversible capacity of ~200 mAh g^{-1} for 50 cycles at 55 °C from 4.6 to 2.0 V vs Li/Li$^+$. In addition to a high theoretical specific energy (~900 Wh kg^{-1}), the cost benefits of completely removing Co makes this material quite attractive. Unfortunately, this material is still far from commercialization. The initial promises of LMR were met with severe performance challenges. These performance hurdles were due to the LMR's high voltage (up to 4.6-4.8 V vs Li/Li$^+$) charging requirements in order to "activate" the Li_2MnO_3 component and realize its high capacity [94]. There were many ensuing issues that are involved in this process: (1) This voltage range is beyond the stability window of the typically used electrolyte. (2) The transformation of the crystal structure on the surface of LMR particles [95]forms what is known as the surface reconstruction layer whichlimits Li-ion diffusion. This layer was formed during the activation process of Li_2MnO_3 where oxygen atoms are irreversibly removed [96]. Without oxygen atoms, the destabilized Ni-ions irreversibly migrate to vacant Li-ion sites [97, 98].

The result was a spinel-like layer situated between a rock-salt-like outer layer and the original LMR layered structure in the bulk [92].(3) The removed oxygen tends to react continuously with the electrolyte forming an ever-growing SEI layer over LMR. The combination of the surface reconstruction layer and the continuous SEI formation ultimately leads to severe voltage decay. This led to the interesting situation where the capacity might exhibit relatively high stability but both the charge and discharge voltage profile drop significantly as presented by J. Zheng [90] as shown in Fig. 5c. The first cycle Coulombic efficiency was found to be very low which was associated with the high charging voltage activation process. The activation process for LMR contains large portions of irreversible capacity which make it problematic for pairing electrodes in a full-cell [99]. The final problem of LMR cathode stems from their relatively poor conductivity which results in rate performances inferior to that of commercial cathodes (NMC, NCA, etc.) [100].

Proposed mitigation strategies for the problems of the 5 V cathode and LMR were both very similar to the methodologies taken in the Ni-rich field of research. By applying a surface coating of various materials [90, 92,

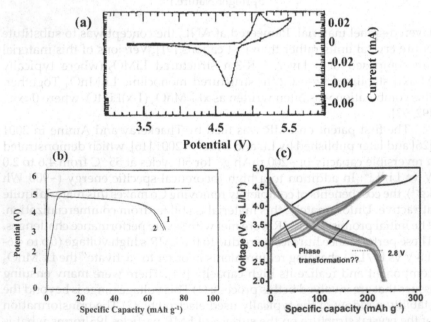

Fig. 5. (a) Cyclic voltammetry and (b) charge discharge profile (1st and 2nd cycle) of the 5 V cathode. (Reproduced with permission [85]. Copyright 2000 Electrochemical Society, Inc.) (c) Charge/discharge profiles of LMR, demonstrating its characteristic voltage fade problem. (Reproduced with permission [90]. Copyright 2015 American Chemical Society)

101, 102], the direct contact between LMR and the electrolyte is limited and so the degree of SEI formation on the cathode as well. Techniques based more on solid-state physics are also very popular. Lattice substitution for Al [103], Mg [104], Ti [105] among other elements can help to prevent phase changes, limiting voltage decay or improve the conductivity [106]. Though scientifically insightful, most of the reported lab scale tests cannot be scaled to the commercial level. Accordingly, the 5 V spinel and LMR class of cathodes have yet to find commercial application. Researchers have even combined the $LiMn_{1.5}Ni_{0.5}O_2$ with the LMR system, known as the integrated layered-spinel structure [107, 108]. First proposed by Park and Thackeray in 2006, this combines the benefit of the high capacity of LMR with the fast lithium diffusion of the spinel structure.

Alloy based anodes [109] such as Ge [110, 111], Sn [112, 113] and especially Si have recently drawn much renewed attention most likely due to the onset of research into nanotechnology. Early on (prior to 1976), these alloy-based anodes have already been researched as a potential replacement to the unstable and dangerous pure Li anode [114]. The problem with alloy based anodes is the enormous volume expansion upon lithiation (>300% for Si) which causes severe electronic disconnections and promotes continuous SEI formation. Strategies adopting nanowire structures [115] and nanoparticles [116] demonstrated enhanced utilization and stability of Si anode. To follow was an explosive research interest into the Si anode technology with focus surrounding the use of breathable conductive networks [117, 118], porous breathable Si morphologies [119], Si dimensional control among other techniques [120]. It is also not necessary to fabricate an anode solely with Si as the active material, since it is possible to blend Si with graphite or hard carbon to achieve a high capacity (700 mAh g^{-1}) without the severe cycle degradation for pure Si [121]. Often neglected in literature, the volume expansion will affect the volumetric capacity at the cell level [122]. With rumors describing Tesla Inc.'s incorporation of small amounts of Si (in the form of SiO_x) into their anodes to boost energy density [123], Si based anodes are arguably the most important next-gen anodes material [71], but probably only certain configurations of Si anodes will be commercially viable. The commonly used nanoparticles in literature will most likely pose severe challenges in achieving appropriate volumetric energy densities.

5. Conclusion (Reproduced with permission [1] Copyright 2018, Wiley-VCH)

Over the years, Ni-based cathodes have become the cornerstone both in terms of academic interest and industrial application. We have provided a brief overview of the development process of both NMC and NCA. The role that Li-based batteries played in revolutionizing consumer electronics and electric vehicles has been clearly indispensable and will probably continue to be so for the foreseeable future. The first commercialized LIB by Sony Co. in 1991 had its energy density as the main design criteria. Since consumer electronics were inherently somewhat of a luxury item, the cost was less important. Modern improvements into the consumer electronics LIB still onto this theme while research into LIB for xEVs mandates cheaper cells. We hope this chapter have presented to the next generation of battery scientists and even veteran battery scientists a comprehensive developmental history of Ni-based Li-ion battery cathodes. For a continued decrease in the cost of xEV, further advances must be made to enable high Ni content or even Co free compositions. While many large companies have achieved decreased costs by economies of scale, it is much more advantageous to solve this problem from a material standpoint.

Acknowledgements

J. Lu and K Amine gratefully acknowledge support from the U.S. Department of Energy (DOE), Office of Energy Efficiency and Renewable Energy, Vehicle Technologies Office. Argonne National Laboratory is operated for DOE Office of Science by UChicago Argonne, LLC, under contract number DE-AC02-06CH11357. Z. Chen and M. Li would like to acknowledge financial supportfrom the Natural Sciences and Engineering Research Council of Canada (NSERC) and the Waterloo Institute for Nanotechnology (WIN).

References

1. Li, M., Lu, J., Chen, Z. and Amine, K. 2018. 30 Years of lithium-ion batteries. Adv. Mater. 1800561.
2. Dyer, L.D., Borie, B.S. and Smith, G.P. 1954. Alkali metal-nickel oxides of the type MNiO$_2$. J. Am. Chem. Soc. 76, 1499-1503.
3. Thomas, M., David, W., Goodenough, J. and Groves, P. 1985. Synthesis and structural characterization of the normal spinel Li [Ni$_2$] O$_4$. Mater. Res. Bull. 20, 1137-1146.
4. Ohzuku, T., Ueda, A. and Nagayama, M. 1993. Electrochemistry and structural chemistry of LiNiO$_2$ (R $\bar{3}$ m) for 4 volt secondary lithium cells. J. Electrochem. Soc. 140, 1862-1870.
5. Liu, W., Oh, P., Liu, X., Lee, M.J., Cho, W., Chae, S., Kim, Y. and Cho, J. 2015. Nickel-rich layered lithium transition-metal oxide for high-energy lithium-ion batteries. Angew. Chem. Int. Ed. 54, 4440-4457.
6. Dahn, J.R., von Sacken, U., Juzkow, M.W. and Al-Janaby, H. 1991. Rechargeable LiNiO$_2$ / carbon cells. J. Electrochem. Soc. 138, 2207-2211.

7. Nitta, N., Wu, F.X., Lee, J.T. and Yushin, G. 2015. Li-ion battery materials: Present and future. Mater Today18, 252-264.
8. Dahn, J.R., Fuller, E.W., Obrovac, M. and von Sacken, U. 1994. Thermal stability of LixCoO$_2$, LixNiO$_2$ and λ-MnO$_2$ and consequences for the safety of Li-ion cells. Solid State Ionics 69, 265-270.
9. Nishi, Y. 2001. Lithium ion secondary batteries: Past 10 years and the future. J. Power Sources 100, 101-106.
10. Choi, J. and Manthiram, A. 2005. Role of chemical and structural stabilities on the electrochemical properties of layered LiNi $_{1/3}$Mn$_{1/3}$Co$_{1/3}$O$_2$ cathodes. J. Electrochem. Soc. 152, A1714-A1718.
11. Arai, H., Okada, S., Sakurai, Y. and Yamaki, J.I. 1998. Thermal behavior of Li$_{1-y}$NiO$_2$ and the decomposition mechanism. Solid State Ionics 109, 295-302.
12. Delmas, C. and Saadoune, I. 1992. Electrochemical and physical properties of the Li$_x$Ni$_{1-y}$Co$_y$O$_2$ phases. Solid State Ionics 53, 370-375.
13. Arai, H., Okada, S., Sakurai, Y. and Yamaki, J.I. 1997. Electrochemical and thermal behavior of LiNi$_{1-z}$M$_z$O$_2$ (M = Co , Mn , Ti) J. Electrochem. Soc. 144, 3117-3125.
14. Delmas, C., Saadoune, I. and Rougier, A. 1993. The cycling properties of the Li$_x$Ni$_{1-y}$Co$_y$O$_2$ electrode. J. Power Sources 44, 595-602.
15. Ohzuku, T., Ueda, A. and Kouguchi, M. 1995. Synthesis and characterization of LiAl$_{1/4}$Ni$_{3/4}$O$_2$ (R3m) for lithium-ion (Shuttlecock) batteries. J. Electrochem. Soc. 142, 4033-4039.
16. Lu, Z., MacNeil, D.D. and Dahn, J.R. 2001. Layered cathode materials [Ni$_x$Li$_{(1/3-2x/3)}$Mn$_{(2/3-x/3)}$]O$_2$ for lithium-ion batteries. Electrochem.Solid-State Lett. 4, A191-A194.
17. Rossen, E., Jones, C.D.W. and Dahn, J.R. 1992. Structure and electrochemistry of Li $_x$Mn$_y$Ni$_{1-y}$O$_2$. Solid State Ionics 57, 311-318.
18. Kang, K., Meng, Y.S., Bréger, J., Grey, C.P. and Ceder, G. 2006. Electrodes with high power and high capacity for rechargeable lithium batteries. Science 311, 977-980.
19. Xu, J., Lin, F., Doeff, M.M. and Tong, W. 2017. A review of Ni-based layered oxides for rechargeable Li-ion batteries. J. Mater. Chem. A5, 874-901.
20. Paulsen, J.M., Thomas, C.L. and Dahn, J.R. 2000. O$_2$ Structure Li$_{2/3}$[Ni$_{1/3}$ Mn$_{2/3}$]O$_2$: A new layered cathode material for rechargeable lithium batteries. I. electrochemical properties. J. Electrochem. Soc. 147, 861-868.
21. Lu, Z. and Dahn, J.R. 2001. Can all the lithium be removed from T$_2$Li$_{2/3}$[Ni$_{1/3}$Mn$_{2/3}$]O$_2$. J. Electrochem. Soc. 148, A710-A715.
22. Paulsen, J.M. and Dahn, J.R. 2000. O$_2$-Type Li$_{2/3}$[Ni$_{1/3}$Mn$_{2/3}$]O$_2$: A new layered cathode material for rechargeable lithium batteries II. structure, composition, and properties. J. Electrochem. Soc. 147, 2478-2485.
23. Guilmard, M., Rougier, A., Grüne, M., Croguennec, L. and Delmas, C. 2003. Effects of aluminum on the structural and electrochemical properties of LiNiO$_2$. J. Power Sources 115, 305-314.
24. Wang, L., Li, J., He, X., Pu, W., Wan, C. and Jiang, C. 2009. Recent advances in layered LiNi$_x$Co$_y$Mn$_{1-x-y}$O$_2$ cathode materials for lithium ion batteries. J. Solid State Electrochem. 13, 1157-1164.

25. Kalyani, P., Kalaiselvi, N., Renganathan, N.G. and Raghavan, M. 2004. Studies on $LiNi_{0.7}Al_{0.3-x}Co_xO_2$ solid solutions as alternative cathode materials for lithium batteries. Mater. Res. Bull. 39, 41-54.

26. Thackeray, M.M., Johnson, C.S., Amine, K. and Kim, J. 2004. Lithium metal oxide electrodes for lithium cells and batteries. US Patent 6677082 B2.

27. Ren, H., Huang, Y., Wang, Y., Li, Z., Cai, P., Peng, Z. and Zhou, Y. 2009. Effects of different carbonate precipitators on $LiNi_{1/3}Co_{1/3}Mn_{1/3}O_2$ morphology and electrochemical performance. Mater. Chem. Phys. 117, 41-45.

28. Hwang, B.J., Tsai, Y.W., Carlier, D. and Ceder, G. 2003. A combined computational/experimental study on $LiNi_{1/3}Co_{1/3}Mn_{1/3}O_2$. Chem. Mater. 15, 3676-3682.

29. Wei, Y., Zheng, J., Cui, S., Song, X., Su, Y., Deng, W., Wu, Z., Wang, X., Wang, W. and Rao, M. 2015. Kinetics tuning of li-ion diffusion in layered Li $(Ni_xMn_yCo_z)O_2$. J. Am. Chem. Soc. 137, 8364-8367.

30. T Ohzuku, Y.M. 2001. Layered lithium insertion material of $LiCo_{1/3}Ni_{1/3}Mn_{1/3}O_2$ for lithium-ion batteries [J]. Chem. Lett. 30, 642-643.

31. Lu, Z., MacNeil, D.D. and Dahn, J.R. 2001. Layered $Li[Ni_xCo_{1-2x}Mn_x]O_2$ cathode materials for lithium-ion batteries. Electrochem. Solid-State Lett. 4, A200-A203.

32. Belharouak, I., Sun, Y.K., Liu, J. and Amine, K. 2003. $Li(Ni_{1/3}Co_{1/3}Mn_{1/3})$ O_2 as a suitable cathode for high power applications. J. Power Sources 123, 247-252.

33. Xiao, Q., Li, B., Dai, F., Yang, L. and Cai, M. 2015. Application of lithium ion battery for vehicle electrification. *In*: Electrochemical Energy: Advanced Materials and Technologies. CRC Press https://www.routledge.com/ Electrochemical-Energy-Advanced-Materials-and-Technologies-1st-Edition/Shen-Wang-Jiang-Sun-Zhang/p/book/9781138748927

34. Kim, J.W., Travis, J.J., Hu, E., Nam, K.W., Kim, S.C., Kang, C.S., Woo, J.H., Yang, X.Q., George, S.M., Oh, K.H., Cho, S.J. and Lee, S.H. 2014. Unexpected high power performance of atomic layer deposition coated $Li[Ni_{1/3}Mn_{1/3}Co_{1/3}]O_2$ cathodes. J. Power Sources 254, 190-197.

35. Mao, Z., Farkhondeh, M., Pritzker, M., Fowler, M. and Chen, Z. 2016. Multi-particle model for a commercial blended lithium-ion electrode. J. Electrochem. Soc. 163, A458-A469.

36. Li, Z., Chernova, N.A., Roppolo, M., Upreti, S., Petersburg, C., Alamgir, F.M. and Whittingham, M.S. 2011. Comparative study of the capacity and rate capability of $LiNi_yMn_yCo_{1-2y}O_2$ (y = 0.5, 0.45, 0.4, 0.33). J. Electrochem. Soc. 158, A516-A522.

37. Jung, R., Metzger, M., Maglia, F., Stinner, C. and Gasteiger, H.A. 2017. Oxygen release and its effect on the cycling stability of $LiNi_xMn_yCo_zO_2$ (NMC) cathode materials for Li-Ion batteries. J. Electrochem. Soc. 164, A1361-A1377.

38. Bak, S.M., Hu, E., Zhou, Y., Yu, X., Senanayake, S.D., Cho, S.J., Kim, K.B., Chung, K.Y., Yang, X.Q. and Nam, K.W. 2014 Structural changes and thermal stability of charged $LiNi_xMn_yCo_zO_2$ cathode materials studied by combined in situ time-resolved XRD and mass spectroscopy. ACS Appl. Mater. Interfaces 6, 22594-22601.

39. Zhang, X., Jiang, W.J., Mauger, A., Qilu, Gendron, F. and Julien, C.M. 2010. Minimization of the cation mixing in $Li_{1+x}(NMC)_{1-x}O_2$ as cathode material. J. Power Sources 195, 1292-1301.

40. Kang, K. and Ceder, G. 2006. Factors that affect Li mobility in layered lithium transition metal oxides. Phys. Rev. B74, 094105.

41. Dixit, M., Markovsky, B., Schipper, F., Aurbach, D. and Major, D.T. 2017. Origin of structural degradation during cycling and low thermal stability of Ni-rich layered transition metal-based electrode materials. J. Phys. Chem. 121, 22628-22636.

42. Noh, H.J., Youn, S., Yoon, C.S. and Sun, Y.K. 2013. Comparison of the structural and electrochemical properties of layered $Li[Ni_xCo_yMn_z]O_2$ (x = 1/3, 0.5, 0.6, 0.7, 0.8 and 0.85) cathode material for lithium-ion batteries. J. Power Sources 233, 121-130.

43. Lee, K.S., Myung, S.T., Amine, K., Yashiro, H. and Sun, Y.K. 2007. Structural and electrochemical properties of layered $Li[Ni_{1-2x}Co_xMn_x]O_2$ (x = 0.1-0.3) positive electrode materials for Li-ion batteries. J. Electrochem. Soc. 154, A971-A977.

44. Hou, P., Zhang, H., Deng, X., Xu, X. and Zhang, L. 2017. Stabilizing the electrode/electrolyte interface of $LiNi_{0.8}Co_{0.15}Al_{0.05}O_2$ through tailoring aluminum distribution in microspheres as long-life, high-rate, and safe cathode for Lithium-ion batteries. ACS Appl. Mater. Interfaces 9, 29643-29653.

45. Yang, J. and Xia, Y. 2016. Suppressing the phase transition of the layered Ni-rich oxide cathode during high-voltage cycling by introducing low-content Li_2MnO_3. ACS Appl. Mater. Interfaces 8, 1297-1308.

46. Chen, Z., Qin, Y., Amine, K. and Sun, Y.K. 2010. Role of surface coating on cathode materials for lithium-ion batteries. J. Mater. Chem. 20, 7606-7612.

47. Sun, Y.K., Myung, S.T., Park, B.C. and Amine, K. 2006. Synthesis of spherical nano-to-microscale core-shell particles $Li[(Ni_{0.8}Co_{0.1}Mn_{0.1})_{1-x} (Ni_{0.5}Mn_{0.5})_x]O_2$ and their applications to lithium batteries. Chem. Mater. 18, 5159-5163.

48. Li, W., Dolocan, A., Oh, P., Celio, H., Park, S., Cho, J. and Manthiram, A. 2017. Dynamic behaviour of interphases and its implication on high-energy density cathode materials in lithium-ion batteries. Nat. Commun 8, 14589.

49. Sun, Y.K., Myung, S.T., Park, B.C., Prakash, J., Belharouak, I. and Amine, K. 2009. High-energy cathode material for long-life and safe lithium batteries. Nat. Mater. 8, 320-324.

50. Hou, P., Zhang, H., Zi, Z., Zhang, L. and Xu, X. 2017. Core-shell and concentration-gradient cathodes prepared via co-precipitation reaction for advanced lithium-ion batteries. J. Mater. Chem. A5, 4254-4279.

51. Lim, B.B., Yoon, S.J., Park, K.J., Yoon, C.S., Kim, S.J., Lee, J.J. and Sun, Y.K. 2015. Advanced concentration gradient cathode material with two-slope for high-energy and safe lithium batteries. Adv. Funct. Mater. 25, 4673-4680.

52. Lee, J.H., Yoon, C.S., Hwang, J.Y., Kim, S.J., Maglia, F., Lamp, P., Myung, S.T. and Sun, Y.K. 2016. High-energy-density lithium-ion battery using a carbon-nanotube–Si composite anode and a compositionally graded $Li[Ni_{0.85}Co_{0.05}Mn_{0.10}]O_2$ cathode. Energy Environ. Sci. 9, 2152-2158.

53. Pillot, C. 2017. Presented at the 31st International Battery Seminar & Exhibit, Fort Lauderdale, FL, USA, March 2017.
54. Olivetti, E.A., Ceder, G., Gaustad, G.G. and Fu, X. 2017. Lithium-ion battery supply chain considerations: Analysis of potential bottlenecks in critical metals. Joule 1, 229-243.
55. Inoue, T. and Mukai, K. 2017. Roles of positive or negative electrodes in the thermal runaway of lithium-ion batteries: Accelerating rate calorimetry analyses with an all-inclusive microcell. Electrochem. Commun. 77, 28-31.
56. Lee, K.K., Yoon, W.S., Kim, K.B., Lee, K.Y. and Hong, S.T. 2001. Characterization of $LiNi_{0.85}Co_{0.10}M_{0.05}O_2$ (M = Al, Fe) as a cathode material for lithium secondary batteries. J. Power Sources 97, 308-312.
57. Anderman, M. 2013. Technical report: Assessing the future of hybrid and electric vehicles: The 2014 xEV Industry Insider Report. Advanced Automotive Batteries, Needham, Ma, USA 2013.
58. Julien, C. 2015. Cathode materials with two-dimensional structure. Springer, Cham, pp. 119-162.
59. Lai, Y.Q., Xu, M., Zhang, Z.A., Gao, C.H., Wang, P. and Yu, Z.Y. 2016. Optimized structure stability and electrochemical performance of $LiNi_{0.8}Co_{0.15}Al_{0.05}O_2$ by sputtering nanoscale ZnO film. J. Power Sources 309, 20-26.
60. Oh, P., Song, B., Li, W. and Manthiram, A. 2016. Overcoming the chemical instability on exposure to air of Ni-rich layered oxide cathodes by coating with spinel $LiMn_{1.9}Al_{0.1}O_4$. J. Mater. Chem. A4, 5839-5841.
61. Xia, S., Li, F., Chen, F. and Guo, H. 2017. Preparation of $FePO_4$ by liquid-phase method and modification on the surface of $LiNi_{0.80}Co_{0.15}Al_{0.05}O_2$ cathode material. J. Alloys Compd. 731, 428-436.
62. Tang, Z.F., Wu, R., Huang, P.F., Wang, Q.S. and Chen, C.H. 2017. Improving the electrochemical performance of Ni-rich cathode material $LiNi_{0.815}Co_{0.15}Al_{0.035}O_2$ by removing the lithium residues and forming Li_3PO_4 coating layer. J. Alloys Compd. 693, 1157-1163.
63. Qi, R., Shi, J.L., Zhang, X.D., Zeng, X.X., Yin, Y.X., Xu, J., Chen, L., Fu, W.G., Guo, Y.G. and Wan, L.J. 2017. Improving the stability of $LiNi_{0.80}Co_{0.15}Al_{0.05}O_2$ by AlPO4 nanocoating for lithium-ion batteries. Sci. China Chem. 60, 1230-1235.
64. Duan, J., Wu, C., Cao, Y., Du, K., Peng, Z. and Hu, G. 2016. Enhanced electrochemical performance and thermal stability of $LiNi_{0.80}Co_{0.15}Al_{0.05}O_2$ via nano-sized LiMnPO4 coating. Electrochim. Acta 221, 14-22.
65. Albrecht, S., Kümpers, J., Kruft, M., Malcus, S., Vogler, C., Wahl, M. and Wohlfahrt-Mehrens, M. 2003. Electrochemical and thermal behavior of aluminum- and magnesium-doped spherical lithium nickel cobalt mixed oxides $Li_{1-x}(Ni_{1-y-z}Co_yMz)O_2$ (M = Al, Mg). J. Power Sources 119, 178-183.
66. Kim, J., Lee, H., Cha, H., Yoon, M., Park, M. and Cho, J. 2017. Prospect and reality of Ni-rich cathode for commercialization. Adv. Energy Mater. 8, 1702028,
67. Grunditz, E.A. and Thiringer, T. 2016. Performance analysis of current BEVs based on a comprehensive review of specifications. IEEE Transactions on Transportation Electrification 2, 270-289.

68. Ewing, J. 2017. Volvo, betting on electric, moves to phase out conventional engines. Energy & Environment [Online early access]. Published Online July 5 2017. https://www.nytimes.com/2017/07/05/business/energy-environment/volvo-hybrid-electric-car.html.

69. UK to ban new ICE vehicles from 2040. https://chargedevs.com/newswire/uk-to-ban-new-ice-vehicles-from-2040/ (accessed Novmber 7th 2017 2017).

70. Chung, D., James, T., Elgqvist, E., Goodrich, A. and Santhanagopalan, S. 2015. Automotive lithium-ion battery (LIB) supply chain and US competitiveness considerations. Clean Energy Manufacturing Analysis Center (CMAC), NREL (National Renewable Energy Laboratory) (No. NREL/PR-7A40-63354). National Renewable Energy Lab.(NREL), Golden, CO (United States).

71. Howell, D., Cunningham, B., Duong, T. and Faguy, P. 2016. Overview of the DOE VTO Advanced Battery R&D Program. Annual Merit Review June, 6.

72. Choi, J.W. and Aurbach, D. 2016. Promise and reality of post-lithium-ion batteries with high energy densities. Nat. Rev. Mater. 1, 16013.

73. Needell, Z.A., McNerney, J., Chang, M.T. and Trancik, J.E. 2016. Potential for widespread electrification of personal vehicle travel in the United States. Nat. Energy 1, 16112.

74. Nykvist, B. and Nilsson, M. 2015. Rapidly falling costs of battery packs for electric vehicles. Nat. Clim. Change 5, 329-332.

75. Chu, S., Cui, Y. and Liu, N. 2017. The path towards sustainable energy. Nat. Mater. 16, 16-22.

76. Kempton, W. 2016. Electric vehicles: Driving range. Nat. Energy 1, 16131.

77. Howell, D. 2016. Electrochemical Energy Storage R&D Overview [Online early access] 2016 (accessed October 25 2017).

78. Technology. 2017. Technology Page on Microvast Website, Microvastc, http://www.microvast.com/index.php/solution/solution_t (accessed: November 2017).

79. Jain, S. 2017. Emerging trends in battery technology. Auto Tech Review 6, 52-55.

80. Nikkei. 2016. Sony pulls plug on pioneering battery business, eyes Murata deal [Online early access] https://asia.nikkei.com/Business/Deals/Sony-pulls-plug-on-pioneering-battery-business-eyes-Murata-deal.

81. Nikkei. Murata to pour $450m into Sony battery ops after purchase [Online early access]. Published Online September 1 2017, 207. https://asia.nikkei.com/Business/Companies/Murata-to-pour-450m-into-Sony-battery-ops-after-purchase (accessed October 18 2017).

82. Zhong, Q., Bonakdarpour, A., Zhang, M., Gao, Y. and Dahn, J.R. 1997. Synthesis and electrochemistry of $LiNi_xMn_{2-x}O_4$. J. Electrochem. Soc. 144, 205-213.

83. Amine, K., Yasuda, H. and Fujita, Y. Positive electrode for lithium battery. US Patent 6420069 B2.

84. Fey, G.T.K., Li, W. and Dahn, J. 1994. $LiNiVO_4$: A 4.8 volt electrode material for lithium cells. J. Electrochem. Soc. 141, 2279-2282.

85. Amine, K., Yasuda, H. and Yamachi, M. 2000. Olivine $LiCoPO_4$ as 4.8 V

electrode material for lithium batteries. Electrochem. Solid-State Lett. 3, 178-179.

86. Tan, S., Ji, Y.J., Zhang, Z.R. and Yang, Y. 2014. Recent progress in research on high-voltage electrolytes for lithium-ion batteries. ChemPhysChem 15, 1956-1969.

87. West, A. 1998. A novel cathode $Li_2CoMn_3O_8$ for lithium ion batteries operating over 5 volts. J. Mater. Chem. 8, 837-839.

88. Truong, Q.D., Devaraju, M.K., Ganbe, Y., Tomai, T. and Honma, I. 2014. Controlling the shape of $LiCoPO_4$ nanocrystals by supercritical fluid process for enhanced energy storage properties. Sci Rep 4.

89. Doan, T.N.L. and Taniguchi, I. 2011. Preparation of $LiCoPO_4$/C nanocomposite cathode of lithium batteries with high rate performance. J. Power Sources 196, 5679-5684.

90. Zheng, J., Xu, P., Gu, M., Xiao, J., Browning, N.D., Yan, P., Wang, C., Zhang, J.G. 2015. Structural and chemical evolution of Li- and Mn-rich layered cathode material. Chem. Mater. 27, 1381-1390.

91. Thackeray, M.M., Kang, S.H., Johnson, C.S., Vaughey, J.T., Benedek, R. and Hackney, S.A. 2007. Li_2MnO_3-stabilized $LiMO_2$ (M = Mn, Ni, Co) electrodes for lithium-ion batteries. J. Mater. Chem. 17, 3112-3125.

92. Zheng, J., Myeong, S., Cho, W., Yan, P., Xiao, J., Wang, C., Cho, J. and Zhang, J.G. 2017. Li- and Mn-Rich cathode materials: Challenges to commercialization. Adv. Energy Mater. 7, 1601284-n/a.

93. Johnson, C., Kim, J., Lefief, C., Li, N., Vaughey, J. and Thackeray, M. 2003. The significance of the Li_2MnO_3 component in 'composite' $xLi_2MnO_{3\cdot(1-x)}$ $LiMn_{0.5}Ni_{0.5}O_2$ electrodes. Electrochem. Commun. 6, 1085-1091.

94. Thackeray, M.M., Kang, S.H., Johnson, C.S., Vaughey, J.T., Benedek, R. and Hackney, S. 2007. Li_2MnO_3-stabilized $LiMO_2$ (M= Mn, Ni, Co) electrodes for lithium-ion batteries. J. Mater. Chem. 17, 3112-3125.

95. Boulineau, A., Simonin, L., Colin, J.F., Bourbon, C. and Patoux, S. 2013. First evidence of manganese–nickel segregation and densification upon cycling in Li-rich layered oxides for lithium batteries. Nano Lett. 13, 3857-3863.

96. Lu, Z. and Dahn, J.R. 2002. Understanding the anomalous capacity of Li/$Li[Ni_xLi_{(1/3-2x/3)}Mn_{(2/3-x/3)}]O_2$ cells using in situ X-ray diffraction and electrochemical studies. J. Electrochem. Soc. 149, A815-A822.

97. Xu, B., Fell, C.R., Chi, M. and Meng, Y.S. 2011. Identifying surface structural changes in layered Li-excess nickel manganese oxides in high voltage lithium-ion batteries: A joint experimental and theoretical study. Energy Environ. Sci. 4, 2223-2233.

98. Yan, P., Nie, A., Zheng, J., Zhou, Y., Lu, D., Zhang, X., Xu, R., Belharouak, I., Zu, X. and Xiao, J. 2014. Evolution of lattice structure and chemical composition of the surface reconstruction layer in $Li_{1.2}Ni_{0.2}Mn_{0.6}O_2$ cathode material for lithium ion batteries. Nano Lett. 15, 514-522.

99. Armstrong, A.R., Holzapfel, M., Novák, P., Johnson, C.S., Kang, S.H., Thackeray, M.M. and Bruce, P.G. 2006. Demonstrating oxygen loss and associated structural reorganization in the lithium battery cathode $Li[Ni_{0.2}Li_{0.2}Mn_{0.6}]O_2$. J. Am. Chem. Soc. 128, 8694-8698.

100. Yu, C., Wang, H., Guan, X., Zheng, J. and Li, L. 2013. Conductivity and electrochemical performance of cathode $xLi_2MnO_{3\cdot(1-x)}LiMn_{1/3}Ni_{1/3}Co_{1/3}$

O_2 (x = 0.1, 0.2, 0.3, 0.4) at different temperatures. J. Alloys Compd. 546, 239-245.

101. Shi, S., Tu, J., Tang, Y., Liu, X., Zhang, Y., Wang, X. and Gu, C. 2013. Enhanced cycling stability of $Li[Li_{0.2}Mn_{0.54}Ni_{0.13}Co_{0.13}]O_2$ by surface modification of MgO with melting impregnation method. Electrochim. Acta 88, 671-679.

102. West, W., Soler, J., Smart, M., Ratnakumar, B., Firdosy, S., Ravi, V., Anderson, M., Hrbacek, J., Lee, E. and Manthiram, A. 2011. Electrochemical behavior of layered solid solution Li_2MnO_3–$LiMO_2$ (M=Ni, Mn, Co) Li-ion cathodes with and without alumina coatings. J. Electrochem. Soc. 158, A883-A889.

103. Nayak, P.K., Grinblat, J., Levi, M., Levi, E., Kim, S., Choi, J.W. and Aurbach, D. 2016. Al doping for mitigating the capacity fading and voltage decay of layered Li and Mn-rich cathodes for Li-ion batteries. Adv. Energy Mater. 6.

104. Yu, R., Wang, X., Fu, Y., Wang, L., Cai, S., Liu, M., Lu, B., Wang, G., Wang, D. and Ren, Q. 2016. Effect of magnesium doping on properties of lithium-rich layered oxide cathodes based on a one-step co-precipitation strategy. J. Mater. Chem. A4, 4941-4951.

105. Yu, Z., Shang, S.L., Gordin, M.L., Mousharraf, A., Liu, Z.K. and Wang, D. 2015. Ti-substituted $Li[Li_{0.26}Mn_{0.6-x}Ti_xNi_{0.07}Co_{0.07}]O_2$ layered cathode material with improved structural stability and suppressed voltage fading. J. Mater. Chem. A3, 17376-17384.

106. Yuan, B., Liao, S.X., Xin, Y., Zhong, Y., Shi, X., Li, L. and Guo, X. 2015. Cobalt-doped lithium-rich cathode with superior electrochemical performance for lithium-ion batteries. RSC Advances 5, 2947-2951.

107. Park, S.H., Kang, S.H., Johnson, C.S., Amine, K., Thackeray, M.M. 2007. Lithium–manganese–nickel-oxide electrodes with integrated layered–spinel structures for lithium batteries. Electrochem. Commun. 9, 262-268.

108. He, H., Cong, H., Sun, Y., Zan, L. and Zhang, Y. 2017. Spinel-layered integrate structured nanorods with both high capacity and superior high-rate capability as cathode material for lithium-ion batteries. Nano Res. 10, 556-569.

109. Zhang, W.J. 2011. A review of the electrochemical performance of alloy anodes for lithium-ion batteries. J. Power Sources 196, 13-24.

110. Mo, R., Rooney, D., Sun, K. and Yang, H.Y. 2017. 3D nitrogen-doped graphene foam with encapsulated germanium/nitrogen-doped graphene yolk-shell nanoarchitecture for high-performance flexible Li-ion battery. Nat. Commun. 8.

111. Kennedy, T., Mullane, E., Geaney, H., Osiak, M., O'Dwyer, C. and Ryan, K.M. 2014. High-performance germanium nanowire-based lithium-ion battery anodes extending over 1000 cycles through in situ formation of a continuous porous network. Nano Lett. 14, 716-723.

112. Zhang, W.M., Hu, J.S., Guo, Y.G., Zheng, S.F., Zhong, L.S., Song, W.G. and Wan, L.J. 2008. Tin-nanoparticles encapsulated in elastic hollow carbon spheres for high-performance anode material in lithium-ion batteries. Adv. Mater. 20, 1160-1165.

113. Besenhard, J.O., Yang, J. and Winter, M. 1997. Will advanced lithium-alloy anodes have a chance in lithium-ion batteries? J. Power Sources 68, 87-90.

114. Sharma, R.A. and Seefurth, R.N. 1976. Thermodynamic properties of the lithium-silicon system. J. Electrochem. Soc. 123, 1763-1768.

115. Chan, C.K., Peng, H.L., Liu, G., McIlwrath, K., Zhang, X.F., Huggins, R.A. and Cui, Y. 2008. High-performance lithium battery anodes using silicon nanowires. Nat Nanotechnol. 3, 31-35.

116. Li, H., Huang, X., Chen, L., Wu, Z. and Liang, Y. 1999. A high capacity nano Si composite anode material for lithium rechargeable batteries. Electrochem. Solid-State Lett. 2, 547-549.

117. Feng, K., Ahn, W., Lui, G., Park, H.W., Kashkooli, A.G., Jiang, G., Wang, X., Xiao, X. and Chen, Z. 2016. Implementing an in-situ carbon network in Si/ reduced graphene oxide for high performance lithium-ion battery anodes. Nano Energy 19, 187-197.

118. Hassan, F.M., Batmaz, R., Li, J., Wang, X., Xiao, X., Yu, A. and Chen, Z. 2015. Evidence of covalent synergy in silicon-sulfur-graphene yielding highly efficient and long-life lithium-ion batteries. Nat Commun 6, 8597.

119. Kim, H., Han, B., Choo, J. and Cho, J. 2008. Three-dimensional porous silicon particles for use in high-performance lithium secondary batteries. Angew. Chem. Int. Edi. 47, 10151-10154.

120. Feng, K., Li, M., Liu, W., Kashkooli, A.G., Xiao, X., Cai, M. and Chen, Z. 2018. Silicon-based anodes for lithium-ion batteries: From fundamentals to practical applications. Small 14, 1702737-n/a.

121. Datta, M.K. and Kumta, P.N. 2007. Silicon, graphite and resin based hard carbon nanocomposite anodes for lithium-ion batteries. J. Power Sources 165, 368-378.

122. Chae, S., Ko, M., Kim, K., Ahn, K. and Cho, J. 2017. Confronting issues of the practical implementation of Si anode in high-energy lithium-ion batteries. Joule 1, 47-60.

123. Tesla tweaks its battery chemistry: A closer look at silicon anode development. https://chargedevs.com/features/tesla-tweaks-its-battery-chemistry-a-closer-look-at-silicon-anode-development/.

Graphene-based Composites as Electrode Materials for Lithium Ion Batteries

Cuiping Han[1#], Shiqiang Zhao[2#], Aurelia Wang[2], Huile Jin[3], Shun Wang[3], Baohua Li[4], Junqin Li[1] and Zhiqun Lin[2*]

[1] College of Materials Science and Engineering, Shenzhen University and Shenzhen Key Laboratory of Special Functional Materials, Shenzhen 518060, China

[2] School of Materials Science and Engineering, Georgia Institute of Technology, Atlanta, GA 30332

[3] College of Chemistry and Materials Engineering, Wenzhou University, Wenzhou, Zhejiang, 325035, China

[4] Engineering Laboratory for Next Generation Power and Energy Storage Batteries, Engineering Laboratory for Functionalized Carbon Materials, Graduate School at Shenzhen, Tsinghua University, Shenzhen 518055, China

1. Introduction

The ever-growing energy demands and the depletion of finite fossil-fuel resources have necessitated the development of advanced renewable energy technology across the world, including both exploring renewable energy resources and searching for effective energy storage technologies [1]. Certain renewable energy resources, for instance, wind, water, sunlight, tidal and geothermal heat, may contribute a great proportion of the total energy supply in the near future. To this end, Denmark has proposed to switch their total energy supply (electricity, mobility and heating/cooling) to 100% renewable energy by 2050, creating a precedent for the rest of the world [2]. In contrast to non-renewable resources that are concentrated to a limited number of countries, renewable energy resources exist over

*Corresponding author: Zhiqun.lin@mse.gatech.edu
#: These two authors contributed equally.

a broad range of geographical areas and are all inherently intermittent. To facilitate rapid deployment of renewable energy for our future power needs, we need optimal energy-storage systems to store the harvested energy from these sustainable sources.

Batteries represent an excellent energy storage solution in the integration of renewable resources due to their pollution-free operation, high round-trip efficiency, long cycle life, low maintenance, compact size, to say the least [3]. Among them, lithium ion batteries (LIBs) have dominated the secondary battery market due to their merits such as high voltage, high energy density, long cycle life, no memory effects and low self-discharge. They are now widely used in mobile phones, laptops and other portable electronic devices, and are also considered as promising candidates for next-generation electric vehicles and grid storage stations [3]. The huge demand for optimized power storage for sustainable energy solutions has motivated scientific and technological efforts dedicated to developing LIBs with superior performance factors, such as higher energy density, higher power density, longer cycle life, lower cost and enhanced safety properties. Since these characteristics of LIBs are highly dependent on the properties of the electrode material, electrodes with improved design through manipulating composition, structure, morphology and reaction kinetics are essential [4].

Graphene is a monolayer of graphite, consisting of sp^2 hybridized carbon atoms arranged in a honeycomb crystal. It has drawn worldwide attention after Andre Geim and Kostya Novoselov won the Nobel prize in physics for their discovery of it in 2010. Thanks to its extraordinary physical and chemical properties, including high electron mobility of $2.5 \times 10^5 \, cm^2 V^{-1} s^{-1}$ [5], excellent thermal conductivity ($3000 \, W \, mK^{-1}$) [6], huge theoretical specific surface area of $2630 \, m^2 \, g^{-1}$ [7], superior mechanical properties and broad electrochemical window, graphene has been studied for use in numerous fields, such as LIBs [8], supercapacitors [9], solar cells [10], fuel cells [11], water purification [12], electronics [13], etc.

Specifically, with respect to the LIB field, graphene can be used as a bare active material or component in hybrids, and displays remarkable results in many cases. Cathode materials commonly used in LIBs are lithium intercalation compounds such as $LiCoO_2$, $LiMn_2O_4$, $LiFePO_4$ and $Li_3V_2(PO_4)_3$. Unfortunately, these materials are semiconductors with limited electronic conductivities. Since the electrochemical behavior of an electrode material is directly related with its electronic conductivity, conductive additives such as carbon black, acetylene black and carbon nanotubes [14] are added to such materials to improve their electrochemical properties. When graphene is used as conductive additive, a comparatively miniscule quantity to other materials can greatly improve the electronic conductivity of electrodes through the "plane-to-point" conducting mode [15, 16]. With regards to anode materials, graphite is the dominant anode material for commercial LIBs, whose specific capacity is 372 mAh g^{-1} based on the intercalation of Li ions that form LiC$_6$. In the past decades, material systems based on the insertion reaction, conversion reaction and alloying reaction mechanism have gained enormous interest due to their largely increased specific capacity [17]. However, these metal or metal oxide anodes have irreversible high capacity loss in the first cycle and rapid capacity fading due to severe volume expansion and contraction during repeated Li insertions and extractions. Since graphene is a perfect 2D building block with superior mechanical and surface properties, rational designs of graphene-based composites have emerged as promising candidates for electrode materials in LIBs. In these hybrids, graphene avoids the issues of poor conductivity and volume changes of the anode materials during the cycling process. Meanwhile, the anode materials serve as a spacer between graphene layers to suppress the restacking of graphene sheets. Therefore, graphene-based hybrids have shown significantly improved electrochemical properties such as high capacity, high rate capability and excellent cycling stability in comparison to their constituents.

In this chapter, most recent advances in graphene-based composites as cathodes and anodes for LIBs are reviewed. Graphene-based composites with unique structural configurations such as anchored, wrapped, encapsulated, sandwiched, layered and mixed models have been prepared [18]. The beneficial role of graphene on the electrochemical properties of the composites is highlighted. Finally, future challenges and potential research directions in the development of high-performance graphene-based hybrids are discussed.

2. Graphene-based Composites as Cathodes

Currently, the biggest challenge for the development of high-energy and high-power LIBs is the limited specific capacity of commercial cathode materials [19], which are primarily lithium intercalation compounds such as LiCoO$_2$ (140 mAh g^{-1}), LiMn$_2$O$_4$ (148 mAh g^{-1}), LiFePO$_4$ (170 mAh g^{-1}) and Li$_3$V$_2$(PO$_4$)$_3$ (197 mAh g^{-1}). To date, many endeavors have been devoted to enhancing the electrochemical performance of these cathode materials, and one of the major points of concern is their low electronic conductivity. The electronic conductivities are 10^{-3} S cm^{-1} for LiCoO$_2$ [20], 10^{-4} S cm^{-1} for LiMn$_2$O$_4$ [21] and 10^{-9} S cm^{-1} for LiFePO$_4$ [22], which play a significant role in imparting the corresponding battery performance. On one hand, electron conducting additives are frequently added to such materials during the electrode preparation process, which include zero-dimensional (0D) carbon black and one-dimensional (1D) carbon nanotubes. On the other hand, as a perfect 2D building block with superior mechanical and surface properties, graphene is hybridized with cathode materials to obtain graphene/cathode composites during the initial materials preparation process. Using either method, the electrochemical performances of cathode materials are improved due to enhanced electronic conductivity.

2.1 Graphene as Conductive Additive for Cathode Materials

As a typical 2D material, graphene is characterized by excellent electronic and thermal conductivity, as well as much larger specific surface area compared to 0D carbon black and 1D carbon nanotubes. Moreover, it has a very low apparent density and good flexibility, making it an ideal candidate for a conductive additive in cathode systems. Ideally, if all the carbon atoms of graphene layers could contribute to the formation of the conductive path, the electronic conductivity of cathode materials would be significantly improved by using a small amount of graphene.

The mass fraction of traditional conductive carbon additives in the cathode system is about 3~5 wt% and even as high as 6~10 wt% in the case of high-rate batteries. These conductive additives do not contribute to the capacity and thus will substantially decrease the energy density. $LiCoO_2$ is one of the main LIB cathode materials used on an industrial scale. Tang et al. [16] demonstrated that only a 0.2 wt% graphene nanosheet combined with 1 wt% Super P can be used to construct an efficient conducting network, guaranteed to match the performance of micro-sized $LiCoO_2$. In this conductive network, the trace amount of graphene sheets provides a long-range electron pathway. Meanwhile, the round, small Super P particles form short-range electron pathways as a supplement. This $LiCoO_2$ with graphene additive demonstrates a specific capacity of 146 mAh g^{-1} at 1C with retention of 96.4% after 50 cycles, which is much better or more compared to batteries containing 3 wt% Super P.

Su et al. [15] further elucidated that the fully-exfoliated graphene sheet bridges the active $LiFePO_4$ particles more effectively through a "plane-to-point" conductive mode. The weight fraction of graphene conductive additive is much lower than that of commercial carbon black conductive additive, in which a "point-to-point" conducting mode is employed. However, it was also found that an excess of graphene additive in a $LiFePO_4$ cathode would block fast Li ion transport, resulting in heavy polarization at a relatively high charge/discharge rate [23, 24]. The steric hindrance for Li ion diffusion is more obvious in thick electrodes due to a longer Li ion diffusion path [25]. This indicates that a proper balance should be achieved between increased electron transport and steric hindrance for Li ion diffusion when using graphene as conductive additive.

2.2 Graphene-based Composites as Cathode Materials

$LiCoO_2$ is one of the main LIB cathode materials used on an industrial scale. However, it suffers from limited cobalt resources, environmental pollution and overcharge during usage causing potential safety hazards. Therefore, replaceable cathode materials such as $LiFePO_4$ have been explored. Olivine type $LiFePO_4$ is a 3.5 V cathode and is generally considered a strong cathode material owing to its high redox potential, large theoretical capacity (170 mAh g^{-1}), excellent stability, low cost and environmental benignity. In fact, $LiFePO_4$ has been commercially employed in numerous commercial batteries for the electric vehicle market. However, critical obstacles for $LiFePO_4$ as a cathode material are its low electronic conductivity (about 10^{-9} S cm^{-1}) [22] and poor Li ion diffusion coefficient ($10^{-14} \sim 10^{-16}$ cm^2 s^{-1}) [26], which exert negative effects on the performance of the resulting LIB, especially in the case of high-rate batteries. To address these issues, conductive graphene nanosheets were used as a scaffold to grow and/or anchor the $LiFePO_4$ materials by various synthesis routes, including hydrothermal, solvothermal, solid state routes, etc. Fan et al. [27] reported the directed growth of $FePO_4$ and $LiFePO_4$ on exfoliated graphene-assembled scaffolds. Rhodanineacetic acid-pyrene (RAAP) was initially functionalized on graphene nanosheets to induce the growth of $FePO_4$ particles, and subsequent thermal reduction converted the $FePO_4$/graphene to $LiFePO_4$/graphene hybrids. The $LiFePO_4$/graphene hybrids delivered 165~163 mAh g^{-1} at 0.2 C and 131~129 mAh g^{-1} at 10 C, whereas the graphene-free sample only showed 102~94 mAh g^{-1} at 10 C. Luo et al. [28] synthesized graphene-encapsulated $LiFePO_4$ composites by self-assembly of surface modified $LiFePO_4$ and graphene oxide with peptide bonds, followed by high temperature thermal reduction. The incorporated graphene forms a continuous conductive coating network connecting the $LiFePO_4$ nanoparticles to facilitate electron transportation. The hybrid material demonstrates 70% capacity retention at 50 C rate and less than 8.6% capacity fading after 950 cycles at 10 C. Thus, the flexible mesh structure of graphene improved the electrical conductivity and rate performance of the $LiFePO_4$/graphene composite material.

$Li_3V_2(PO_4)_3$ is another attractive phosphate cathode material for LIBs with a monoclinic structure due to its high operating voltage and large theoretical specific capacity. The structure of $Li_3V_2(PO_4)_3$ consists of a three-dimensional framework of slightly distorted VO_6 octahedral and PO_4 tetrahedral sharing oxygen vertices, which host Li ions in relatively large interstitial sites [29]. All three Li ions in $Li_3V_2(PO_4)_3$ can be removed and reversibly intercalated between 3.0 and 4.8 V, resulting in a high capacity of 197 mAh g^{-1} [30]. Compared to 1D channel-structured $LiFePO_4$, $Li_3V_2(PO_4)_3$ exhibits a much higher Li-ion diffusion efficiency (from 10^{-9} to 10^{-10} cm^2 s^{-1}) [30] and higher intercalation potentials, which indicate that $Li_3V_2(PO_4)_3$ has great potential to achieve a higher rate capability than $LiFePO_4$. However, one of the key drawbacks of $Li_3V_2(PO_4)_3$ is that its low intrinsic electronic conductivity of about 2.4×10^{-7} S cm^{-1} at room temperature largely limits its rate performance [29]. Therefore, graphene may be introduced to improve its electrochemical performance. Cui et al. [31] reported a nitrogen-doped graphene nanosheet decorated $Li_3V_2(PO_4)_3$/C composite for LIBs. The $Li_3V_2(PO_4)_3$/C nanocrystals with typical sizes of 50-100 nm are uniformly anchored on the nitrogen-doped graphene nanosheets through a microwave-assisted hydrothermal method. The composite material delivered a specific discharge capacity of 191.5 mAh g^{-1} at 0.1 C, 172.6 mAh g^{-1} and 160.5 mAh g^{-1} at higher rates of 5 C and 10 C, respectively. In addition, it showed a stable cycling performance with 87.3% capacity retention after 1000 cycles at 20 C. Kim successfully obtained a $Li_3V_2(PO_4)_3$/reduced graphene oxide (rGO) multilayer composite with a high tap density of 0.6 g cm^{-3}. In detail, the electrostatic interactions between anionic species of $LiH_{2-n}PO_4^{n-}$ and VO_3^- and the cetyltrimethylammonium bromide (CTAB) functionalized graphene oxide (GO) flakes produce a self-assembled multilayer precursor. The following heat treatment at 800 °C converts the $Li_3V_2(PO_4)_3$ precursor into $Li_3V_2(PO_4)_3$ particles and GO into highly conductive rGO sheets. The multilayer structure has open channels between the highly conductive rGO layers, which could effectively improve the rate capability. Specifically, the $Li_3V_2(PO_4)_3$/rGO multilayer composites exhibit a specific capacity of 131 mAh g^{-1} at 0.1 C, 88% capacity retention at 60 C, and 97% capacity retention after 500 cycles at 10 C within 3.0-4.3 V.

Spinel $LiMn_2O_4$ is another type of attractive cathode thanks to its three-dimensional Li$^+$ diffusion paths, along with low cost, environmental friendliness, good structural stability, and greatly improved safety. On one hand, $LiMn_2O_4$ suffers from Jahn-Teller distortion and Mn dissolution into the electrolyte at elevated temperatures (>55 °C). To suppress these problems, various cations such as Ni and Co have been substituted onto the Mn-site in $LiMn_2O_4$ to improve its structural stability. On the other hand, $LiMn_2O_4$ and its derivatives suffer from poor rate capability due to sluggish Li$^+$ ion diffusion and limited electrical conductivity. Hybrids of $LiMn_2O_4$ combined with graphene have shown great improvement in cathode performance [32, 33]. Nam reports a well-crystallized and nano-sized spinel $LiMn_2O_4$/rGO hybrid prepared by a microwave-assisted hydrothermal method at 200 °C for 30 min. [32]. The $LiMn_2O_4$ particles sized from 10-40 nm were evenly dispersed on the rGO template without agglomeration. The high active surface area of individual $LiMn_2O_4$ nanoparticles and the high conductivity of rGO enabled the hybrid to exhibit remarkable specific capacities of 117 and 101 mAh g^{-1} at a rapid charge/discharge rate of 50 C and 100 C, respectively. Moreover, additional promising cathode materials (e.g. V_2O_5) hybridized with graphene were also investigated and demonstrated improved electrochemical performance [34].

3. Graphene-based Materials as Anodes

Initially, graphene was studied as a bare active material for LIBs [35]. It was proposed that lithium ions could be adsorbed on both sides of a single-layered graphene sheet by the formation of Li_2C_6 with a theoretical capacity of 744 mAh g^{-1}, which is much higher than 372 mAh g^{-1} of a graphite anode [36, 37]. Intensive research has been devoted to achieving high Li storage performance of graphene anodes through reducing the layer number, expanding the interlayer spacing, tuning the surface chemistry, or increasing edge sites and defects [38, 39]. However, several challenges still remain, such as high irreversible capacity loss during the initial several cycles, formation of a solid electrolyte interface (SEI) layer at lower potentials (below 1.0 V vs. Li/Li$^+$), limited rate capability and significant capacity decay during cycling. Therefore, further effort has been given towards exploring graphene as a 2D building block for nanostructured composite anodes, which will be discussed presently.

3.1 Graphene-based Insertion Anodes

Generally, an electrochemical reaction using lithium involves three different mechanisms: Li insertion/extraction reaction, conversion (redox) reaction, and alloying/de-alloying reaction. The insertion type anodes for LIBs are metal oxides that contain transition metal ions and empty sites in the host crystal lattice, in the form of 2D layers (van der Waals gap) or as channels in the 3D network, which include TiO_2, $Li_4Ti_5O_{12}$, Nb_2O_5, V_xO_y, $TiNb_2O_7$, layered-MoO, etc. Such insertion type anodes usually undergo a "topotactic reaction" process which involves the insertion and extraction of guest Li ions into and out of the host lattice with minor modification of the crystal structure, such as small volume expansion and contraction. Therefore, these type of anodes typically exhibit long cycling life and high Coulombic efficiency. However, their electrochemical performance is greatly hindered by their poor intrinsic electronic conductivity and low lithium ion diffusion coefficient. To facilitate the insertion of Li ions, graphene is introduced to create a transport path for both Li ions and electrons.

TiO_2 is a promising candidate as an anode for LIBs due to its low cost, wide availability, and environmental friendliness. To date, TiO_2 polymorphs including anatase, rutile and TiO_2-B (bronze), have been reported for lithium electrochemical reactivity [40]. Generally, Li ion intercalation/deintercalation with TiO_2 is expressed as the following (Eq. 1):

$$TiO_2 + xLi^+ + xe^- \leftrightarrow Li_xTiO_2 \ (x \leq 1)$$ (1)

where x is varied with different TiO_2 polymorphs, morphologies and crystallographic orientations, and the maximum theoretical capacity of TiO_2 is 335 mA·h g^{-1} for 1 mol of Li per mole of TiO_2 when cycled to the lower cutoff voltage of 1.0 V versus Li/Li$^+$. Unfortunately, sluggish lithium-ion diffusion and low electrical conductivity of TiO_2 constitute the major obstacles for its practical application in LIBs. Graphene, as a new type of 2D carbon nanomaterial, appears particularly promising to improve the electrochemical performance of TiO_2 owing to its superior electrical conductivity, large surface area, and excellent structural flexibility. Consequently, many endeavors to modify TiO_2 anode material with graphene have been reported. On one hand, attempts to precisely control the particle size and nanostructure of TiO_2 on graphene nanosheets have been reported. Zhou et al. reported that, with the assistance of sodium dodecylbenzene sulfonate (SDBS) as a surfactant, hydroxyl titanium oxalate (HTO) can condense as a flower-like nanostructure on GO sheets [41]. By calcination, the HTO/GO intermediate can be converted to a TiO_2/graphene nanocomposite that still retains the flower-like nanostructure. The TiO_2/graphene nanocomposite is able to deliver a capacity of 230 mAh g^{-1} at 17 mA g^{-1} as well as superior rate and cycling stability. Ding reported the growth of TiO_2 nanosheets with exposed (001) high-energy facets on graphene sheets using a solvothermal treatment [42]. The highly active facets endow the composite with a reversible capacity of 107 mAh g^{-1} at a current rate of 20 C.

On the other hand, assembling TiO_2 and 2D graphene into specially designed architectures has been realized. These architectures provide graphene materials with high specific surface area, good mechanical properties, flexibility, and fast mass and electron transport kinetics. Yang et al. [43] designed a TiO_2/graphene composite with a sandwich-like structure using graphene and mesoporous silica as templates. In this architecture, the graphene layer within each nanosheet acts as a mini-current collector, which is favorable for fast electron transport in the electrode. As a consequence, the composites can reach reversible capacities of 162, 123 and 80 mAh g^{-1} at current rates of 1 C, 10 C, and 50 C, respectively. Zhu reported a 3D nitrogen-doped graphene foam embedded with ultrafine TiO_2 nanoparticles prepared by a hydrothermal self-assembly and a subsequent freeze-drying method [44]. In this architecture, the N-doped graphene sheets were randomly overlapped to form an interconnected conducting 3D network, which provides a multidimensional electronic network that facilitates charge transfer, enlarges the contact area between electrolyte and electrode, and provides numerous open channels for the access of electrolyte, thus favoring diffusion kinetics for both electrons and lithium ions. The as-obtained composite delivers good rate capacity up to 96 mAh g^{-1} at 20 C and remains at a specific capacity of 165 mAh g^{-1} after 200 cycles at 1 C. Both strategies result in significant improvement in the electrochemical performance of TiO_2/graphene composites.

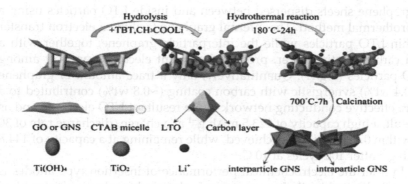

Fig. 1. Schematic illustration of the synthesis and structure of carbon coated LTO/ graphene (C-LTO/G) architecture [51]. (Copyright © 2014 Elsevier)

Spinel $Li_4Ti_5O_{12}$ (LTO) is considered a competitive anode material for high-power LIB applications due to its reliable safety, excellent high rate capability and extremely long cycling stability [45]. The spinel structure of LTO can accommodate three Li ions with a theoretical specific capacity of 175 mAh g^{-1} at a flat potential of 1.55 V, corresponding to a chemical formula of $Li_7Ti_5O_{12}$. A negligible volume change of <1% is observed during lithium insertion/extraction, which ensures an extremely long cycling life for this "zero-strain" material [46]. Moreover, the spinel structure of LTO provides 3D transportation pathways for lithium ions, thereby providing outstanding lithium ion mobility that is conducive for high rate applications [47]. Unfortunately, the insulating properties arising from the empty Ti 3d state, with a calculated band energy of ca. 2 eV, remain a critical issue [48], as they lead to poor electrochemical performance of LTO electrodes. Alternatively, a LTO/graphene hybrid created through simple mechanical mixing exhibited improved conductivity and reduced polarization. As a consequence, a specific capacity of 122 mAh g^{-1} at 30C and a capacity loss of less than 6% over 300 cycles at 20 C were achieved with 5 wt.% grapheme [49]. To achieve a more uniform hybridization, flower-like LTO hollow microspheres consisting of nanosheets are prepared via a hydrothermal process, and then wrapped by graphene through electrostatic interactions [50]. Lin also found that electron transfer at the LTO/graphene heterojunction interface reduces the localized work function of these composites, decreases energy required for electrons to escape and consequently results in lower charge-transfer resistances. The graphene-wrapped LTO hollow spheres show a specific capacity of 272.7 mAh g^{-1} at 750 mA g^{-1} after 200 cycles, which is higher than the 235.6 mAh g^{-1} of pristine LTO. In comparison with the *ex-situ* method, *in-situ* hybridization can achieve more uniform hybridization. Dong et al. [51] demonstrated a LTO/graphene composite architecture consisting of graphene sheets dispersed between and inside LTO particles using a hydrothermal method. The internal graphene facilitated electron transfer within LTO particles while the interparticle graphene, together with a thin carbon coating layer, promoted efficient electron transport among LTO particles (Fig. 1). Quantitatively, only a trace amount of graphene (~ 0.4 wt%) synergistic with carbon coating (~0.8 wt%) contributed to a more effective conducting network in the resultant LTO electrode and as a result, a high capacity of 123.5 mAh g^{-1} at a charge/discharge rate of 30 C (within 0.01-3.0 V) was achieved, while remaining at a capacity of 144.8 mAh g^{-1} after 100 cycles at 10 C.

In brief the electrochemical performance of insertion type anodes is generally limited by their poor intrinsic electrical conductivities and low lithium ion diffusion coefficients. Hybridization of these materials with graphene could noticeably enhance the rate property due to improvement in conductivity and/or enhanced electrode/electrolyte interfaces. However, graphene appears to exert a limited influence on improving the specific capacity of hybrids owing to limited available vacancies present in the host lattice. Moreover, the redox potentials of Ti^{4+}/Ti^{3+}, Nb^{5+}/Nb^{4+}, and Nb^{4+}/Nb^{3+} ranging from 1.0 to 1.6 V are slightly higher, which will lower the energy density of the resulting full cells.

3.2 Graphene-based Alloying Anodes

The elements Si, Sb, Ge, Sn, In, Zn and Cd can store lithium through alloying reactions with high theoretical lithium storage capacities (Table 1) [52]. The low charge-discharge voltage plateaus of the alloying-dealloying reactions make them the most ideal alternative anode materials for LIBs. Representatively, as the second-most abundant element in the earth's crust, Si can deliver an ultrahigh theoretical lithium storage capacity of ~4200 mAh g^{-1} *via* alloying with Li and generating Li$_{4.4}$Si, which is 11.3 times of that of graphite (372 mAh g^{-1}). Similarly, the theoretical capacities for Ge, Sn and Sb are also very attractive [53]. However, their

Table 1. Alloying reactions and theoretical capacities of alloying anodes

Element	Alloying reaction	Theoretical capacity/mAh g^{-1}
Silicon (Si)	$Si + 4.4Li^+ + 4.4e^- \leftrightarrow Li_{4.4}Si$	4198.8
Germanium (Ge)	$Ge + 4.4Li^+ + 4.4e^- \leftrightarrow Li_{4.4}Ge$	1624.1
Tin (Sn)	$Sn + 4.4Li^+ + 4.4e^- \leftrightarrow Li_{4.4}Sn$	993.4
Stibium (Sb)	$Sb + 3Li^+ + 3e^- \leftrightarrow Li_3Sb$	660.3
Indium (In)	$In + 3Li^+ + 3e^- \leftrightarrow Li_3In$	700.3
Zinc (Zn)	$Zn + Li^+ + e^- \leftrightarrow LiZn$	409.9
Cadmium (Cd)	$Cd + Li^+ + e^- \leftrightarrow LiCd$	238.4

applications in LIBs are severely hindered by rapid capacity decay due to pulverization of electrode materials, disintegration of electrodes and instability of the SEI layer, originating from the large volume expansion upon lithium insertion (~400% for Si, 370% for Ge, 260% for Sn, 200% for Sb) and volume contraction upon lithium extraction [53-56].

Encapsulation of the alloying anode materials into a graphene matrix has been proven to be an effective strategy for solving the aforementioned problems. In graphene/alloy composites, flexible graphene nanosheets could encapsulate the electrode materials to prevent their pulverization, provide abundant void spaces to buffer volume expansion, construct a layered self-assembly structure to enhance the structural stability of electrode and SEI layer, and form an electrically conductive framework to facilitate electron transfer [57]. Benefiting from the positive effects of graphene, a graphene/alloy composite electrode could express a high Li storage capacity, good cycling stability and enhanced rate capability as listed in Table 2 [58, 59].

Among all the alloying anode materials, Si and Sn are the most widely investigated. The construction of uniformly dispersed graphene/Si hybrids is challenging yet critical for performance enhancement. The hydrophobic surface of commercial Si nanoparticles results in their poor dispersion in aqueous solutions of graphene. In addition, Si is negatively charged due to being easily oxidized to form a surface layer of silicon oxide, but GO also exhibits a negative charge owing to ionization of the carboxylic acid groups and the presence of phenolic hydroxyl groups. Various approaches have been devoted to achieve uniform anchoring of Si nanoparticles on the surfaces of graphene nanosheets. One frequently used strategy is modifying the surface electrical properties of Si or graphene by polymers. Yang et al. [61] grafted Si nanoparticles on the surface of electropositive p-phenylenediamine ($PhNH_2$) functionalized graphene (Fig. 2a) and the resultant Si-Ph-G nanocomposite delivered initial and 50th charge capacities of 1079 and 828 mAh g^{-1} at 300 mA g^{-1}, respectively. Zhou et al. [62] modified Si nanoparticles with positively charged poly(diallydimethylammonium chloride) (PDDA) and then mixed the Si-PDDA with negatively charged graphene oxides in order to fabricate a Si-NP@G nanocomposite that retained a high capacity of 1205 mAh g^{-1} after 150 cycles at 100 mA g^{-1}.

Another effective strategy is the *in-situ* generation of Si particles on graphene or graphene on Si particles to fabricate a Si/G composite. Ko et al. [63] prepared a graphene/Si nanocomposite with amorphous Si particles (<10 nm) uniformly anchored on a graphene backbone by decomposition of SiH_4 gas through a chemical vapor deposition (CVD) reaction (Fig. 2b). The graphene/Si nanocomposite exhibited a remarkable initial Coulombic efficiency (ICE) of 92.5% with a first reversible charge capacity of 2858 mAh g^{-1} at 56 mA g^{-1} and an average charge capacity of 1103 mAh g^{-1}

Table 2. Electrochemical performances of the reported graphene/alloying anode material (G/M, M=Si, Ge, Sn...) for LIBs

G/M hybrids (M=Si, Ge, Sn...)	M content (wt%)	Voltage range (V vs Li/Li+)	Current density (A g⁻¹)	Specific capacity (mAh g⁻¹)	ICE (%)	Cycles	References
G/Si	77.8	0.05-1.2	0.2	880	~70	30	[60]
Si-Ph-G	91.7	0.01-3	0.3	828	59.9	50	[61]
Si-NP@G	80.1	0.05-1	0.1	1205	58.9	150	[62]
G/amorphous Si	82	0.01-1.5	14	1103	92.5	1000	[63]
rGO-Si	75.8	0.005-1.5	2.1	1162	64.5	100	[64]
Si@Gr	91	0.01-1	2.1	2800	93.2	300	[65]
G/Sn@C	60.1	0.05-2	0.05	590	66.5	60	[66]
Sn@C-GNs	77.96	0.01-3	0.075	566	70.1	100	[67]
Sn@G-PGNWs	46.8	0.005-3	2	656	69	1000	[68]
Si NWs-rGO	90.9	0.01-1.5	1.2	2230	78	100	[69]
Graphene/Sn-nanopillar	70	0.02-3	0.05	700	71.2	30	[70]
rGO-Sn@C nanocables	61	0.005-2.5	0.05	630	66.9	50	[71]
Graphene/Ge nanowires	–	0-2.9	4.8	1059	68	200	[72]
3D porous Si/G	82.9	0-1.2	0.21	1299.6	/	25	[73]
SnO₂ NC@N-rGO	70	0.005-3	0.5	1346	61.3	500	[75]
Graphene/GeO₂ microtubes	89.6	0.01-1.5	110	919	33	100	[76]

ICE: Initial Coulombic efficiency.

after 1000 cycles at 14 A g^{-1}. Kannan et al. [64] synthesized rGO-Si from GO-SiO$_2$ by a NaCl-assisted magnesiothermic reduction method without generating silicon carbide (SiC) impurities (Fig. 2c), which retained a high capacity of 1162 mAh g^{-1} after 100 cycles at 2.1 A g^{-1} (a 87.1% capacity retention).

In graphene/Sn composites, the metallic Sn prefers to grow into large particles during electrochemical reaction processes due to its low melting point (232°C), thus the construction of an additional surface carbon coating layer around Sn in graphene/Sn composites is widely implemented to buffer the volume change and suppress the coarsening of Sn to maintain cycling stability. The Sn nanoparticles are typically fabricated by reducing SnO$_2$ nanoparticles. Luo et al. [66] coated a glucose-derived carbon layer on the surface of a graphene/SnO$_2$ precursor and then obtained a carbon-coated graphene/Sn (G/Sn@C) nanocomposite by carbothermal reduction of SnO$_2$ into Sn during calcinations. The G/Sn@C nanocomposite exhibited a stable capacity of over 590 mAh g^{-1} after 60 cycles at 50 mA g^{-1}. Qin et al. [68] developed a one-step CVD technique to fabricate graphene shell-encapsulated Sn nanoparticles (5-30 nm) anchored in three-dimensional (3D) porous graphene networks (Sn@G-PGNWs) with a SnCl$_2$ precursor

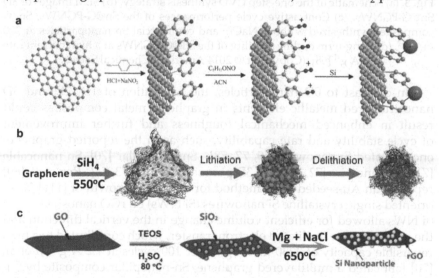

Fig. 2. Schematics of synthesis strategies of (a) Grafting Si nanoparticles on surface of electropositive p-phenylenediamine (PhNH$_2$) functionalized graphene for generating a Si-Ph-G nanocomposite [61]. (Copyright © 2012 Royal Society of Chemistry) (b) Anchoring amorphous Si particles on a graphene backbone by decomposition of SiH$_4$ for synthesizing a G/Si nanocomposite [63]. (Copyright © 2014 American Chemical Society) (c) Preparing a rGO-Si nanocomposite from GO-SiO$_2$ by a NaCl-assisted magnesiothermic reduction method [64]. (Copyright © 2016 Royal Society of Chemistry)

as a catalyst and NaCl particles as a template (Fig. 3). The Sn@G-PGNWs exhibited a reversible capacity of 682 mAh g^{-1} even at a high current density of 2 A g^{-1} with a high capacity retention of 96.3% after 1000 cycles.

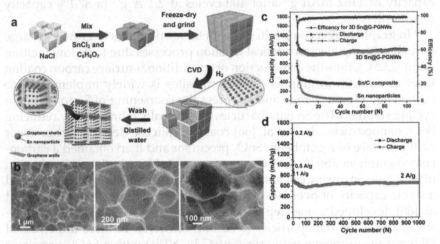

Fig. 3. (a) Schematic of the one-step CVD synthesis strategy. (b) SEM images of 3D Sn@G-PGNWs. (c) Contrastive cycle performances of the Sn@G-PGNWs, Sn/C composite (synthesized without NaCl), and commercial Sn nanoparticles at 200 mA g^{-1}. (d) Long-term cycling stability of the Sn@G-PGNWs at a high current rate of 2 A g^{-1} [68]. (Copyright © 2014 American Chemical Society)

In contrast to 0D nanoparticles, the fabrication of 1D, 2D and 3D nanostructured metallic elements in graphene/metal composites could result in enhanced mechanical toughness and further improvement of cycle stability and rate capability, such as in the reported graphene-encapsulated Si nanowire [69, 77, 78], Sn nanopillar [70], Sn nanocable [71], Ge nanowire [72, 79], and 3D porous Si sphere [73, 74] . Ren et al. [69] reported an Au-seeded CVD method for the *in-situ* growth of [111] facet-oriented single crystalline Si nanowires (Si NWs) on rGO nanosheets. The Si NWs allowed for efficient volume change in the vertical direction and the rGO substrate facilitated electron transfer, which contributed to a high reversible capacity of 2230 mA h g^{-1} after 100 cycles at 1.2 A g^{-1}. Ji et al. [70] fabricated a multilayered graphene/Sn-nanopillar composite by Sn-Graphene film alternating deposition and an annealing process, which retained a capacity of 700 mAh g^{-1} after 30 cycles at 50 mA g^{-1}. Kim et al. [72] reported a CVD process for the *in-situ* growth of a single to a few layers of graphene on Ge nanowires in a CH_4 and H_2 atmosphere. The graphene/Ge nanowire composite exhibited a high capacity of 1059 mAh g^{-1} after 200 cycles at a high current rate of 4.8 A g^{-1}.

Compositing graphene with the oxides (i.e. SnO_2 and GeO_2) of metallic alloys could also enhance the lithium storage capability by improving conductivity and forming highly reactive nanosized particles [80, 81] Normally, there are certain two-step reactions, the conversion reaction (Eq. 2) and alloying reaction (Eq. 3), involved in these metallic oxide composites. During the lithiation process, the metal oxides are firstly reduced to their metallic states followed by the alloying reaction with Li. The conversion reaction is widely reported to be irreversible for bulk particles, but the generated Li_2O in the first discharge process could serve as a buffering domain to hinder the coarsening of metal nanoparticles and contribute to the stable Li storage capacity of the alloying reaction. In recent years, it has been proved that the conversion reaction can be reversible for SnO_2 and GeO_2 nanoparticles [82]. However, the low conductivity and easy aggregation of oxide nanoparticles inhibit their electrochemical performance. Zhou et al. [75] reported a SnO_2 nanocrystal/nitrogen-doped reduced graphene oxide hybrid material (denoted SnO_2 NC@N-rGO) by an *in-situ* hydrazine monohydrate vapor reduction method. The homogeneous confinement of small SnO_2 nanocrystals (4-5 nm) in the N-rGO nanosheets endowed SnO_2 NC@N-rGO with good conductivity, and the Sn-N bonds formed between rGO and SnO_2 nanocrystals restricted the aggregation of SnO_2 or Sn nanoparticles in the cycling process. The 500th discharge capacity of SnO_2 NC@N-rGO was 1346 mAh g^{-1}, nearly equal to the theoretical capacity of SnO_2 (1494 mAh g^{-1}). Chen et al. [76] fabricated a multilayer graphene/GeO_2 tubular nanoarchitecture by a strain-driven strategy after dissolving the substrate. Due to their excellent conductivity and structure stability, the graphene/GeO_2 microtubes expressed a reversible capacity of 919 mAh g^{-1} after 100 cycles at 110 mA g^{-1} with a higher rate performance than bare GeO_2 microtubes.

Conversion-alloying reaction of metal oxides anode materials:

$$\text{Conversion reaction: } MO_2 + 4Li^+ + 4e^- \leftrightarrow M + 2Li_2O \qquad (2)$$
$$\text{Alloying reaction: } M + 4.4Li^+ + 4.4e^- \leftrightarrow Li_{4.4}M \text{ (M = Sn or Ge)} \qquad (3)$$

Although graphene/alloy composites exhibit many unique synergistic effects in contrast to bare alloy materials, challenges remain in decreasing the costs of raw materials (i.e. Si nanoparticles) and simplifying the preparation process. Moreover, it is important to increase the low ICE by reducing the irreversible capacity loss that is caused by the formation of a SEI layer, as well as limiting the consumption of lithium by its reactions with surface functional groups of graphene.

3.3 Graphene-based Conversion Anodes

The conversion lithium storage reaction is normally characterized by the reversible conversion reaction from metal oxides and lithium to metal nanoparticles and lithium oxides *via* a reduction-oxidation reaction of the metal elements (Eq. 4). Conversion type anodes generally deliver a high theoretical specific capacity because multiple electrons are involved in the reduction-oxidation reaction between transition metal oxides and metal nanoparticles. In addition, conversion anode materials are cheaper and more easily fabricated than metallic alloying electrodes. As a consequence, conversion anodes are the most widely investigated electrode materials for LIBs. In addition to transition metal oxides, a wide range of transition metal sulfides and carbonates have also been investigated as conversion type anodes [83, 84].

Conversion reaction of transition metal oxides:

$$M_xO_y + 2yLi^+ + 2ye^- \leftrightarrow xM + yLi_2O$$
$$(M = Mn, Fe, Co, Ni, Cu, Cr, Mo \text{ or } Ru) \qquad (4)$$

Unfortunately, the drastic volume change of these materials during lithium insertion and extraction usually leads to the pulverization of particles and eventually the disintegration of electrodes, resulting in a short cycle life. Moreover, these materials also possess low electrical conductivities, large voltage polarization and insufficient Coulombic efficiency [83]. Similar to the functions of graphene in the above discussed graphene-based alloying type anodes, graphene can improve the electrochemical properties of conversion type anode materials by increasing conductivity and buffering volume changes. Generally, graphene is added *in-situ* in the synthesis system of conversion materials, which could serve as a template for the generation of nanostructured materials with short lithium and electron transfer distance. As a result, graphene-based conversion anodes typically express higher capacity, longer cycling life and better rate performance than their bare counterparts.

3.3.1 Graphene-based Transition Metal Oxides and Sulfides

Tarascon et al. [85] reported transition metal oxides (MOs) of CoO, Co_3O_4, NiO and Cu_2O as lithium storage materials by a conversion reaction of $MO + 2Li \leftrightarrow Li_2O + M$ with theoretical capacities of ~1000 mAh g^{-1}. Since then, various transition MOs, including monocomponent (i.e. M_xO_y, M = Mn, Fe, Co, Ni, Cu, et al.) and multicomponent (i.e. MFe_2O_4, MCo_2O_4 and MMn_2O_4) variants, have been investigated as anodes of LIBs [86, 87]. In these composites, graphene affords not only a buffering matrix for large volume variation, but also a conductive scaffold for electron and lithium ion transport. Thereby, graphene/MO hybrids exhibit improved specific capacity, rate capability, and cycle stability as summarized in Table 3.

The oxygen-containing functional groups on GO electrostatically adsorb metal ions in the reaction solution, which guides the *in situ* crystallization of nanostructured metal compounds on GO and generates a metal compound/GO composite. Sun et al. [118] fabricated a MnO/G nanocomposite by thermally annealing the Mn-based precursor/GO composite. The shortened lithium and electron transportation length

Table 3. Electrochemical performances of the reported graphene/transition metal oxide/sulfide hybrids for LIBs

G/M_xO_y or G/M_xS_y hybrids	M_xO_y content (wt%)	Voltage range (V vs Li/Li$^+$)	Current density (A g^{-1})	Specific capacity (mAh g^{-1})	ICE (%)	Cycles	References
MnO nanosheets/G	82.6	0.01-3.0	3	625.8	–	400	[118]
N-doped MnO/G	91.1	0.01-3.0	5	202	69.5	90	[88]
Mn_3O_4/G platelet	65.7	0.01-3.0	0.075	720	–	100	[89]
Mn_3O_4/rGO	90	0.1-3.0	1.6	390	–	–	[90]
MnO_2 nanorods/PEDOT/G	–	0.2-3.0	0.4	698	–	15	[91]
MnO_2 nanoflakes/G	~50	1.8-4.0	0.2	230	–	–	[92]
MnO_2 nanotubes/G film	~50	0.01-3.0	1.6	208	–	–	[93]
α-Fe_2O_3/G	72	0.01-3.0	0.8	620	66	–	[94]
α-Fe_2O_3 nanorods/G	58.3	0.005-3.0	1	210.7	64.7	30	[95]
α-Fe_2O_3 nanodisks/rGO	52	0.005-3.0	0.2	931	–	50	[96]
α-Fe_2O_3 nanospindles/rGO	75.6	0.005-3.0	0.1	969	65	100	[97]
Hollow Fe_3O_4/G	90.2	0.005-3.0	0.1	~900	70	50	[98]
Fe_3O_4 nanorods/G	75	0.01-3.0	4.864	569	65.8	–	[99]
Fe_3O_4/G nanoscroll	50	0.01-3.0	5	300	73	50	[100]
Fe_3O_4/G foam	80	0.01-3.0	60C	190	65.1	500	[101]
Co_3O_4/G	75.4	0.01-3.0	0.05	~935	68.6	30	[102]
Co_3O_4 hollow sphere/G	76.2	0.01-3.0	5	259	69.3	100	[103]
Co_3O_4 nanorods/G	80	0.01-3.0	1	1090	70.4	40	[104]

(Contd.)

Table 3. (*Contd.*)

G/M_xO_y or G/M_xS_y hybrids	M_xO_y content (wt%)	Voltage range (V vs Li/Li$^+$)	Current density (A g^{-1})	Specific capacity (mAh g^{-1})	ICE (%)	Cycles	References
Co_3O_4 nanosheets/G	–	0.05-3.0	5C	130	76.6	50	[105]
Ni/NiO-G	<96	0.001-3.0	3	700	73	300	[106]
NiO NPs/G	68.5	0.01-3.0	4	152	65.1	100	[107]
CuO nanorods/G	–	0.001-3.0	5C	262	64.7	50	[108]
CuO nanosheets/G	50	0.01-3.0	0.067	736.8	91.6	50	[109]
Hollow CuO/Cu_2O-G	–	0.005-3.0	5	183	65	60	[110]
Cu_2O/G	~75	0.01-3.0	0.1	599.8	70	50	[111]
Cu_2O@rGO	~83	0.01-3.0	0.1	1097	65.8	200	[112]
Al_2O_3 coated ZnO/G	53	0.01-3.0	1	415	53.1	100	[113]
C/ZnO/G	47	0.01-3.0	0.978	420	62.8	–	[114]
$NiFe_2O_4$/rGO	70	0.01-3.0	0.2	489	–	50	[115]
$MnFe_2O_4$-GNS	90	0.01-3.0	0.1	1017	65	90	[116]
$CoFe_2O_4$-G	80	0.01-3.0	0.1	1082	65	50	[117]
G@FeS-GNRs	69	0.01-3.0	0.4	536	–	100	[120]
Ni_3S_4/NG	86	0.01-3.0	0.141	1323.2	–	100	[121]
CoS/G	76.5	0.01-3.0	0.062	749	–	40	[122]
FeS_2/G	90.4	0.01-3.0	0.87	970	–	300	[123]
MnS/G	75.4	0.01-2.6	0.5	830	–	100	[124]

of MnO nanosheets and the good electrical conductivity of graphene contribute to the excellent rate capability of the MnO/G nanocomposite, with an ultrahigh capacity of 2014 mAh g^{-1} after 150 cycles at 200 mA g^{-1}. Similarly, Jiang et al. [119] synthesized a core-void-shell structured Fe$_3$O$_4$@ rGO composite with SiO$_2$ as a sacrificial layer and (3-Ami-nopropyl) triethoxysilane (APTES) as an electrostatic absorbent between SiO$_2$ and GO. The void space between the Fe$_3$O$_4$ core and rGO shell accommodated the volume changes of the Fe$_3$O$_4$ particles during the lithium insertion and extraction process, and resulted in an enhanced cycling stability in contrast to mechanically mixed Fe$_3$O$_4$/rGO and bare Fe$_3$O$_4$.

Conversion reaction of cobalt ferrite (CoFe$_2$O$_4$):

First discharge: $CoFe_2O_4 + 8Li^+ + 8e^- \leftrightarrow Co + 2Fe + 4Li_2O$ (5)

Later cycles: $Co + Li_2O \leftrightarrow CoO + 2Li^+ + 2e^-$;   (6)

$$2Fe + 2Li_2O \leftrightarrow Fe_2O_3 + 6Li^+ + 6e^-$$ (7)

In contrast to monocomponent MOs, multicomponent MOs exhibit enhanced cycling stability due to the synergistic effect between different metallic components. As shown in the lithium storage reaction mechanism of CoFe$_2$O$_4$ (Eqs 5-7), in the first discharge process, the CoFe$_2$O$_4$ is reduced into metallic Co and Fe, which are oxidized into separate CoO and Fe$_2$O$_3$ in the following charge process, rather than recovering into CoFe$_2$O$_4$. In the subsequent cycles, there are reversible reactions in CoO and Fe$_2$O$_3$. The uniform coexistence of CoO and Fe$_2$O$_3$ nanoparticles in the cycled CoFe$_2$O$_4$ electrode can prevent the coarsening of same-phase nanoparticles and result in enhanced cycling stability [86]. Zhu et al. [115] reported a one-pot hydrothermal method for the *in-situ* growth of NiFe$_2$O$_4$ nanoparticles (5-10 nm) on a rGO nanocomposite, which exhibited higher rate performance than bare NiFe$_2$O$_4$ nanoparticles. Xiao et al. [116] prepared MnFe$_2$O$_4$-graphene nanocomposites (MnFe$_2$O$_4$-GNSs) via ultrasonic mixing of MnFe$_2$O$_4$ nanoparticles and GO followed by a hydrazine hydrate reduction, which retained a high capacity of 1017 mAh g^{-1} after 90 cycles at 100 mA g^{-1} (Figs 4a and 4b). Xia et al. [117] synthesized CoFe$_2$O$_4$-G nanocomposites by an ethanol-water solvothermal reaction, which delivered a reversible capacity of 1082 mAh g^{-1} after 50 cycles at 100 mA g^{-1}.

Similar to transition MOs, the hybridization of graphene with various transition metal sulfides (M$_x$S$_y$, M = Mn, Fe, Co or Ni) also exhibits advanced electrochemical properties as anodes of LIBs via conversion reactions (Eq. 8). Li et al. [120] fabricated sandwich structured graphene-wrapped FeS-graphene nanoribbons (G@FeS-GNRs) with GNRs as an inner layer for the loading of FeS nanoparticles and with graphene as an outer layer for encapsulating the FeS nanoparticles (Fig. 4c). Cationic polydiallyldimethylammonium chloride (PDADMAC) served as a

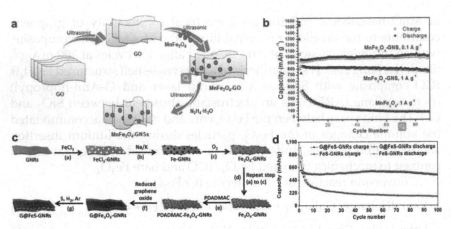

Fig. 4. (a, c) Schematics of the synthesis methods and (b, d) cycling performances of (a, b) MnFe$_2$O$_4$-GNSs [116] Copyright © 2014 American Chemical Society and (c, d) graphene-wrapped FeS-graphene nanoribbons (G@FeS-GNRs) [120] Copyright © 2016 Tsinghua University Press and Springer-Verlag Berlin Heidelberg.

medium to combine graphene and FeS through electrostatic interactions in order to construct the sandwich structure. The G@FeS-GNRs exhibited improved cycling stability compared to FeS-GNRs and maintained a high capacity of 536 mAh g^{-1} after 100 cycles at 400 mA g^{-1} (Fig. 4d). With N doped graphene (NG) nanosheets as templates, Mahmood et al. [121] prepared a Ni$_3$S$_4$/NG composite that expressed excellent cycling stability with an ultrahigh capacity of 1323.2 mAh g^{-1} after 100 cycles at a current rate of 141 mA g^{-1}. Moreover, the reported graphene-wrapped CoS nanoparticles [122], FeS$_2$ microspheres [123] and mesoporous MnS clusters [124] maintained reversible capacities of 1056, 970 and 980 mA h g^{-1} after long term cycles, respectively.

Conversion reaction of transition metal sulfides (M$_x$S$_y$):

$$M_xS_y + 2yLi^+ + 2ye^- \leftrightarrow xM + yLi_2S \quad (M = Mn, Fe, Co, Ni \text{ or } Cu) \quad (8)$$

3.3.2 Graphene-based Transition Metal Carbonates

In recent years, transition metal carbonates (MCO$_3$, M = Mn, Fe, Co or Ni) have attracted increasing attention as a new kind of high-capacity anode material for LIBs [125-136]. According to the two-step conversion reactions (Eqs 9 and 10), the theoretical capacity of MCO$_3$ is as high as ~1600 mAh g^{-1}. However, the reported practical capacities of bare MCO$_3$ are usually much lower than the proposed theoretical capacities. The low electrical conductivity and poor structural stability of MCO$_3$ are blamed as the main obstacles. Fortunately, compositing graphene has been proven as an effective strategy for increasing both the conductivity and structural stability and then remarkably improving the lithium storage capability of MCO$_3$ [130, 134, 137-140].

Conversion reaction of transition metal carbonates:

$$MCO_3 + 2Li \leftrightarrow Li_2CO_3 + M \text{ (M = Mn, Fe, Co or Ni)} \tag{9}$$

$$Li_2CO_3 + (4+0.5x)Li \leftrightarrow 3Li_2O + 0.5Li_xC_2 \text{ (}x = 0, 1 \text{ or } 2) \tag{10}$$

MCO_3 can be easily fabricated by a precipitation reaction between M^{2+} and CO_3^{2-} ions in aqueous solution. Thus, it is easy to obtain graphene-encapsulated MCO_3 composites by *in situ* addition of GO in the reaction solution. Zhong et al. [141] prepared large-area and small-area graphene (LG and SG) by a modified Hummers method using natural graphite flakes with 80 and 3600 mesh sizes, respectively. Then, $MnCO_3$-LG and $MnCO_3$-SG composites were fabricated after the reaction of MnOOH with $NaHCO_3$ in the presence of LG and SG, respectively. Due to the efficient ion-transport path along with the uninterrupted electron-conducting networks of the LG, $MnCO_3$-LG exhibited superior lithium storage performances with a stable capacity of 1050 mAh g^{-1} after 1100 cycles at a current rate of 2 A g^{-1}. Gao et al. [142] synthesized graphene-wrapped $MnCO_3$ mesoporous single crystals by a simple dynamic floating electrodeposition method, which delivered a high reversible capacity of over 1000 mAh g^{-1} after 450 cycles at 100 mA g^{-1} (Figs 5a and 5c). Zhao et al. [143] reported that the *in-situ* addition of GO in the synthesis solution of $MnCO_3$ flower-

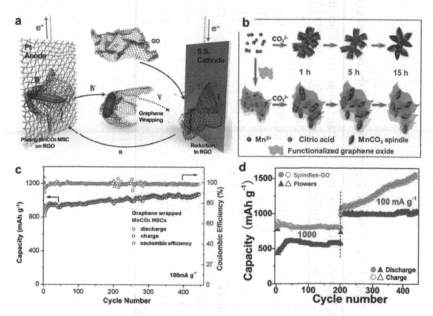

Fig. 5. (a, b) Schematics of the synthesis methods and (c, d) cycling performances of (a, c) graphene-wrapped mesoporous $MnCO_3$ [142]. (Copyright © 2015 Royal Society of Chemistry) and (b, d) $MnCO_3$ spindle-GO composite [143]. (Copyright © 2015 Royal Society of Chemistry)

Table 4. Electrochemical performances of the reported graphene/transition metal carbonate hybrids for LIBs

G/MCO₃ hybrids M = Mn, Fe, Co	M_xO_y content (wt%)	Voltage range (V vs Li/Li⁺)	Current density (A g⁻¹)	Specific capacity (mAh g⁻¹)	ICE (%)	Cycles	References
MnCO₃-LG	–	0.01-3	2	1050	–	1100	[141]
MnCO₃ spindle-GO	91.7	0.01-3	0.1	1474	71	400	[143]
MnCO₃-G	90.3	0.01-3	0.1	>1000	67.8	450	[142]
MnCO₃-RGO	97	0.01-3	0.1	849.1	54	200	[144]
MnCO₃@RGO	68	0.01-3	0.12	857	–	100	[145]
CoCO₃/GNS	90	0.01-3	0.05	930	–	40	[130]
CoCO₃/GA	42.3	0.01-3	0.1	1102	–	80	[140]
FeCO₃/RGO	83.9	0.01-3	0.1	1224	–	50	[139]
FeCO₃/GNS	68.25	0.01-3	0.1	1166	–	255	[146]

like architectures could induce a flower-to-petal structural conversion and generate a well-dispersed $MnCO_3$ spindle-GO composite (Fig. 5b). The $MnCO_3$ spindle-GO composite exhibited enhanced conductivity and structural stability in contrast to bare $MnCO_3$ flowers, and delivered a reversible capacity of 1474 mAh g^{-1} after 400 cycles at 100 mA g^{-1} (Fig. 5d). The lithium storage performances of other reported MCO_3-G composites are summarized in Table 4. The facile synthesis methods and ultrahigh capacities of MCO_3-G hybrids make them promising for mass production and practical applications as anodes of LIBs under the condition that their low ICE can be improved.

4. Conclusion and Outlooks

In this chapter, the most recent advances in the preparation of graphene and graphene-based composite electrodes and their Li storage properties are summarized. Graphene can be used as a bare active material or as a component in hybrids, and in many cases displays significantly improved electrochemical properties such as high capacity, high rate capability and excellent cycling stability due to its large specific surface area, superior electrical conductivity and excellent mechanical and thermal properties. Regarding cathode materials, commonly used cathodes in LIBs are lithium intercalation compounds with limited electronic conductivities, such as $LiCoO_2$, $LiMn_2O_4$, and $LiFePO_4$. With a miniscule amount of graphene, the electronic conductivity of cathode materials can be greatly improved through a "plane-to-point" conducting mode. With great attention to anodes, a number of reports have provided a framework for graphene-containing material categories, such as metal oxides, metal sulfides, metal carbonates, alloys, etc. In these hybrids, graphene serves as an ideal 2D support for growing or assembling electrode materials with well-defined structures, including anchored, wrapped, encapsulated, sandwich-like, layered and mixed models. The inclusion of graphene solves the problem of poor conductivity and buffers the volume changes of anode materials during cycling processes. Moreover, anode materials serve as a spacer between graphene layers to suppress the restacking of graphene sheets. Due to the synergistic effect between graphene and electrode materials and the unique properties of graphene, these graphene-based hybrid anodes show excellent electrochemical properties.

Although great endeavors have been made to improve them, great challenges for graphene-based electrode materials still exist. Firstly, although graphene-based composites with various nanostructures have been reported, it is hard to identify which structural configuration enables better performance over others. Therefore, more understanding should be gained on the correlation between the architecture and the electrochemical performance for graphene-based composite electrodes, which will provide guidelines for future material design and synthesis. Second, it remains a challenge to increase the low initial Coulombic efficiency and reduce the large irreversible capacity loss of graphene-based composite electrodes. Third, various green and cost-effective production approaches towards graphene-based composites need to be developed for industrial applications. Finally, mass production of high-quality graphene oxide and graphene with reduced cost is the key for their widespread applications in LIBs.

References

1. Larcher, D. and Tarascon, J.M. 2015. Towards greener and more sustainable batteries for electrical energy storage. Nat. Chem. 7(1), 19-29.
2. Mathiesen, B.V., Lund, H., Connolly, D., Wenzel, H., Østergaard, P.A., Möller, B., et al. 2015. Smart energy systems for coherent 100% renewable energy and transport solutions. Appl. Energy 145, 139-154.
3. Dunn, B., Kamath, H. and Tarascon, J.-M. 2011. Electrical energy storage for the grid: A battery of choices. Science 334(6058), 928-935.
4. Luo, B. and Zhi, L. 2015. Design and construction of three dimensional graphene-based composites for lithium ion battery applications. Energy Environ. Sci. 8(2), 456-477.
5. Mayorov, A.S., Gorbachev, R.V., Morozov, S.V., Britnell, L., Jalil, R., Ponomarenko, L.A., et al. 2011. Micrometer-scale ballistic transport in encapsulated graphene at room temperature. Nano Lett. 11(6), 2396-2399.
6. Balandin, A.A. 2011. Thermal properties of graphene and nanostructured carbon materials. Nat. Mater. 10, 569.
7. Stoller, M.D., Park, S., Zhu, Y., An, J. and Ruoff, R.S. 2008. Graphene-based ultracapacitors. Nano Lett. 8(10), 3498-3502.
8. Yao, X. and Zhao, Y. 2017. Three-dimensional porous graphene networks and hybrids for lithium-ion batteries and supercapacitors. Chem 2(2), 171-200.
9. Xu, L., Shi, R., Li, H., Han, C., Wu, M. and Wong, C.-P. 2018. Pseudocapacitive anthraquinone modified with reduced graphene oxide for flexible symmetric all-solid-state supercapacitors. Carbon 127, 459-468.
10. Yin, Z., Zhu, J., He, Q., Cao, X., Tan, C., Chen, H. et al. 2014. Graphene-based materials for solar cell applications. Adv. Energy Mater. 4(1), 1300574.
11. Wang, R., Yan, M., Li, H., Zhang, L., Peng, B. and Sun, J. 2018. FeS$_2$ nanoparticles decorated graphene as microbial-fuel-cell anode achieving high power density. Adv. Mater. 30(22), 1800618.
12. Pan, L., Liu, S., Oderinde, O., Li, K., Yao, F. and Fu, G. 2018. Facile fabrication of graphene-based aerogel with rare earth metal oxide for water purification. Appl. Surf. Sci. 427, 779-786.
13. Ito, Y., Tanabe, Y., Sugawara, K., Koshino, M., Takahashi, T. and Tanigaki, K. 2018. Three-dimensional porous graphene networks expand graphene-based electronic device applications. Phys. Chem. Chem. Phys. 20(9), 6024-6033.

14. Landi, B.J., Ganter, M.J., Cress, C.D., DiLeo, R.A. and Raffaelle, R.P. 2009. Carbon nanotubes for lithium ion batteries. Energy Environ. Sci. 2(6), 638-654.

15. Su, F.-Y., You, C., He, Y.-B., Lv, W., Cui, W. and Jin, F. 2010. Flexible and planar graphene conductive additives for lithium-ion batteries. J. Mater. Chem. 20(43), 9644-9650.

16. Tang, R., Yun, Q., Lv, W., He, Y.-B., You, C. and Su, F. 2016. How a very trace amount of graphene additive works for constructing an efficient conductive network in $LiCoO_2$-based lithium-ion batteries. Carbon 103, 356-362.

17. Reddy, M.V., Subba Rao, G.V. and Chowdari, B.V.R. 2013. Metal oxides and oxysalts as anode materials for Li ion batteries. Chem. Rev. 113(7), 5364-5457.

18. Wu, Z.-S., Zhou, G., Yin, L.-C., Ren, W., Li, F. and Cheng, H.-M. 2012. Graphene/metal oxide composite electrode materials for energy storage. Nano Energy 1(1), 107-131.

19. Wu, S., Xu, R., Lu, M., Ge, R., Iocozzia, J. and Han, C. 2015. Graphene-containing nanomaterials for lithium-ion batteries. Adv. Energy Mater. 5(21), 1500400.

20. Molenda, J., Stokłosa, A. and Bąk, T. 1989. Modification in the electronic structure of cobalt bronze Li_xCoO_2 and the resulting electrochemical properties. Solid State Ionics 36(1), 53-58.

21. Shimakawa, Y., Numata, T. and Tabuchi, J. 1997. Verwey-type transition and magnetic properties of the $LiMn_2O_4$ spinels. J. Solid State Chem. 131(1), 138-143.

22. Chung, S.-Y. and Chiang, Y.-M. 2003. Microscale measurements of the electrical conductivity of doped $LiFePO_4$. Electrochem. Solid-State Lett. 6(12), A278-A281.

23. Wei, W., Lv, W., Wu, M.-B., Su, F.-Y., He, Y.-B. and Li, B. 2013. The effect of graphene wrapping on the performance of $LiFePO_4$ for a lithium ion battery. Carbon 57, 530-533.

24. Su, F.-Y., He, Y.-B., Li, B., Chen, X.-C., You, C.-H. and Wei, W. 2012. Could graphene construct an effective conducting network in a high-power lithium ion battery? Nano Energy 1(3), 429-439.

25. Ke, L., Lv, W., Su, F.-Y., He, Y.-B., You, C.-H. and Li, B. 2015. Electrode thickness control: Precondition for quite different functions of graphene conductive additives in $LiFePO_4$ electrode. Carbon 92, 311-317.

26. Prosini, P.P., Lisi, M., Zane, D. and Pasquali, M. 2002. Determination of the chemical diffusion coefficient of lithium in $LiFePO_4$. Solid State Ionics 148(1-2), 45-51.

27. Fan, Q., Lei, L., Xu, X., Yin, G. and Sun, Y. 2014. Direct growth of $FePO_4$/graphene and $LiFePO_4$/graphene hybrids for high rate Li-ion batteries. J. Power Sources 257, 65-69.

28. Luo, W.-B., Chou, S.-L., Zhai, Y.-C. and Liu, H.-K. 2014. Self-assembled graphene and $LiFePO_4$ composites with superior high rate capability for lithium ion batteries. J. Mater. Chem. A 2(14), 4927-4931.

29. Yin, S.C., Strobel, P.S., Grondey, H. and Nazar, L.F. 2004. $Li_{2.5}V_2(PO_4)_3$: A room-temperature analogue to the fast-ion conducting high-temperature γ-Phase of $Li_3V_2(PO_4)_3$. Chem. Mater. 16(8), 1456-1465.

30. Rui, X.H., Ding, N., Liu, J., Li, C. and Chen, C.H. 2010. Analysis of the chemical diffusion coefficient of lithium ions in $Li_3V_2(PO_4)_3$ cathode material. Electrochim. Acta 55(7), 2384-2390.

31. Cui, K., Hu, S. and Li, Y. 2016. Nitrogen-doped graphene nanosheets decorated $Li_3V_2(PO_4)_3$/C nanocrystals as high-rate and ultralong cycle-life cathode for lithium-ion batteries. Electrochim. Acta 210, 45-52.

32. Bak, S.-M., Nam, K.-W., Lee, C.-W., Kim, K.-H., Jung, H.-C. and Yang, X.-Q. 2011. Spinel $LiMn_2O_4$/reduced graphene oxide hybrid for high rate lithium ion batteries. J. Mater. Chem. 21(43), 17309-17315.

33. Lin, B., Yin, Q., Hu, H., Lu, F. and Xia, H. 2014. $LiMn_2O_4$ nanoparticles anchored on graphene nanosheets as high-performance cathode material for lithium-ion batteries. J. Solid State Chem. 209, 23-28.

34. Han, C., Yan, M., Mai, L., Tian, X., Xu, L., Xu, X., et al. 2013. V_2O_5 quantum dots/graphene hybrid nanocomposite with stable cyclability for advanced lithium batteries. Nano Energy 2(5), 916-922.

35. Zhu, Y., Murali, S., Cai, W., Li, X., Suk, J.W., Potts, J.R., et al. 2010. Graphene and graphene oxide: Synthesis, properties, and applications. Adv. Mater. 22(35), 3906-3924.

36. Wang, G., Shen, X., Yao, J. and Park, J. 2009. Graphene nanosheets for enhanced lithium storage in lithium ion batteries. Carbon. 47(8), 2049-2053.

37. Dahn, J.R., Zheng, T., Liu, Y. and Xue, J.S. 1995. Mechanisms for lithium insertion in carbonaceous materials. Science 270(5236), 590-593.

38. Vinayan, B.P., Nagar, R., Raman, V., Rajalakshmi, N., Dhathathreyan, K.S. and Ramaprabhu, S. 2012. Synthesis of graphene-multiwalled carbon nanotubes hybrid nanostructure by strengthened electrostatic interaction and its lithium ion battery application. J. Mater. Chem. 22(19), 9949-9956.

39. Xu, C., Xu, B., Gu, Y., Xiong, Z., Sun, J. and Zhao, X.S. 2013. Graphene-based electrodes for electrochemical energy storage. Energy Environ. Sci. 6(5), 1388-1414.

40. Guan-Nan Zhu, Y.-GW. and Yong-Yao, Xia. 2012. Ti-based compounds as anode materials for Li-ion batteries. Energy Environ. Sci. 5, 6652-6667.

41. Xin, X., Zhou, X., Wu, J., Yao, X. and Liu, Z. 2012. Scalable synthesis of TiO_2/graphene nanostructured composite with high-rate performance for lithium ion batteries. ACS Nano. 6(12), 11035-11043.

42. Ding, S., Chen, J.S., Luan, D., Boey, F.Y.C., Madhavi, S. and Lou, X.W. 2011. Graphene-supported anatase TiO_2 nanosheets for fast lithium storage. Chem. Commun. 47(20), 5780-5782.

43. Yang, S., Feng, X., Muellen, K. 2011. Sandwich-like, graphene-based titania nanosheets with high surface area for fast lithium storage. Adv. Mater. 23(31), 3575.

44. Jiang, X., Yang, X., Zhu, Y., Jiang, H., Yao, Y. and Zhao, P. 2014. 3D nitrogen-doped graphene foams embedded with ultrafine TiO_2 nanoparticles for high-performance lithium-ion batteries. J. Mater. Chem. A 2(29), 11124-11133.

45. Li, B., Han, C., He, Y.-B., Yang, C., Du, H. and Yang, Q.-H. 2012. Facile synthesis of $Li_4Ti_5O_{12}$/C composite with super rate performance. Energy Environ. Sci. 5(11), 9595-9602.

46. Han, C., He, Y.-B., Wang, S., Wang, C., Du, H. and Qin, X. 2016. Large polarization of $Li_4Ti_5O_{12}$ lithiated to 0 V at large charge/discharge rates. ACS Appl. Mater. Interfaces 8(29), 18788-18796.

47. Wagemaker, M., van Eck, E.R.H., Kentgens, A.P.M. and Mulder, F.M. 2008. Li-ion diffusion in the equilibrium nanomorphology of spinel $Li_{4+x}Ti_5O_{12}$. J. Phys. Chem. B 113(1), 224-230.

48. Ouyang, C.Y., Zhong, Z.Y. and Lei, M.S. 2007. Ab initio studies of structural and electronic properties of $Li_4Ti_5O_{12}$ spinel. Electrochem. Commun. 9(5), 1107-1112.

49. Shi, Y., Wen, L., Li, F. and Cheng, H.-M. 2011. Nanosized $Li_4Ti_5O_{12}$/graphene hybrid materials with low polarization for high rate lithium ion batteries. J. Power Sources 196(20), 8610-8617.

50. Lin, Z., Yang, Y., Jin, J., Wei, L., Chen, W. and Lin, Y., 2017. Graphene-wrapped $Li_4Ti_5O_{12}$ hollow spheres consisting of nanosheets as novel anode material for lithium-ion batteries. Electrochim. Acta 254, 287-298.

51. Dong, H.-Y., He, Y.-B., Li, B., Zhang, C., Liu, M. and Su, F. 2014. Lithium titanate hybridized with trace amount of graphene used as an anode for a high rate lithium ion battery. Electrochim. Acta 142(0), 247-253.

52. Palacin, M.R. 2009. Recent advances in rechargeable battery materials: A chemist's perspective. Chem. Soc. Rev. 38(9), 2565-2575.

53. Chan, C.K., Zhang, X.F. and Cui, Y. 2007. High capacity Li ion battery anodes using Ge nanowires. Nano Lett. 8(1), 307-309.

54. Beaulieu, L.Y., Eberman, K.W., Turner, R.L., Krause, L.J. and Dahn, J.R. 2001. Colossal reversible volume changes in lithium alloys. Electrochem. Solid-State Lett. 4(9), A137-A140.

55. Kasavajjula, U., Wang, C. and Appleby, A.J. 2007. Nano- and bulk-silicon-based insertion anodes for lithium-ion secondary cells. J. Power Sources 163(2), 1003-1039.

56. Wu, H. and Cui, Y. 2012. Designing nanostructured Si anodes for high energy lithium ion batteries. Nano Today 7(5), 414-429.

57. He, Y.-S., Gao, P., Chen, J., Yang, X., Liao, X.-Z., Yang, J., et al. 2011. A novel bath lily-like graphene sheet-wrapped nano-Si composite as a high performance anode material for Li-ion batteries. Rsc Adv. 1(6), 958-960.

58. Zhou, X., Bao, J., Dai, Z. and Guo, Y.-G. 2013. Tin nanoparticles impregnated in nitrogen-doped graphene for lithium-ion battery anodes. J. Phys. Chem. C 117(48), 25367- 25373.

59. Liu, X.H., Zhong, L., Huang, S., Mao, S.X., Zhu, T. and Huang, J.Y. 2012. Size-dependent fracture of silicon nanoparticles during lithiation. ACS Nano 6(2), 1522-1531.

60. Wong, D.P., Tseng, H.-P., Chen, Y.-T., Hwang, B.-J., Chen, L.-C. and Chen, K.-H. 2013. A stable silicon/graphene composite using solvent exchange method as anode material for lithium ion batteries. Carbon 63, 397-403.

61. Yang, S., Li, G., Zhu, Q. and Pan, Q. 2012. Covalent binding of Si nanoparticles to graphene sheets and its influence on lithium storage properties of Si negative electrode. J. Mater. Chem. 22(8), 3420-3425.

62. Zhou, X., Yin, Y.-X., Wan, L.-J. and Guo, Y.-G. 2012. Self-assembled nanocomposite of silicon nanoparticles encapsulated in graphene through electrostatic attraction for lithium-ion batteries. Adv. Energy Mater. 2(9), 1086-1090.

63. Ko, M., Chae, S., Jeong, S., Oh, P. and Cho, J. 2014. Elastic a-silicon nanoparticle backboned graphene hybrid as a self-compacting anode for high-rate lithium ion batteries. ACS Nano. 8(8), 8591-8599.

64. Kannan, A.G., Kim, S.H., Yang, H.S. and Kim, D.W. 2016. Silicon nanoparticles grown on a reduced graphene oxide surface as high-performance anode materials for lithium-ion batteries. Rsc Adv. 6(30), 25159-25166.

65. Li, Y., Yan, K., Lee, H.-W., Lu, Z., Liu, N. and Cui, Y. 2016. Growth of conformal graphene cages on micrometre-sized silicon particles as stable battery anodes. Nat. Energy 1, 15029.

66. Luo, B., Wang, B., Li, X., Jia, Y., Liang, M. and Zhi, L. 2012. Graphene-confined Sn nanosheets with enhanced lithium storage capability. Adv. Mater. 24(26), 3538-3543.

67. Wang, D., Li, X., Yang, J., Wang, J., Geng, D. and Li, R. 2013. Hierarchical nanostructured core-shell Sn@C nanoparticles embedded in graphene nanosheets: Spectroscopic view and their application in lithium ion batteries. Phys. Chem. Chem. Phys. 15(10), 3535-3542.

68. Qin, J., He, C.N., Zhao, N.Q., Wang, Z.Y., Shi, C.S. and Liu, E.Z. 2014. Graphene networks anchored with Sn@Graphene as lithium ion battery anode. ACS Nano. 8(2), 1728-1738.

69. Ren, J.-G., Wang, C., Wu, Q.-H., Liu, X., Yang, Y. and He, L. 2014. A silicon nanowire-reduced graphene oxide composite as a high-performance lithium ion battery anode material. Nanoscale 6(6), 3353-3360.

70. Ji, L., Tan, Z., Kuykendall, T., An, E.J., Fu, Y. and Battaglia, V. 2011. Multilayer nanoassembly of Sn-nanopillar arrays sandwiched between graphene layers for high-capacity lithium storage. Energy Environ. Sci. 4(9), 3611-3616.

71. Luo, B., Wang, B., Liang, M.H., Ning, J., Li, X.L. and Zhi, L.J. 2012. Reduced graphene oxide-mediated growth of uniform tin-core/carbon-sheath coaxial nanocables with enhanced lithium ion storage properties. Adv. Mater. 24(11), 1405-1409.

72. Kim, H., Son, Y., Park, C., Cho, J. and Choi, H.C. 2013. Catalyst-free direct growth of a single to a few layers of graphene on a germanium nanowire for the anode material of a lithium battery. Angew. Chem. Int. Ed. 52(23), 5997-6001.

73. Wu, P., Wang, H., Tang, Y., Zhou, Y. and Lu, T. 2014. Three-dimensional interconnected network of graphene-wrapped porous silicon spheres: In situ magnesiothermic-reduction synthesis and enhanced lithium-storage capabilities. ACS Appl. Mater. Interfaces 6(5), 3546-3552.

74. Ge, M., Rong, J., Fang, X., Zhang, A., Lu, Y. and Zhou, C. 2013. Scalable preparation of porous silicon nanoparticles and their application for lithium-ion battery anodes. Nano Res. 6(3), 174-181.

75. Zhou, X., Wan, L. and Guo, Y. 2013. Binding SnO_2 nanocrystals in nitrogen-doped graphene sheets as anode materials for lithium-ion batteries. Adv. Mater. 25(15), 2152-2157.

76. Chen, Y., Yan, C.L. and Schmidt, O.G. 2013. Strain-driven formation of multilayer graphene/GeO_2 tubular nanostructures as high-capacity and very long-life anodes for lithium-ion batteries. Adv. Energy Mater. 3(10), 1269-1274.

77. Lu, Z., Zhu, J., Sim, D., Shi, W., Tay, Y.Y. and Ma, J. 2012. In situ growth of Si nanowires on graphene sheets for Li-ion storage. Electrochim. Acta 74, 176-181.

78. Wang, X.-L. and Han, W.-Q. 2010. Graphene enhances Li storage capacity of porous single-crystalline silicon nanowires. ACS Appl. Mater. Interfaces 2(12), 3709-3713.

79. Chockla, A.M., Panthani, M.G., Holmberg, V.C., Hessel, C.M., Reid, D.K., Bogart, T.D., et al. 2012. Electrochemical lithiation of graphene-supported silicon and germanium for rechargeable batteries. J. Phys. Chem. C 116(22), 11917-11923.

80. Chen, J.S., Archer, L.A. and Wen Lou, X. 2011. SnO_2 hollow structures and TiO_2 nanosheets for lithium-ion batteries. J. Mater. Chem. 21(27), 9912-9924.

81. Wu, S., Han, C., Iocozzia, J., Lu, M., Ge, R. and Xu, R. 2016. Germanium-based nanomaterials for rechargeable batteries. Angew. Chem. Int. Ed. 55(28), 7898-7922.

82. Jiang, B.B., He, Y.J., Li, B., Zhao, S.Q., Wang, S., He, Y.B., et al. 2017. Polymer-templated formation of polydopamine-coated SnO_2 nanocrystals: Anodes for cyclable lithium-ion batteries. Angew. Chem. Int. Ed. 56(7), 1869-1872.

83. Cabana, J., Monconduit, L., Larcher, D. and Palacín, M.R. 2010. Beyond intercalation-based Li-ion batteries: The state of the art and challenges of electrode materials reacting through conversion reactions. Adv. Mater. 22(35), E170-E192.

84. Deng, Y., Wan, L., Xie, Y., Qin, X. and Chen, G. 2014. Recent advances in Mn-based oxides as anode materials for lithium ion batteries. RSC Adv. 4(45), 23914-23935.

85. Poizot, P., Laruelle, S., Grugeon, S., Dupont, L. and Tarascon, J. 2000. Nano-sized transition-metaloxides as negative-electrode materials for lithium-ion batteries. Nature 407(6803), 496-499.

86. Jiang, B.B., Han, C.P., Li, B., He, Y.J. and Lin, Z.Q. 2016. In-situ crafting of $ZnFe_2O_4$ nanoparticles impregnated within continuous carbon network as advanced anode materials. ACS Nano 10(2), 2728-2735.

87. Zhao, S., Sewell, C.D., Liu, R., Jia, S., Wang, Z., He, Y., Yuan, K., Jin, H., Wang, S., Liu, X.Q. and Lin, Z. 2020. SnO_2 as advanced anode of alkali-ion batteries: Inhibiting Sn coarsening by crafting robust physical barriers, void boundaries, and heterophase interfaces for superior electrochemical reaction reversibility. Adv. Energy Mater. 10(6), 1902657.

88. Zhang, K., Han, P., Gu, L., Zhang, L., Liu, Z. and Kong, Q. 2012. Synthesis of nitrogen-doped MnO/graphene nanosheets hybrid material for lithium ion batteries. ACS Appl. Mater. Interfaces 4(2), 658-664.

89. Lavoie, N., Malenfant, P.R.L., Courtel, F.M., Abu-Lebdeh, Y. and Davidson, I.J. 2012. High gravimetric capacity and long cycle life in Mn_3O_4/graphene platelet/LiCMC composite lithium-ion battery anodes. J. Power Sources 213, 249-254.

90. Wang, H., Cui, L.-F., Yang, Y., Casalongue, H.S., Robinson, J.T. and Liang, Y. 2010. Mn_3O_4-graphene hybrid as a high-capacity anode material for lithium ion batteries. J. Am. Chem. Soc. 132(40), 13978-13980.

91. Guo, C.X., Wang, M., Chen, T., Lou, X.W. and Li, C.M. 2011. A hierarchically nanostructured composite of MnO_2/conjugated polymer/graphene for high-performance lithium ion batteries. Adv. Energy Mater. 1(5), 736-741.

92. Li, J., Zhao, Y., Wang, N., Ding, Y. and Guan, L. 2012. Enhanced performance of a MnO_2-graphene sheet cathode for lithium ion batteries using sodium alginate as a binder. J. Mater. Chem. 22(26), 13002-13004.

93. Yu, A., Park, H.W., Davies, A., Higgins, D.C., Chen, Z. and Xiao, X. 2011. Free-standing layer-by-layer hybrid thin film of graphene-MnO_2 nanotube as anode for lithium ion batteries. J. Phys. Chem. Lett. 2(15), 1855-1860.

94. Wei, D., Liang, J., Zhu, Y., Yuan, Z., Li, N. and Qian, Y. 2013. Formation of graphene-wrapped nanocrystals at room temperature through the colloidal coagulation effect. Part. Part. Syst. Charact. 30(2), 143-147.

95. Zhao, B., Liu, R., Cai, X., Jiao, Z., Wu, M. and Ling, X. 2014. Nanorod-like Fe_2O_3/graphene composite as a high-performance anode material for lithium ion batteries. J. Appl. Electrochem. 44(1), 53-60.

96. Qu, J., Yin, Y.-X., Wang, Y.-Q., Yan, Y., Guo, Y.-G. and Song, W.-G. 2013. Layer structured alpha-Fe_2O_3 nanodisk/reduced graphene oxide composites as high-performance anode materials for lithium-ion batteries. ACS Appl. Mater. Interfaces 5(9), 3932-3936.

97. Bai, S., Chen, S., Shen, X., Zhu, G. and Wang, G. 2012. Nanocomposites of hematite (alpha-Fe_2O_3) nanospindles with crumpled reduced graphene oxide nanosheets as high-performance anode material for lithium-ion batteries. Rsc Adv. 2(29), 10977-10984.

98. Chen, D., Ji, G., Ma, Y., Lee, J.Y. and Lu, J. 2011. Graphene-encapsulated hollow Fe_3O_4 nanoparticle aggregates as a high-performance anode material for lithium ion batteries. ACS Appl. Mater. Interfaces 3(8), 3078-3083.

99. Hu, A., Chen, X., Tang, Y., Tang, Q., Yang, L. and Zhang, S. 2013. Self-assembly of Fe_3O_4 nanorods on graphene for lithium ion batteries with high rate capacity and cycle stability. Electrochem. Commun. 28, 139-142.

100. Zhao, J., Yang, B., Zheng, Z., Yang, J., Yang, Z. and Zhang, P. 2014. Facile preparation of one-dimensional wrapping structure: Graphene nanoscroll-wrapped of Fe_3O_4 nanoparticles and its application for lithium-ion battery. ACS Appl. Mater. Interfaces 6(12), 9890-9896.

101. Luo, J., Liu, J., Zeng, Z., Ng, C.F., Ma, L. and Zhang, H. 2013. Three-dimensional graphene foam supported Fe_3O_4 lithium battery anodes with long cycle life and high rate capability. Nano Lett. 13(12), 6136-6143.

102. Wu, Z.-S., Ren, W., Wen, L., Gao, L., Zhao, J. and Chen, Z. 2010. Graphene anchored with Co_3O_4 nanoparticles as anode of lithium ion batteries with enhanced reversible capacity and cyclic performance. ACS Nano 4(6), 3187-3194.

103. Sun, H., Sun, X., Hu, T., Yu, M., Lu, F. and Lian, J. 2014. Graphene-wrapped mesoporous cobalt oxide hollow spheres anode for high-rate and long-life lithium ion batteries. J. Phys. Chem. C 118(5), 2263-2272.

104. Tao, L., Zai, J., Wang, K., Zhang, H., Xu, M. and Shen, J. 2012. Co_3O_4 nanorods/graphene nanosheets nanocomposites for lithium ion batteries with improved reversible capacity and cycle stability. J. Power Sources 202, 230-235.

105. Sun, H., Liu, Y., Yu, Y., Ahmad, M., Nan, D. and Zhu, J. 2014. Mesoporous Co_3O_4 nanosheets-3D graphene networks hybrid materials for high-performance lithium ion batteries. Electrochim. Acta 118, 1-9.

106. Choi, S.H., Ko, Y.N., Lee, J.-K. and Kang, Y.C. 2014. Rapid continuous synthesis of spherical reduced graphene ball-nickel oxide composite for lithium ion batteries. Sci. Rep. 4, 5786.

107. Zhuo, L., Wu, Y., Zhou, W., Wang, L., Yu, Y. and Zhang, X. 2013. Trace amounts of water-induced distinct growth behaviors of NiO nanostructures on graphene in CO_2-expanded ethanol and their applications in lithium-ion batteries. ACS Appl. Mater. Interfaces 5(15), 7065-7071.

108. Wang, Q., Zhao, J., Shan, W., Xia, X., Xing, L. and Xue, X. 2014. CuO nanorods/graphene nanocomposites for high-performance lithium-ion battery anodes. J. Alloys Compd. 590, 424-427.

109. Liu, Y., Wang, W., Gu, L., Wang, Y., Ying, Y. and Mao, Y. 2013. Flexible CuO nanosheets/reduced-graphene oxide composite paper: Binder-free anode for high-performance lithium-ion batteries. ACS Appl. Mater. Interfaces 5(19), 9850-9855.

110. Zhou, X., Shi, J., Liu, Y., Su, Q., Zhang, J. and Du, G. 2014. Microwave-assisted synthesis of hollow CuO-Cu_2O nanosphere/graphene composite as anode for lithium-ion battery. J. Alloys Compd. 615, 390-394.

111. Xu, Y.T., Guo, Y., Song, L.X., Zhang, K., Yuen, M.M.F. and Xu, J.B. 2015. Co-reduction self-assembly of reduced graphene oxide nanosheets coated Cu_2O sub-microspheres core-shell composites as lithium ion battery anode materials. Electrochim. Acta 176, 434-441.

112. Xu, Y.T., Guo, Y., Li, C., Zhou, X.Y., Tucker, M.C., Fu, X.Z., et al. 2015. Graphene oxide nano-sheets wrapped Cu_2O microspheres as improved performance anode materials for lithium ion batteries. Nano Energy 11, 38-47.

113. Yu, M., Wang, A., Wang, Y., Li, C. and Shi, G. 2014. An alumina stabilized ZnO-graphene anode for lithium ion batteries via atomic layer deposition. Nanoscale 6(19), 11419-11424.

114. Hsieh, C.-T., Lin, C.-Y., Chen, Y.-F. and Lin, J.-S. 2013. Synthesis of ZnO@Graphene composites as anode materials for lithium ion batteries. Electrochim. Acta 111, 359-365.

115. Zhu, P.Y., Liu, S.Y., Xie, J., Zhang, S.C., Cao, G.S. and Zhao, X.B. 2014. Facile Synthesis of $NiFe_2O_4$/reduced graphene oxide hybrid with enhanced electrochemical lithium storage performance. J. Mater. Sci. Technol. 30(11), 1078-1083.

116. Xiao, Y.L., Zai, J.T., Tao, L.Q., Li, B., Han, Q.Y. and Yu, C. 2013. $MnFe_2O_4$-graphene nanocomposites with enhanced performances as anode materials for Li-ion batteries. Phys. Chem. Chem. Phys. 15(11), 3939-3945.

117. Xia, H., Zhu, D.D., Fu, Y.S. and Wang, X. 2012. $CoFe_2O_4$-graphene nanocomposite as a high-capacity anode material for lithium-ion batteries. Electrochim. Acta 83, 166-174.

118. Sun, Y.M., Hu, X.L., Luo, W., Xia, F.F. and Huang, Y.H. 2013. Reconstruction of conformal nanoscale MnO on graphene as a high-capacity and long-life anode material for lithium ion batteries. Adv. Funct. Mater. 23(19), 2436-2444.

119. Jiang, Y., Jiang, Z.J., Yang, L.F., Cheng, S. and Liu, M.L. 2015. A high-performance anode for lithium ion batteries: Fe_3O_4 microspheres encapsulated in hollow graphene shells. J. Mater. Chem. A 3(22), 11847-11856.

120. Li, L., Gao, C.T., Kovalchuk, A., Peng, Z.W., Ruan, G.D. and Yang, Y. 2016. Sandwich structured graphene-wrapped FeS-graphene nanoribbons with

improved cycling stability for lithium ion batteries. Nano Res. 9(10), 2904-2911.

121. Mahmood, N., Zhang, C.Z. and Hou, Y.L. 2013. Nickel sulfide/nitrogen-doped graphene composites: Phase-controlled synthesis and high performance anode materials for lithium ion batteries. Small 9(8), 1321-1328.

122. Gu, Y., Xu, Y. and Wang, Y. 2013. Graphene-wrapped CoS nanoparticles for high-capacity lithium-ion storage. ACS App. Mater. Inter. 5(3), 801-806.

123. Xue, H.T., Yu, D.Y.W., Qing, J., Yang, X., Xu, J. and Li, Z.P. 2015. Pyrite FeS$_2$ microspheres wrapped by reduced graphene oxide as high-performance lithium-ion battery anodes. J. Mater. Chem. A 3(15), 7945-7949.

124. Chen, D.Z., Quan, H.Y., Luo, X.B. and Luo, S.L. 2014. 3-D graphene cross-linked with mesoporous MnS clusters with high lithium storage capability. Scripta Mater. 76, 1-4.

125. Zhao, S., Yu, Y., Wei, S., Wang, Y., Zhao, C. and Liu, R. 2014. Hydrothermal synthesis and potential applicability of rhombohedral siderite as a high-capacity anode material for lithium ion batteries. J. Power Sources 253, 251-255.

126. Zhao, S., Wang, Z., He, Y., Jiang, B., Harn, Y. and Liu, X. 2017. Interconnected Ni(HCO$_3$)$_2$ hollow spheres enabled by self-sacrificial templating with enhanced lithium storage properties. ACS Energy Lett. 2(1), 111-116.

127. Zhao, S., Wang, Y., Liu, R., Yu, Y., Wei, S., Yu, F., et al. 2015. Full-molar-ratio synthesis and enhanced lithium storage properties of Co$_x$Fe$_{1-x}$CO$_3$ composites with an integrated lattice structure and an atomic-scale synergistic effect. J. Mater. Chem. A 3(33), 17181-17189.

128. Zhao, S., Wei, S., Liu, R., Wang, Y., Yu, Y. and Shen, Q. 2015. Cobalt carbonate dumbbells for high-capacity lithium storage: A slight doping of ascorbic acid and an enhancement in electrochemical performances. J. Power Sources 284, 154-161.

129. Su, L., Hei, J., Wu, X., Wang, L. and Zhou, Z. 2017. Ultrathin layered hydroxide cobalt acetate nanoplates face-to-face anchored to graphene nanosheets for high-efficiency lithium storage. Adv. Funct. Mater. 27(10), 1605544.

130. Su, L., Zhou, Z., Qin, X., Tang, Q., Wu, D. and Shen, P. 2013. CoCO$_3$ submicrocube/graphene composites with high lithium storage capability. Nano Energy 2(2), 276-282.

131. Wang, L., Tang, W., Jing, Y., Su, L. and Zhou, Z. 2014. Do transition metal carbonates have greater lithium storage capability than oxides? A case study of monodisperse CoCO$_3$ and CoO microspindles. ACS App. Mater. Interfaces 6(15), 12346-12352.

132. Zhong, Y., Su, L., Yang, M., Wei, J. and Zhou, Z. 2013. Rambutan-like FeCO$_3$ hollow microspheres: Facile preparation and superior lithium storage performances. ACS App. Mater. Interfaces 5(21), 11212-11217.

133. Zhou, X., Zhong, Y., Yang, M., Zhang, Q., Wei, J. and Zhou, Z. 2015. Co$_2$(OH)$_2$CO$_3$ nanosheets and CoO nanonets with tailored pore sizes as anodes for lithium ion batteries. ACS App. Mater. Interfaces 7(22), 12022-12029.

134. Zhou, J., Cheng, S., Jiang, Y., Zheng, F., Yang, L. and Rong, H. 2017. High rate and high capacity lithiation of rGO-coated Co$_2$(OH)$_2$CO$_3$ nanosheet arrays

for lithium-ion batteries through the involvement of CO_3^{2-}. Electrochim. Acta 235, 98-106.

135. Zhao, J. and Wang, Y. 2014. High-capacity full lithium-ion cells based on nanoarchitectured ternary manganese-nickel-cobalt carbonate and its lithiated derivative. J. Mater. Chem. A 2(36), 14947-14956.

136. Li, Q., Hu, Y., Li, L. and Feng, C. 2016. Synthesis and electrochemical performances of $Mn_xCo_yNi_zCO_3$. J. Mater. Sci.: Mater. Electron. 27(2), 1700-1707.

137. Zhao, S., Wang, Z., He, Y., Jiang, H., Harn, Y.W., Liu, X., Su, C., Jin, H., Li, Y., Wang, S., Shen, Q. and Lin, Z. 2019. A robust route to $Co_2(OH)_2CO_3$ ultrathin nanosheets with superior lithium storage capability templated by aspartic acid-functionalized graphene oxide. Adv. Energy Mater. 9(26), 1901093.

138. Li, H., Tseng, C., Yang, C., Lee, T., Su, C. and Hsieh, C. 2017. Eco-efficient synthesis of highly porous $CoCO_3$ anodes from supercritical CO_2 for Li^+ and Na^+ storage. Chemsuschem. 10(11), 2464-2472.

139. Zhang, F., Zhang, R., Feng, J., Ci, L., Xiong, S. and Yang, J. 2015. One-pot solvothermal synthesis of graphene wrapped rice-like ferrous carbonate nanoparticles as anode materials for high energy lithium-ion batteries. Nanoscale 7(1), 232-239.

140. Garakani, M., Abouali, S., Zhang, B., Takagi, C., Xu, Z. and Huang, J. 2014. Cobalt carbonate/and cobalt oxide/graphene aerogel composite anodes for high performance Li-ion batteries. ACS App. Mater. Interfaces 6(21), 18971-18980.

141. Zhong, Y.R., Yang, M., Zhou, X.L., Luo, Y.T., Wei, J.P. and Zhou, Z. 2015. Orderly packed anodes for high-power lithium-ion batteries with super-long cycle life: Rational design of $MnCO_3$/large-area graphene composites. Adv. Mater. 27(5), 806-812.

142. Gao, M.W., Cui, X.W., Wang, R.F., Wang, T.F. and Chen, W.X. 2015. Graphene-wrapped mesoporous $MnCO_3$ single crystals synthesized by a dynamic floating electrodeposition method for high performance lithium-ion storage. J. Mater. Chem. A 3(27), 14126-14133.

143. Zhao, S.Q., Feng, F., Yu, F.Q. and Shen. Q. 2015. Flower-to-petal structural conversion and enhanced interfacial storage capability of hydrothermally crystallized $MnCO_3$ via the in situ mixing of graphene oxide. J. Mater. Chem. A 3(47), 24095-25102.

144. Wang, K., Shi, Y.H., Li, H.H., Wang, H.F., Li, X.Y. and Sun, H.Z. 2016. Assembly of $MnCO_3$ nanoplatelets synthesized at low temperature on graphene to achieve anode materials with high rate performance for lithium-ion batteries. Electrochim. Acta 215, 267-275.

145. Zhou, L.K., Kong, X.H., Gao, M., Lian, F., Li, B.J. and Zhou, Z.F. 2014. Hydrothermal fabrication of $MnCO_3@rGO$ composite as an anode material for high-performance lithium ion batteries. Inorg. Chem. 53(17), 9228-9234.

146. Yao, B., Ding, Z.J., Feng, X.Y., Yin, L.W., Shen, Q. and Shi, Y.C. 2014. Enhanced rate and cycling performance of $FeCO_3$/graphene composite for high energy Li ion battery anodes. Electrochim. Acta 148, 283-290.

Polymer Electrolytes for Lithium Ion Batteries

A. Saxena[1], N. Gnanaseelan[2], S.K. Kamaraj[3]* and F. Caballero-Briones[2]*

[1] Department of Physics, Hindustan College of Science and Technology, Mathura, India
[2] Instituto Politécnico Nacional, Materials and Technologies for Energy Health and Environment (GESMAT), CICATA Altamira, 89600 Altamira, México
[3] InstitutoTecnológico de El Llano, Km. 18 Carretera Aguascalientes-San Luis Potosi, 20330 El Llano, Aguascalientes, México

1. Introduction

Batteries are the components which reversibly convert electrical current into chemical stored energy. Batteries are not only a simple and convenient way to make electricity portable, they also provide energy independent from the utility grids in a variety of applications from entertainment to emergency situations. While the first battery was demonstrated nearly 200 years ago by Alessandro Volta, battery development continues on as a result of the growing need for higher and more reliable portable power in applications ranging from localization, remote sensing, internet of things, gadgets and to a plethora of applications. Lots of innovations in the field have allowed increased power and energy densities, as well as reduced size and weight, and one of the substantial changes in battery technology, in the recent years, has been the rechargeable Li ion battery (LIB from herein) which due to the mentioned characteristics, has reached production and sales volumes of millions of units/month [1-5]. The technological maturity and outstanding properties of LIBs have enhanced the market penetration of Hybrid-Electric Vehicles (HEVs) and complete Electric Vehicles (EVs) which are currently the most sustainable ways of transportation [6]. However, despite their market penetration and effectiveness for several

*Corresponding authors: fcaballero@ipn.mx; sathish.bot@gmail.com

applications, there are limitations in their performance and efficiency, which will be discussed later in this chapter.

The common configuration of a commercial lithium ion battery, shown in Fig. 1, consists mostly of a graphite negative electrode, a positive electrode built of a transition metal oxide or phosphate and an electrolyte composed of an organic solvent containing a lithium salt or a mixture of lithium salts as well as additives to enhance stability and safety. In the basic configuration, the electrolyte is soaked into a porous polymer that acts as separator between the electrodes, while allowing the transport of ionic charge carriers and preventing short circuit. The battery charging implies that Li ions are transported through the electrolyte and intercalated into the graphite electrode, and the battery discharging implies the opposite. Both charging and discharging reactions imply volume changes, heating, and an overpotential. During the first charge process, an interfacial layer is formed onto the graphite surface, so-called SEI (Solid Electrolyte Interphase) that protects the graphite electrode from degradation during the battery cycling. The battery performance is measured with the battery voltage and current (power), as well as by its cycling stability.

Li-ion batteries (LIBs) are now dominating the battery industry for a wide range of applications from microelectronics to the automobile industry due to low self-discharge rate, no memory effect, long cycle life, high volumetric and gravimetric energy density. However, some serious impediments limit the application of LIBs, such as low power density compared with Ni-Cd batteries, slow Li ion diffusion rates and poor electrical and thermal conductivity between electrode and electrolyte. Consequently, several materials have had to be developed to achieve efficient electrodes, electrolytes and separators, in order to improve

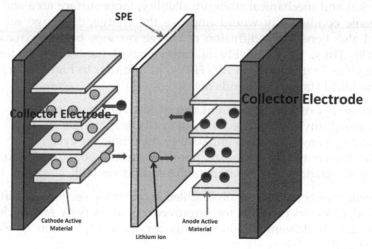

Fig. 1. Li ion rechargeable battery.

energy density, safety features and reliability [7]. Nanomaterials have been gaining importance for electrode materials in LIBs because of the following advantages:

1. There is a higher activity of the electrode reaction in the nanosized electrodes, compared to bulk materials.
2. The higher concentration of reaction centers with respect to the electrode volume, increases the rate of lithium intercalation/deintercalation at the electrode surface.
3. The electron mobility can be improved by tuning the nanomaterials.
4. The electrolyte interaction at the electrode surface is higher because it increases active areas [8].

Between the nanomaterials of potential interest for LIBs, carbon based nanomaterials such as carbon nanotubes, nanofibers, hollow nanospheres, graphene, porous carbon and their variants are considered as good electrode materials due to their high lithiation capability, conductivity, surface area, availability, and chemical and thermal stabilities. The nanoporosity often found in these structures reduces the transport distance for Li^+ ions, allowing larger interfaces between electrode and electrolyte for charge storage and the charge-transfer reaction, while maintaining the electrode stability.

For application in electric and hybrid-electric cars, there are important requirements which LIBs must fulfill, such as thermal safety, affordable price, high specific energy and high specific power. Another important issue that has to be considered for Li based batteries in EV/HEVs is the problem of capacity deterioration during faster charge/discharge rates in the duty cycle that occurs when the vehicle accelerates or moves uphill. Chemical and mechanical strength, stability, large surface area and high electronic conductivity would improve the charge/discharge rate and would also benefit the diffusion of charge carriers, both electrons and ions [9]. Thus, the suitability of carbon based nanomaterials for LIB applications, particularly in EV/HEVs, in addition to high surface area and chemical stability, depends on:

1. Possibility of having different shapes, sizes and aspect ratios.
2. Accessibility and processability of available materials.
3. Reduced environmental impact thru the whole life cycle.
4. Texture control, i.e. pore size, pore volume and pore distribution.
5. Operational ranges for temperature and power [10].

Some pioneer works are highlighted in following sections on different types of Li based batteries for improved power density and cycle life, which are fundamental requirements for their application in electric vehicles.

The different types of Li based batteries are shown and briefed in Fig. 2.

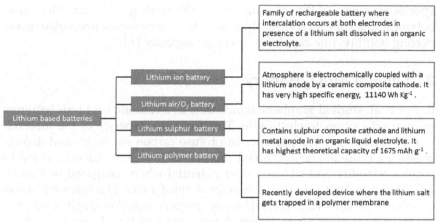

Fig. 2. Types of lithium based batteries.

2. Li Ion Batteries

In 2012, Qie et al. fabricated carbon nanofiber webs (CNFWs) doped with nitrogen from polypyrrole (PPy) nanofiber webs with 16 wt% nitrogen content and KO as activation agent. Resulting CNFWs with porous nano structure contributed to a high reversible capacity of 943 mAh g^{-1} at current density of 2 A g^{-1} after 600 cycles. The superior performance was attributed to the large number of nanopores, which accommodated a larger number of Li^+ ions. The availability of larger surface area (2381 m^2 g^{-1}) led to the creation of adequate electrode/electrolyte interface, which facilitated absorption of Li^+ ions and boosted faster charge-transfer reactions [11]. S. Lee synthesized carbon-coated spinel $LiMnO_4$. The material had achieved high rate capability and high energy density. The power density of the active material was 45 kW/kg. Fast chemical reactions, fast ion pathways, and high electron conductivity were other merits of the material [12]. Z. Chen stressed the importance of diligence in carbon coating method, which has an impact on capacity, rate capability and trap density; they synthesized $LiFePO_4$/C composite (3.5 wt % carbon) using sugar during the preliminary stage of heating of the reactants, which resulted in uniform particle size and proficient carbon coating, which promoted good rate capability [8]. K.T. Lee et al. synthesized three dimensional ordered macroporous (3DOM) material with well-interconnected pores. This material could be directly used as an electrode material without additional binders or conducting agents and a well-interconnected wall provided a continuous electronic pathway, which improved electrical conductivity. The specific discharge of material was 299 mAh g^{-1} at the specific current of 15.2 mA g^{-1} [13]. H. Wang et al. adopted a two-step solution-phase method for growing Mn_3O_4 nanoparticles on graphene oxide (GO). The specific capacity of the material was ~900 mAh g^{-1}. Strong interaction between substrate graphene and Mn_3O_4 nanoparticles promoted good cycling stability, rate capability and high capacity [14].

3. Lithium Air/O$_2$ Battery

B. Sun et al. studied graphene nanosheets as a cathode for Li-air batteries in an alkyl carbonate electrolyte. The discharge capacity of the material was 2332 mAh g^{-1}. The presence of profuse carbon vacancies and defects ensured a more oxygen involved reaction in the Li-air reaction. It led to better cyclability and reduced over potential when compared to Vulcan XC-72 carbon electrode [15]. J. Xiao et al. used a colloidal microemulsion approach to construct hierarchically porous functionalized graphene sheets, which incorporate lattice defects and functional groups such as hydroxyl, epoxy and carboxyl groups. The specific capacity was as high as 15,000 mAh/g. The reason for this amazing performance is the unique morphology of the material, which has interconnected pore channels in both micro and nanosize. This results in faster oxygen diffusion and enhanced catalytic reactions of Li-O$_2$ [16]. C. Sun synthesized graphene (G)-Co$_3$O$_4$ nanocomposite through the hydrothermal route followed by thermal treatment process, which ensured good cycle performance. The catalyst had good charge-discharge curves at a current density of 160 mA g^{-1}. The material had higher round trip efficiency, robust long-term stability and cost effectiveness, when compared to noble metal catalysts [17]. L. Wang et al. reported bifunctional CoMn$_2$O$_4$-graphene composites for higher catalytic activity for oxygen reduction reaction (ORR) and oxygen evolution reaction (OER). The material exhibited a high discharge capacity around 3000 mAh g^{-1} in the first cycle because of the material's high specific area and 2D structure. The material also exhibited good rate capability [18].

4. Lithium Sulphur Battery

N. Jayaprakash et al. synthesized C@S carbon-sulphur nanocomposite capsules to address the problem of limited rate capability and long-term cycle life in Li-S batteries due to high interfacial impendence at the Li metal electrode. The composite material attributed the promising electrochemical performance with initial specific discharge capacity of 1071 mAh g^{-1} and reversible capacity of 974 mAh g^{-1} (at rate of 0.5 C) after 100 cycles with retention capacity of 91%. Authors hypothesized that minimum loss of lithium polysulfides to the electrolyte was due to presence of confined sulphur in pores and interior void space in the composite and it also hindered shuttling. Partial graphitic character was believed to be contributed mechanical and electrochemical stability [19]. H. Wang et al.

developed a novel graphene-sulphur composite material. In addition, the material was integrated with polymeric (polyethyleneglycol) 'cushions', which facilitated minimum dissolution and diffusion of polysulfides and showed volume expansion during discharge. The composite had shown a specific capacity of ~600 mAh g^{-1} with good cycling stability [20]. G. Zhou et al. fabricated sandwich structure with pure sulphur placed between two graphene membranes. One side of graphene membrane was coated with sulphur on it as the active material, which acted as current collector and another side of graphene membrane was coated on the commercial polymer separator. Presence of sulphur on both sides promoted excellent electric conductivity and large volumetric expansion of sulphur during lithiation was possible due to sandwich structure. The discharge capacity was 1345 mAh g^{-1} and showed excellent capacity retention [21]. L. Ji used economic and environmental friendly chemical reaction method to synthesis graphene oxide-sulfur (GO-S) nanocomposite cathode for Li/S batteries in ionic liquid electrolyte. The partially reduced graphene oxide possessed huge surface area, which located cavities everywhere and established strong electronic contact with S. This had effectively avoided aggregation and loss of electrical contact. The material exhibited a high reversible capacity of 950-1400 mAh g^{-1} and stable cycling for more than 50 deep cycles at 167.5 mA g^{-1} [22].

5. Lithium Polymer Battery

B. Jin et al. added multi walled carbon nanotubes (MWCNTs) to improve conductivity of pure $LiFePO_4$ by eight orders of magnitude higher than pure $LiFePO_4$. It also enhanced lithium-ion diffusion co-efficient, reduced crystalline size and charge transfer efficiency. The material showed more stable discharge capacity retention, which was 111 mAh g^{-1} and decreased to 96 mAh g^{-1} after 30 cycles. However, for pure $LiFePO_4$, capacity falls from 110 mAh g^{-1} to 99 mAh g^{-1} during the first cycle to the seventh cycle [23].

6. Overview of Polymer Electrolytes

Polymer electrolytes correspond to a relatively new class of conductors which are ionically conductive in their solid state. They are, in general, solutions of salts that are dissociated within a polymer host, thus they have better conductivity above their T_g. Polymer electrolytes are different as compared to common solid ionic materials which are supported by glasses, ceramics or inorganic compounds in terms of the charge transport property, although the scale of the ionic conductivity is of the order of 10^2 to 10^3 times lesser than the inorganic materials. This disadvantage can be overcome by the high flexibility of the polymer material which permits polymer electrolytes to be processed into very thin films of high surface areas to keep a high-power level. The inherent flexibility allows to stand volume changes due to cyclic lithium transport [23]. Because their low conductivity, ca. 10^{-2} S/m, polymer electrolytes are often processed in thin and large-area elements, thus keeping an acceptable internal resistance. Summarizing, the main advantageous features of solid polymer electrolytes (also abbreviated as SPEs) are [24, 25]:

1. They can be processed as thin films, allowing an increase in the energy density.
2. They serve as spacers, thus eliminating the need of another separator within the battery.
3. They are used as binders, enhancing the electrical contact with the electrodes during charging and discharging.

Additionally, corrosion problems in the containers and battery seals, are avoided by substituting liquid electrolytes by a plastic material. Also, the absence of a liquid phase allows low-pressure packaging, as no gases are expected to evolve, thus opening the way for manufacturing the cells in almost any size and shape. Roll-to-roll technologies of the plastic industry can be used for the manufacturing process.

7. Materials Science Aspects of Polymer Electrolytes

The SPEs must fulfil some important properties in order to accomplish their function as spacer and electrolyte in solid state batteries:

1. As electrolyte, the SPEs must have large enough ionic conductivity for reasonable current densities. Acceptable values lie around 10^{-2}-10^{-3} S/cm.
2. SPEs must be stable within the range of electrochemical reactions at the electrodes, including the overpotentials in charge and discharge.
3. The electrode surfaces and the SPEs must be chemically compatible to allow degradation reactions and decomposition in equilibrium and during the charge-discharge cycling.
4. The SPEs must have good thermal stability within the operation temperature ranges.
5. For scaling from laboratory to pilot, to production, issues like mechanical stability and polymer availability and affordable price have to be taken into account. Despite novel polymer suitability at laboratory scale, they could not be affordable at full production.
6. As a solvent for the salt, the electron donor power of functional groups needs to form coordinate bonds with the cation salt.
7. The energy barriers to allow bond rotation in the polymer chains, must be low.

8. The polymer structure and ion sizes need to be optimized as there is evidence of an optimal distance between the coordinating centers for the formation of intra-polymer ionic bonds.

8. Classification of Polymer Electrolytes

The first ion conducting polymer was introduced in 1973 by Fenton; thereafter, several types of polymer electrolytes have been investigated to obtain better electrolytic properties. A brief classification is shown in Fig. 3, attending to the synthesis methods, and the structural and physical properties of the polymer electrolytes.

8.1 Solvent-free Solid Polymer Electrolytes

This is the oldest class of polymer electrolytes which shows the presence of macromolecular systems which are competent to dissolve salts/acids. To prepare films and membranes of these polymers, the technique of choice has been solution casting. Other approaches for film casting also include hot-press [26, 27]. Solid polymer electrolytes are prepared by dispersing ion donating salts/acids (such as NaI, NaCl, KI) in high molecular weight polymers, i.e. PVP, PPO and PEO. The suitable polymers have polar groups (NH-, -O-, CN-) that allow complex formation with the dispersed salts, while maintaining the phase of the whole system [24, 28, 29]. The ionic conductivity of such

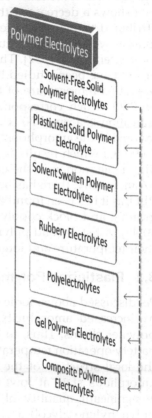

Fig. 3. Classification of polymer electrolytes.

polymer electrolytes arises from the dispersed inorganic salts/acids. The structure of the complexed system can be explained on the basis of lattice and solvation energies of the polymers as well as inorganic salts. The stability of electrolyte complex is related to the activation energies of the polymeric and the inorganic part. Polyethylene oxide is one of the most promising polymeric hosts for SPEs, in which the presence of oxygen on the ether moieties of the chain, making them suitable for solvating several inorganic acids and salts, providing good conductivity. It has been observed that there is increased ion conductivity in polymer electrolytes with low crystallinity (i.e. "high amorphicity"), and the origin of this behavior is related with the degree of motion and local relaxation in the polymer chains out of crystalline order [30]. Another factor that affects ionic conductivity in polymer electrolytes based in PEOs is the salt/acid concentration dispersed within the polymer host; the observations indicate the maximum conductivity with the salt/acid concentration, that shows a decrease with further acid/salt increments; this behavior is attributed to the generation of ion pairs that reduce the number of free ions, as well as to the reduced mobility of the polymer segments at high ionic strengths [31, 32]. The ion-pair generation at higher concentrations of salt have been studied by many workers [33-39]. Another factor is the operation temperature of the electrolyte; for example, the melting point of PEO which corresponds to its semicrystalline-to-amorphous phase transition is around 70 °C. As explained above, the ionic conductivity is favored in the amorphous state and PEO mostly exists in this state around 70-90 °C, which results in good segmental motion, which in turn, gives high ionic motion in the electrolyte and conductivity values around 10^{-4} S/cm. Finally, there has been a strong effort in structural modification of PEO, as it has been found that the conductivity increases by cross-linking, grafting and block copolymerization between other approaches [24, 28, 40-42]. A list of solid polymer electrolytes is given in Table 1 along with their temperature and ionic conductivity.

8.2 Plasticized Polymer Electrolytes

As discussed above, ionic conductivity in solvent-free SPEs is supported by increased amorphicity and higher order segmental motion of the polymer chains. Thus, to increase the ionic conductivity of SPEs at the battery operation temperatures, strong efforts have been made to modify the local structure of the polymer matrix to keep the polymers in the amorphous state at lower temperatures. To achieve this modification, a significant quantity of additives of low molecular weight such as poly(ethylene glycol) and/or aprotic organic solvents such as ethylene carbonate (EC), propylene carbonate (PC), diethylene carbonate (DEC), and dimethylsulfoxide (DMSO) are added to the solvent-free SPE matrix to act as "plasticizers". These additives enhance the amorphicity of the system by decreasing the glass transition temperature and supporting segmental motion of the polymer chains which, in turn, increase the conductivity of the plasticized polymer electrolytes at ambient conditions. A further contribution of the plasticizer to the conductivity, comes from the high dielectric constant of organic plasticizers, like EC and PC, that results in a higher ion dissociation. Hariharan et al. [67] studied the effect of PEG400 as plasticizer in $(PEO)_6:NaPO_3$ for conductivity modification. Bhattacharya et al. [68] reported an enhancement in conductivity for PEO/PDMS blend complexed with LiI, close to 10^{-3} S/cm at 303 K.

Some selective polymer electrolyte systems with plasticizers are listed in Table 2.

Table 1. Some solvent-free Solid Polymer Electrolytes (SPEs) along with their temperature and conductivity

Solid polymer electrolyte (SPE)	Temp (°C)	Conductivity (S/cm)	References
(PEO)x-NaSCN	20	1.0×10^{-7}	[43]
(PEO)x-NH$_4$SCN	60	1.0×10^{-5}	[44]
PVAc-LiSCN	100	1.0×10^{-3}	[45]
PEO-LiBr	256	5.3×10^{-8}	[46]
PEO-Cu(ClO$_4$)$_2$	25	1.0×10^{-6}	[47]
PEO-(NH$_4$)$_2$SO$_4$	30	1.0×10^{-7}	[48]
PEO-AgNO$_3$	30	4.0×10^{-7}	[49]
PEO-NaPF$_6$	30	5.7×10^{-6}	[50]
PEO-(NH$_4$)$_2$SO$_4$	30	2.7×10^{-6}	[51]
PEO-LiClO$_4$	27	1×10^{-4}	[52]
PEO-LiCF$_3$SO$_4$	40	1×10^{-4}	[53]
PEO-LiBF$_4$	25	1×10^{-6}	[54]
PEO-LiN(CF$_3$SO$_3$)$_2$	25	1×10^{-6}	[55]
PEO-NH$_4$I	-	$\sim10^{-5}$	[56]
MEEP-LiN(CF$_3$SO$_3$)$_2$	20	6.5×10^{-5}	[57]
MEEP-NaCF$_3$SO$_4$	25	1×10^{-5}	[58]
PPO-NaCF$_3$SO$_4$	45	1×10^{-5}	[59]
PEO-NaI	60	1×10^{-5}	[60]
PEO-AgNO$_3$	27	4×10^{-6}	[49]
PEO-KIO$_3$	27	4.4×10^{-7}	[61]
PEO-NH$_4$ClO$_4$	-	1.05×10^{-5}	[62]
PEO: AgNO$_3$	27	4.0×10^{-6}	[63]
PEO-LiPF$_6$	RT	1.5×10^{-4}	[64]
PEO-KI	27	6.30×10^{-7}	[65]
PEO-KI/I$_2$-MB (2-Mercapto benzimidazole)	26	1.8×10^{-5}	[66]

8.3 Solvent Swollen Polymer Electrolytes

There exist certain polymers like poly(vinyl alcohol) (PVA), poly(acrylic acid) (PAA), poly(vinyl pyrrolidone) (PVP) etc, which swell-up by the addition of some aqueous/nonaqueous solvents. In this swollen

Table 2. Some selective plasticized polymer electrolytes with
temperature and conductivity

Plasticized polymer electrolytes	Temp (°C)	Conductivity (S/cm)	References
$(PEO)_8$-$LiClO_4$ (EC:PC, 20 mol%)	20	1×10^{-3}	[29]
PEO-$LiCF_3SO_3$- PEG	25	$\sim 10^{-3}$	[69]
$(PEO)_8$-$LiClO_4$(PC, 50 mol%)	20	8×10^{-4}	[29]
PEO-$LiBF_4$ with 12-crown-4	RT	7×10^{-4}	[70]
PEO-$LiCF_3SO_3$-EC	23	1.4×10^{-6}	[71]
PEO-$LiCF_3SO_3$-PC	23	5.4×10^{-8}	[71]
PVC- x PEMA–PC–$LiClO_4$	30	3.45×10^{-3}	[72]
PEO-PDMA-LiI	27	$\sim 10^{-3}$	[68]
PAN-EC-$LiCF_3SO_3$	RT	1.32×10^{-3}	[73]
PAN-EC-$LiCF_3SO_3$	RT	5.49×10^{-3}	[73]
PVA-H_3PO_4-PEG	30	5.3×10^{-4}	[74]
PVA-NH_4Cl-DMSO	30	1×10^{-5}	[75]
PVA-NH_4-DMSO	30	5.7×10^{-4}	[75]
PVA-NH_4I-DMSO	30	2.5×10^{-3}	[75]
PVA-NH_4NO_3-DMSO	30	7.5×10^{-3}	[76]
Chitosan-PVA-NH_4NO_3-EC	RT	1.6×10^{-3}	[77]

polymer matrix different types of dopant solutes like H_3PO_4, NH_4SCN
etc, are accommodated and are responsible for electrical conductivity.
Electrical conductivity of such types of unstable systems depends on the
concentration of the dopant solute present in the swollen matrix, relative
humidity, temperature, ambient conditions etc. as studied by Awadhia
and Agrawal (2007) [78], Polak et al. (1986) [79], Daniel et al. (1988) [80],
Rekukawa et al. (1996) [81], Gupta et al. (1989) [82].

8.4 Rubbery Electrolytes

A novel kind of polymer electrolyte described as *"rubbery electrolyte"*
was first introduced by Angell et al. [83], in which they reported the
electrolyte system as consisting of PEO and PPO host polymer complexed
with comparatively large amounts of mixtures of different lithium salts.
They stated that a small amount of high molecular weight polymer

soluble in the melt of salt mixture exhibits a rubbery character with low T_g temperature by means of the entanglement mechanism and facilitates high ion conduction due to decoupled cation motion. The rubbery electrolyte, $AlCl_3$-LiBr-$LiClO_4$-PPO, reported by Angell et al. exhibited the conductivity as high as $\sim 2 \times 10^{-2}$ S/cm at room temperature [84]. Feng and Cui reported a rubbery electrolyte system, $LiClO_4$-$LiN(CF_3SO_2)_2$-PEO, and obtained encouraging results with improved conductivity of $\sim 10^{-6}$ S/cm [85]. However, despite the better electrical properties of these electrolytes, very few reports are available in this category.

8.5 Polyelectrolytes

Polyelectrolytes have definite groups of ions connected with the primary chain of the polymer, thus having self-ion generating groups [86]. In solvent-free polyelectrolytes ions like Na^+, H^+, Cl^-, OH^- etc., are strongly attached to the polymer chains with strong coulombic forces between the collections of oppositely charged ions. A wide amount of proton conducting polyelectrolytes have been studied in the literature [87-90]. Since the preparation of these polyelectrolytes involves complex chemical processing, they have not attracted a larger interest than their alternatives.

8.6 Gel Polymer Electrolytes (GPEs)

To achieve a compromise between the dimensional stability inherent to solid polymer electrolytes and the high conductivity of liquid electrolytes, gel polymer electrolyte systems (GPEs) have been developed. GPEs are solvent swollen polymers, which depict a high ionic conductivity comparable with that of SPEs. One of the methods of preparation of GPEs is a two-step process: first, the preparation of a liquid electrolyte by dissolving salt into a polar solvent or ionic liquid and second, incorporation of the liquid electrolyte into an inert polymeric host [26]. The other method for preparation of GPEs is called 'phase inversion technique' in which a separate porous polymer matrix is prepared using some non-solvents or electrospun and dipping it into the liquid electrolyte to soak the liquid electrolyte into its pores [91]. The large amount of liquid present in the polymer matrix, gives rise to both the adhesive characteristic of solids and the dispersive transport characteristic of liquids. A main disadvantage is the reduction in the mechanical integrity upon addition of large amounts of liquid to the polymer; however, this can be overcome by adding components that can be cross-linked and/or thermoset with the GPEs [92].

8.7 Composite Polymer Electrolytes

This category of the polymer electrolytes is prepared by incorporation of nano or micro sized dispersed particles of inert ceramic materials to the polymer electrolytes. Weston and Steele [102] demonstrated for the first

Table 3. Some polymer gel electrolytes along with their temperature and conductivity

Polymer gel electrolyte	Temp (°C)	Conductivity (S/cm)	References
PAN-PC-EC-LiN $(CF_3SO_2)_2$	25	2.0×10^{-3}	[93]
PAN-PC-EC-LiCF$_4$	25	4.5×10^{-3}	[93]
PAN-PC-EC-LiAlF$_6$	20	4.6×10^{-3}	[93]
PMMA-PC-LiClO$_4$	20	2.1×10^{-2}	[94]
PMMA-PC-NaClO$_4$	20	4.3×10^{-2}	[94]
PMMA-PC-ZnClO$_4$	20	4.2×10^{-2}	[94]
PMMA-PC-Mg(ClO$_4$)$_2$	20	1.5×10^{-2}	[94]
PMMA-EC-PC-LiCF$_3$SO$_3$	30	2.3×10^{-3}	[95]
PMMA-EC-PC-LiN(CF$_3$SO$_2$)$_2$	30	7.1×10^{-3}	[95]
PMMA-DEC-LiCF$_3$SO$_3$	20	2.5×10^{-3}	[96]
PMMA-PC-LiClO$_4$	20	4.8×10^{-3}	[96]
PMMA-DMA-LiCF$_3$SO$_3$	20	5.5×10^{-3}	[96]
PVdF(HFP-(oxalic acid) (OA)-DMA	RT	4.2×10^{-5}	[97]
PVdF(HFP)-PC-DEC-LiClO$_4$	30	2.6×10^{-3}	[98]
MG49-NH$_4$CF$_3$SO$_4$	30	1.2×10^{-2}	[99]
PVdF(HFP-PC- Mg(ClO$_4$)$_2$	25	5.0×10^{-3}	[100]
TMPTMA-EMC-DMC-LiPF$_6$	20	7.4×10^{-3}	[101]

time the concept of inclusion of electrochemically unreactive ceramic filler particles of α-alumina in the PEO based SPE and shows significant increase in room temperature conductivity. Since then, several research groups have attempted to disperse a variety of ceramic fillers into different SPE hosts and a large number of CPEs consequently have been reported. It has been observed that dispersion of these ceramic particles not only enhance the ionic conductivity but also improve the other physical, mechanical and electrochemical properties of polymer electrolyte systems [103-105]. The size of the particles and the dispersible nature of the material have significant effects on the electrolyte properties; for example Weiczorek et al. [106, 107] observed a significant increase in the conductivity when smaller sized Al$_2$O$_3$ particles (smaller than 4 μm) were incorporated in the PEO-NaI SPE. They suggested that surface groups in the ceramic particles have a significant effect in promoting the local structural modification in the polymer electrolyte. Croce et al. [108, 109] observed substantial increase in the ionic conductivity and mechanical integrity on incorporation of inert sub-micrometer particles of TiO_2, Al_2O_3 and SiO_2 in PEO based SPEs. They suggested a solid plasticizer-like behavior of these fillers in the polymer matrix, which improves the amorphicity in PEO when annealed at ~70 °C, increasing the electrical properties as discussed above. Panero et al. [110] prepared a dual composite polymer electrolyte by incorporation of calixpyrrole (CP) and super-acid zirconia (S-ZrO_2) in PEO-$LiCF_3SO_3$ solid polymer electrolyte system and suggested that CP enhances the Li^+ transference number and $SZrO_2$ ceramic component

Table 4. Some Composite Polymer Electrolytes with temperature and conductivity values

Composite polymer electrolytes	Temp (°C)	Conductivity (S/cm)	References
PEO-NH_4I-Al_2O_3	70	8×10^{-4}	[114]
PEO-$LiClO_4$-α-Al_2O_3	25	1×10^{-5}	[115]
PEO-NaI-SiO_2	25	5×10^{-4}	[115]
PEO-$LiClO_4$-SiC	30	1×10^{-5}	[116]
PEO-$LiClO_4$-$BaTiO_3$	70	1×10^{-3}	[117]
PEO-$LiClO_4$-TiO_2	30	2×10^{-5}	[118]
PEO-$LiClO_4$-Al_2O_3	30	1×10^{-5}	[118]
PEO-$LiBF_4$-TiO_2	RT	1×10^{-5}	[119]
PEO-$LiBF_4$-ZrO_2	RT	1×10^{-5}	[119]
PMMA-$LiCO_4$-DMP-CeO_2	RT	5.36×10^{-4}	[120]
PVdF(HFP)-EC-DEC-$LiPF_6$-Al_2O_3	RT	1.95×10^{-3}	[121]
PVdF(HFP)-PMMA-$LiCF_3SO_3$-(PC-DEC)-SiO_2	30	1×10^{-3}	[122]
PVdF(HFP)-EC-PC-$LiClO_4$-Al_2O_3	RT	2.11×10^{-3}	[123]
PVdF(HFP)-$LiAlO_2$	RT	8.12×10^{-3}	[124]
PMMA-$LiCF_3SO_3$-SiO_2	30	7.30×10^{-5}	[125]
MG49-TiO_2-$LiBF_4$	RT	1.4×10^{-5}	[126]
PMMA-EC-PC-$NaClO_4$-SiO_2	20	3.4×10^{-3}	[127]
PMMA-PC-DEC-$LiClO_4$-MMT	RT	1.3×10^{-3}	[128]
PMMA-EC-PC-$LiN(CF_3SO_2)_2$-SiO_2	RT	2.0×10^{-3}	[125]
PEO-$LiClO_4$-$LiAlO_2$	RT	9.76×10^{-5}	[129]

promotes the conductivity of the electrolyte system. Falaras et al. [111], analyzed a PEO/titania polymer electrolyte for photoelectrochemical cells manufactured with $Ru(dcbpy)_2(NCS)_2$ as sensitizer and photoanodes made of nanoporous TiO_2 films.

Many other reports are also available on these composite electrolytes where inert filler particles, such as Al_2O_3, SiO_2, TiO_2, $BaTiO_3$, etc., have been incorporated and the prepared composite polymer electrolytes have been found to exhibit improved electrolytic properties. Conducting ceramic particles like, zeolites and ionites, have been used as fillers in composite polymer electrolytes to enhance the ionic conductivity and mechanical strength [112, 113]. Table 4 shows some composite polymer electrolytes from literature.

9. Challenges, Present and Forecasted Solutions

9.1 Electrodes and Selection of Electrode Material

The selection of battery materials significantly depends on the open circuit voltage (OCV) of the cell. The OCV value relies directly on chemical potential of the electrode materials and is described as [130]

$$V_{OC} = \frac{(\mu_A - \mu_C)}{F}$$

where μ_A and μ_C are the chemical potentials of the anode and cathode materials, respectively and F is the Faraday constant. Due to low conductivity of the solid phases and liquid electrolyte phase, as well as slow electrochemical reactions in a battery electrode, the voltage V_{dis} obtained during discharge can be reduced by an amount of η_{dis}, which is called over potential [130]. A similar over potential would appear during charging. The open circuit potential can be tuned according to the above mentioned equation by changing the electrode materials, although the electrode material cannot be randomly chosen as the electrolyte might decompose if the potential is beyond the stability potential window of the electrolyte as discussed above in this chapter.

Materials for battery electrodes in LIBs must satisfy some of the following criteria [131]:

1. The lithium chemical potential at the anode must be high, while at the cathode must be low for maximizing the cell voltage.
2. The cell capacity in Ah/L or Ah/kg is maximized when a large amount of lithium is transported, depending on the host material surface chemistry as well as on the coordination number of lithium in the electrodes.
3. The electrode must be structurally stable in the lithium insertion/ extraction cycles.

4. Charge/discharge rate must be ensured by the high electronic conductivity of the electrode and high ionic conductivity of lithium ion within the electrolyte.
5. The electrode materials must be thermally and chemically stable to avoid reaction with the electrolyte and subsequent performance degradation. To ensure this, the band gap of the electrodes must be within the redox potential window of the electrolyte.
6. Materials must be monetarily affordable, environmentally innocuous, and lightweight.

10. Recent Progress

Though tremendous progress has been made over the last couple of decades, state-of-the-art LIBs still face various problems such as low energy and power density, short lifetime, and high cost. Among them, thermal instability appears as one of the top concerns, to which the anode, cathode, and electrolyte contribute jointly. Depression of the peak heating rate has been investigated in cathode materials which are olivine-based or $LiMn_2O_4$, with the end result of improved cathode stability. Other solutions for heat management will be discussed in the later section of this chapter.

For LIBs to be used in EV/HEVs, the capacity over a long time span and several charge-discharge cycles is critical. Therefore, the challenge is the design of high-power-density and high-energy-density batteries with improved stability and high discharging/charging rates. Employment of nanomaterials in different shapes, such as nanowires, rods, spheres, fibers, tubes, and plates, has been demonstrated to enhance material transport characteristics and, therefore, discharge/charge rate. Particularly, Zhang et al. [132] have developed three-dimensional bi-continuous electrode nano architectures consisting of a thin layer of electrolytically active material between three-dimensionally connected electrolyte and conductive scaffold phases, providing a highly conductive pathway for electrons, a short ion diffusion length in the intercalation compound, and a fast mass transport channel in the liquid electrolyte. A high discharge rate of 1114 W mAh g^{-1} was demonstrated.

Element doping and surface coating in battery nanomaterials have also been employed as strategies to enhance charge carriers in the form of lithium ions and electrons. The pioneering work [133] from the Chiang group at MIT first demonstrated a factor of 108 improvements in conductivity when doping $LiFePO_4$ with cationic elements such as aluminium, niobium, and zirconium, though the conductivity improvement mechanism is still in debate. Coating of cathodes with either ionically or electronically conductive materials has been demonstrated to significantly improve high-rate capacity and capacity retention. AlF_3

coating on $LiMn_2O_4$ showed only 3.4% capacity loss at 55 °C after 50 cycles compared to ~18% decay without the coating [134]. Kang and Ceder at MIT had created a fast ion-conducting surface phase on $LiFePO_4$ through controlled off-stoichiometry [135]. Their materials were able to discharge at ~200 °C and still achieve a capacity more than 120 mAh/g. At this high rate, the time to fully charge a battery would be a matter of seconds rather than hours, and this will make new technological applications possible and induce lifestyle changes. This is particularly attractive when speaking of the EV/HEVs market, where the long charging times are a cause of the poor penetration of these technologies. However, a debate is still going on, because a report showed a carbon amount up to 65%, which poses the question whether the device can be considered a battery or a supercapacitor.

11. Separator Materials in Batteries

Battery separators create a barrier between the anode and cathode; the barrier prevents short-circuiting between the internal components. In order to enable ion flux while also serving as a barrier, battery separator materials must present a porous structure with exceptional mechanical strength [136]. A particularly critical role of the battery separator is the so-called "*shutdown function*" which is the separator material's ability to "seal itself off" if the battery's temperature rises to abnormal levels for any reason. Because of this, the separator material is considered to play an important part in overall battery safety.

12. Transition from Dry to Wet Separators in EV Batteries

Two main classes of separator materials are used, wet and dry separators. While the former have become the most widely used in consumer electronics, they have until now seen little utilization in the EV batteries due to cost and safety considerations.

Wet separators can be made thinner than their dry counterparts due to superior mechanical strength. However, the use of solvents in the manufacturing process of wet separators reduce their advantage by increasing the cost. Furthermore, wet separators have until now had a greater number of safety issues when used in tandem with high-nickel cathode materials in EV battery applications.

Despite this, the price difference between wet and dry separators is being reduced as production size and volume efficiency for wet separator production increases, and the gap is expected to continue shrinking. To solve the safety concerns, research in the recent years has been directed to the addition of ceramic coatings to the wet separator surface, to increase heat resistance. Thus, there are high expectations for an increase in the number of manufacturers using wet separators for automotive applications in the near future [136].

The materials consist of polyolefin separators that are either polypropylene (PP) or polyethylene (PE) based. The battery separator materials contain micro and nano pores that provide ion transportation capabilities within the Li-ion battery. They are also offered in a ceramic-coated variety for enhanced thermal deformation resistance and mechanical strength. There are several industrial suppliers for these kinds of separators. Amongst the reported advantages of PP materials are the high shut down temperature and melting point, porosity distribution, mechanical resistance upon puncture and low shrinkage while for PE, the uniform thickness, ion penetration and puncture resistance are the most prominent characteristics [136].

13. Thermal Management of Li Ion Battery

Boeing 787 Dreamliner is a long range, wide body, twin-engine jet airliner, which started its commercial journey in late 2011. However, in a short span of almost more than a year, on 16 January 2013, all Boeing 787 Dreamliners have permanently landed on the ground without any further take-off. This fateful decision cost the company its reputation, finances, Dreamliner operators and suppliers/vendors. What exactly went wrong? The failure of Li-ion battery in the aircraft contributed to its grounding.

Boeing 787 Dreamliner used two identical lithium ion batteries, which assisted the operation of the power unit when the aircraft landed on the ground and utilized it as a backup for the electronic flight system. As known, the current is generated in the Li-ion batteries by shuttling lithium ions to transport the charges between the electrodes. A polymer membrane is used to separate the electrode to avoid internal short circuit and a lithium salt containing organic electrolyte acts as medium to channelize the ion transport. The volatile nature of electrolyte poses serious safety issues with Li-ion battery. High operating temperature and short circuit could spur exothermic reactions to produce combustible gases, lead to melting of the separator and contribute to thermal runaway. Battery catching fire or exploding is considered to be the worst-case scenario [137].

Victor A. (2014) wrote a detailed review on the formation and stability of solid electrolyte interface on the graphite anode. Reaction between the electrolyte and anode during the initial stage of charging leads to the formation of various species on the electrode surface. This formation of surface species is called solid electrolyte interphase (SEI). Basically it is a Li^+ conductor but an insulator to electron flow and protects the electrolyte from decomposition to enhance life cycle stability. Electrode materials and composition of electrolyte play a major role in effective formation of the SEI layer [138]. The optimum temperature range of operation of lithium ion batteries is –20°C to 60°C [139]. When the battery operates outside this optimal temperature window it contributes to the degradation of the SEI and safety is compromised.

The safety of the Li-ion battery is often compromised by failure of individual components (cathode, anode, electrolytes or current collector) and during the abusive conditions of the battery such as overcharging and thermal runway. There are three stages of the thermal runway which occur when LIBs are heated externally. In this situation deterioration occurs with increase of temperature. Primarily it starts with slow anodic reaction at 90°C and reaction rate become rapid with increase of temperature. Solid electrolyte interface (SEI) breaks at 120°C, which results in a large electrolyte reduction rate at the lithiated graphite anode. At 140°C, exothermic reactions are initiated at the cathode and generate oxygen. This oxygen catalyzes interfacial oxidation of the electrolyte above 180°C, leading to high thermal runaway with temperature increased rapidly at 100°C/min. The other reason for thermal runaway is sustained overcharge, which leads to increase in internal temperature and pressure. This may lead to rupture of the battery [140].

Another problem is confinement of hundreds of lithium ion batteries, which are connected in series and parallel to satisfy the power requirement of electric/hybrid vehicles. This creates the problem of heat dissipation from array of batteries to the external environment [141]. Tesla, one of the leading manufacturers of electric cars, has been in the news since past few years regarding its cars catching fire, owing to Li-ion batteries. On 23 March 2018, a 38-year-old man was killed in an accident when he rammed his Tesla model X into an unshielded highway median at Highway 101 near Mountain View, California and subsequently the car caught fire. It became a nightmare for firefighters to extinguish the fire, as they were not trained to handle battery vehicle fire accidents to deal with electric emission hazards. Lithium-ion fires can emit many types of toxic gases such as carbon monoxide, soot, hydrogen fluoride, and particulates of nickel, aluminum, lithium, copper, and cobalt [142]. Comparatively, although gasoline is a very flammable and risky substance, technology has made gasoline-powered vehicles very safe based on 130 years of research and development. For EV/HEVs, technology is in its early stages and efforts to make LIBs safer are ongoing [143].

Application of phase-change materials (PCM) is widely utilized to deal with thermal problems. The PCM materials reduce temperature inside the battery by storing excess heat as latent heat without increasing temperature and thereby phase change takes place. PCM also acts like a buffer to handle extreme fluctuations in ambient temperature. Melting point and temperature ranges could be modified by changing the chemical composition. Thermal conductivity of PCM paraffin wax (IGI–1260) was improved by two orders by combining it with graphene and latent heat storage ability was preserved. Paraffin waxes are common PCM because of wide availability, chemical stability, durability and cycling. Its latent heat of fusion is as high as 200-250 kJ/kg and melting point range is suitable for greater control of batteries and portable electronics. Melting and boiling points of IGI-1260 are T_M ~70°C and T_B ~90°C respectively. Easier attachment of paraffinic hydrocarbon molecules to graphene flakes might be the reason for increased thermal conductivity [144]. Phase changes do happen in PCMs from solid or liquid to gas at constant temperature during process of latent heat storage. They are cheap, non-corrosive and have large latent heat storage capability.

On the other hand, high porosity, good thermal physical properties and mechanical strength are the properties that make the metal foams, a good option for PCM materials. Hussian et al. proposed a novel material, graphene coated nickel foam saturated with paraffins [145]. This material has higher melting point and lower freezing point when compared to pure paraffins. The latent heat and specific heat of this material are reduced to 30% and 34% respectively. Increase of the surface temperature of the battery is also reduced to 17% when compared to nickel foam under 1.7 A discharge current. Qi et al. synthesized PEG/GO/GNP composite (polyethylene glycol/graphene oxide/graphene nanoplatelets composite) for PCM applications. High phase change enthalpy, chemical and thermal stability, biodegradation, non-toxicity, non-corrosiveness, low vapor pressure and ability to alter melting temperature are the properties of polyethylene glycol, which makes it suitable for PCMs. The presence of GO provided the rigid network structure and strong interaction with PEG, which gave the composite a better shape-stability. The presence of GNB enhanced the thermal conductivity of the composite to 490%, that is 1.72 W/mK when compared to pure PEG, which had 0.29 W/mK. GNB also helped to make the composite a better electrical conductor by forming effective conductive network [146]. Wu et al. developed a copper mesh (CM)-enhanced paraffin (PA)/expanded graphite (EGComposite as PCM material for thermal management applications. The composite offered good mechanical strength and thermal conductivity due to presence of CM. CM also improved heat dissipation performance and uniform distribution of temperature. The PA also absorbed heat generated from cells during melting [147].

14. Summary

In the present chapter, the basic structure of lithium ion batteries (LIBs) has been revised and their applicability in the field of electric vehicles and hybrid-electric vehicles (EV/HEVs) has been stressed on. The current technologies in LIBs, as well as the challenges and current improvements in electrodes, separators, electrolytes and thermal management from the Materials Science point of view, have been pointed out, with particular emphasis on the use of carbon nanomaterials and more specifically, graphene in particular. Although these materials are showing large applicability in this field, further work in scaling up the applications of carbon nanomaterials to real scale EV/HEVs is urgently needed.

Acknowledgements

This work was partially financed by Instituto Politecnico Nacional (Mexico), through the project SIP-20181187.

References

1. Campbell, C.J. 2002. The Assessment and Importnace of Oil Depletion. Paper at the International Workshop on Oil Depletion, Uppsala University.
2. Campbell, C.J. and Laherrere, J.H. 1998. The End of Cheap Oil. Scientific American.
3. Gratzel, M. 2001. Photoelectrochemical cells. Nature, 414, 338-344.
4. Steel, B.C.H. and Heinzel, A. 2001. Materials for fuel-cell technologies. Nature, 414, 345-352.
5. Scrosati, B. 1995. Battery technology: Challenge of portable power. Nature. 373, 557-558.
6. Zeng, Xianlai, Li, Jinhui and Singh, Narendra. 2014. Recycling of spent lithium-ion battery: A critical review. Critical Reviews in Environmental Science and Technology 10, 1129-1165.
7. Warnar, J. 2015. The Handbook of Lithium-Ion Battery Pack Design. Elsevier, USA.
8. Chen, Z. and Dahn, J.R. 2002. Reducing carbon in LiFePO$_4$/C composite electrodes to maximize specific energy, volumetric energy, and tap density. Journal of the Electrochemical Society 149(9), A1184-A1189.
9. Park, C. 2010. Next Generation Lithium Ion Batteries for Electrical Vehicles. In-Tec. Croatia.
10. Frackowiak, E. and Béguin, F. 2001. Carbon materials for the electrochemical storage of energy in capacitors. Carbon. 39, 937-950.
11. Qie, L., Chen, W.M., Wang, Z.H., Shao, Q.G., Li, X., Yuan, L.X., Hu, X.L., Zhang, W.X. and Huang, Y.H. 2012. Nitrogen-doped porous carbon nanofiber webs as anodes for lithium ion batteries with a superhigh capacity and rate capability. Adv. Mater. 24, 2047-2050.
12. Lee, S., Cho, Y., Song, H.K., Lee, K.T. and Cho, J. 2012. Carbon-coated single-crystal LiMn$_2$O$_4$ nanoparticle clusters as cathode material for high-energy and high-power lithium-ion batteries. Angew. Chemie. Int. Ed. 51, 8748-8752.

13. Lee, K.T., Lytle, J.C., Ergang, N.S., Oh, S.M. and Stein, A. 2005. Synthesis and rate performance of monolithic macroporous carbon electrodes for lithium-ion secondary batteries. Adv. Func. Mater. 15, 547-556.

14. Wang, H., Cui, L., Yang ,Y., Casalongue, H.S. and Liang, Y. 2017. Mn_3O_4-graphene hybrid as a high capacity anode material for lithium ion batteries. J. Am. Chem. Soc. 132, 13978-13980.

15. Sun, B., Wang, B., Su, D., Xiao, L., Ahn, H. and Wang, G. 2012. Graphene nanosheets as cathode catalysts for lithium-air batteries with an enhanced electrochemical performance. Carbon. 50, 727-733.

16. Xiao, J., Mei, D., Li, X., Xu, W., Wang, D., Graff, G.L., Bennett, W.D., Nie, Z., Saraf, L.V., Aksay, I.A., Liu, J. and Zhang, J.G. 2011. Hierarchically porous graphene as a lithium-air battery electrode. Nano Lett. 11, 5071-5078.

17. Sun, C., Li, F., Ma, C., Wang, Y., Ren, Y., Yang, W., Ma, Z., Li, J., Chen, Y., Kim, Y. and Chen, L. 2014. Graphene-Co_3O_4 nanocomposite as an efficient bifunctional catalyst for lithium-air batteries. J. Mater. Chem. A 2, 7188-7196.

18. Wang, L., Zhao, X., Lu, Y., Xu, M., Zhang, D., Ruoff, R.S., Stevenson, K.J. and Goodenough, J.B. 2011. $CoMn_2O_4$ spinel nanoparticles grown on graphene as bifunctional catalyst for lithium-air batteries. J. Electrochem. Soc. 158, A1379.

19. Jayaprakash, N., Shen, J., Moganty, S.S., Coron, A. and Archer, L.A. 2011. Porous hollow carbon@sulfur composites for high-power lithium-sulfur batteries. Angew. Chemie. Int. Ed. 50, 5904-5908.

20. Wang, H., Yang, Y., Liang Y., Robinson, J.T., Li, Y., Jackson, A. and Dai, H. 2011. Graphene-wrapped sulfur particles as a rechargeable lithium-sulfur battery cathode material with high capacity and cycling stability. Nano Letters 11(7), 2644-2647.

21. Li, F., Zhou, G., Pei, S., Li, L., Wang, D.W., Wang, S., Huang, K., Yin, L.C. and Cheng, H.M. 2014. A graphene-pure-sulfur sandwich structure for ultrafast, long-life lithium-sulfur batteries. Adv. Mater. 26, 625-631.

22. Ji, L., Rao, M., Zheng, H., Zhang, L., Li, O.Y. and Duan, W. 2011. Graphene oxide as a sulfur immobilizer in high performance lithium/sulfur cells. J. Am. Chem. Soc. 133, 18522-18525.

23. Jin, B., Jin, E.M., Park, K.H. and Gu, H.B. 2008. Electrochemical properties of $LiFePO_4$-multiwalled carbon nanotubes composite cathode materials for lithium polymer battery. Electrochem. Commun. 10, 1537-1540.

24. Gray, F.M. 1991. Solid Polymer Electrolytes: Fundamental and Technological Applications. New York: VCH. Chap. 1, pp. 1-10.

25. Park, M., Zhang, X., Chung, M., Less, G.B. and Sastry, A.M. 2010. A review of conducting phenomenon in Li-ion batteries, J. Power Sources 195, 7904-7929.

26. Gray, F.M., McCallum, J.R. and Vincent, C.A. 1986. Poly(ethylene oxide) – $LiCF_3SO_3$ – polystyrene electrolyte systems. Solid State Ionics, 18-19, 282-286.

27. Agrawal, R.C. and Pandey, G.P. 2008. Solid polymer electrolyte: Materials designing and all-solid state battery applications: An overview. J. Phys. D. Appl. Phys. 41, 223001.

28. MacCullam, J.R. and Vincent, C.A. (Eds.). 1987 & 1989. Polymer Electrolyte Reviews. Vol. 1-2. Elsevier Applied Science, London.

29. Gray, F.M. 1997. Polymer Electrolytes. Royal Society of Chemistry Monographs, Cambridge.
30. Wright, P.V. 1975. Electrical conductivity in ionic complexes of poly(ethylene oxide). Brit. Polym. J. 7, 319-327.
31. Olsen, I., Koksbang, R. and Skou, E. 1995. Transference number measurements on a hybrid polymer electrolyte. Electrochim. Acta 40, 1701-1706.
32. Petersen, G., Jacobsson, P. and Torell, LM. 1992. A Raman study of ion-polymer and ion-ion interactions in low molecular weight polyether-LiCF$_3$SO$_3$ complexes. Electrochim. Acta, 37, 1495-1497.
33. Bruce, P.G. and Vincent, C.A. 1989. Effect of ion association on transport in polymer electrolytes. Faraday Discuss. Chem. Soc. 88, 43-54.
34. Pollock, D.W., Williamson, K.J., Weber, K.S., Lyons, L.S. and Sharpe, L.R. 1994. Ion pairing and ionic conductivity in amorphous polymer electrolytes: A structural investigation employing EXAFS. Chem. Mater. 6, 1912-1914.
35. MacCallum, J.R., Tomlin, A.S. and Vincent, C.A. 1986. An investigation of the conducting species in polymer electrolytes. European Polymer Journal 22, 787-791.
36. Schantz, S. 1991. On the ion association at low salt concentrations in polymer electrolytes: A Raman study of NaCF$_3$SO$_3$ and LiClO$_4$ dissolved in poly(propylene oxide). J. Chem. Phys. 94, 6296.
37. Bakker, A., Gejji, S., Lindgren, J., Hermansson, K. and Probst, M.M. 1995. Contact ion pair formation and ether oxygen coordination in the polymer electrolytes M[N(CF$_3$SO$_2$)$_2$]$_2$PEOn for M = Mg, Ca, Sr and Ba. Polymer. 36, 4371-4378.
38. Reddy, M.J. and Chu, P.P. 2002. Ion pair formation and its effect in PEO:Mg solid polymer electrolyte system. J. Power Sources 109, 340-346.
39. M-Vosshage, D. and Chowdari, B.V.R. 1995. XPS studies on (PEO) nLiCF$_3$SO$_3$ and (PEO)nCu(CF$_3$SO$_3$)$_2$ polymer electrolytes. Electrochim. Acta 40, 2109-2114.
40. Nest, J.F.L., Callens, S., Gandini, A. and Armand, A. 1992. A new polymer network for ionic conduction. Electrochim. Acta 37, 1585-1588.
41. Dias, F.B., Plomp, L. and Veldhuis, J.B.J. 2000. Trends in polymer electrolytes for secondary lithium batteries. J. Power Sources 88, 169-191.
42. Kim, D.W., Park, J.K., Gong, M.S. and Song, H.Y. 1994. Effect of grafting degree and side PEO chain length on the ionic conductivities of NBR-g-PEO based polymer electrolytes. Polym. Eng. Sci. 34, 1305-1313.
43. Gray, F.M., MacCallum, J.R., Vincent, C.A. and Giles, J.R.M. 1988. Novel polymer electrolytes based on ABA block copolymers. Macromolecules 21, 392-397.
44. Ooura, Y., Machida, N., Naito, M. and Shigematsu, T. 2012. Electrochemical properties of the amorphous solid electrolytes in the system Li$_2$S-Al$_2$S$_3$-P$_2$S$_6$. Solid State Ionic 225, 350-353.
45. Wintersgill, M.C., Fontanella, J.J., Calme, J.P., Smith, M.K., Jones, T.B., Greenbaum, S.G., Adamic, K.J., Shetty, A.N. and Andeen, C.G. 1986. Conductivity, DSC, FTIR, and NMR studies of poly(vinyl acetate) complexed with alkali metal salts. Solid State Ionics 18-19, 326-331.
46. Rietman, E.A., Kaplan, M.L. and Cave, R.J. 1987. Alkali metal ion-

poly(ethylene oxide) complexes. II: Effect of cation on conductivity. Solid State Ionics 25, 41-44.

47. Magistris, A., Chiodelli, G., Singh, K. and Ferloni, P. 1990. Electrical and thermal properties of PEO-Cu(ClO$_4$)$_2$ polymer electrolytes. Solid State Ionics 38, 235-240.

48. Maurya, K.K., Hashmi, S.A. and Chandra, S. 1992. Proton conducting polymer electrolytes: Polyethylene oxide + (NH$_4$)$_2$SO$_4$ system. Journal of Physical Society of Japan 61, 1709-1716.

49. Chandra, S., Hashmi, S.A., Saleem, M. and Agrawal, R.C. 1993. Investigations on poly ethylene oxide based polymer electrolyte complexed with AgNO$_3$. Solid State Ionics 67, 1-7.

50. Hashmi, S.A. and Chandra, S. 1995. Experimental investigations on a sodium-ion conducting polymer electrolyte based on poly(ethylene oxide) complexed with NaPF$_6$. Materials Science & Engineering: B 34, 18-26.

51. Ali, A.M.M., Mohamed, N.S. and Arof, A.K. 1998. Polyethylene oxide (PEO)-ammonium sulphate ((NH$_4$)$_2$SO$_4$) complexes and electrochemical cell performance. J. Power Sources 74, 135-141.

52. MacCallum, J.R., Smith, M.J. and Vincent, C.A. 1984. The effects of radiation-induced crosslinking on the conductance of LiClO$_4$·PEO electrolytes. Solid State Ionics 11, 307-312.

53. Mustarelli, P., Quartarone, E., Tomasi, C. and Magistris, A. 2000. New materials for polymer electrolytes. Solid State Ionics 135, 81-86.

54. Chiodelli, G., Ferloni, P., Magistris, A. and Sanesi, M. 1988. Ionic conduction and thermal properties of poly (ethylene oxide)-lithium tetrafluoroborate films. Solid State Ionics 28-30, 1009-1013.

55. Armand, M., Gorecki, W. and Andreani, R. 1991. Proc. 2nd Int. Meeting on Polymer Electrolytes (New York). Scrosati, B. (Ed.). Elsevier. Amsterdam.

56. Maurya, K.K., Srivastava, N., Hashmi, S.A. and Chandra, S. 1992. Proton conducting polymer electrolyte: II poly ethylene oxide + NH$_4$I system. J. Mater. Sci. 27, 6357-6364.

57. Abraham, K.M. and Alamgir, M. 1993. Ambient temperature rechargeable polymer electrolyte batteries. J. Power Sources 43-44, 195-208.

58. Greenbaum, S.G., Adamić, K.J., Pak, Y.S., Wintersgill, M.C. and Fontanella, J.J. 1988. NMR, DSC and electrical conductivity studies of MEEP complexed with NaCF$_3$SO$_3$. Solid State Ionics 28-30, 1042-1046.

59. Watanabe, M. and Ogata, N. 1987. Polymer Electrolyte Review I. MacCallum, J.R. and Vincent, C.A. (eds.). Elsevier Applied Sciences, London.

60. Fauteux, D., Lupien, M.D. and Robitaille, C.D. 1987. Phase diagram, conductivity and transference number of PEO NaI electrolyte. J. Electrochem. Soc. 134, 2761-2767.

61. Agrawal, R.C., Mahipal, Y.K. and Ashrafi, R. 2011. Materials and ion transport property studies on hot-press casted solid polymer electrolyte membranes: [(1 − x) PEO:x KIO$_3$]. Solid State Ionics 92, 6-8.

62. Hashmi, S.A., Kumar, A., Maurya, K.K. and Chandra, S. 1990. Proton-conducting polymer electrolyte. I: The polyethylene oxide+NH$_4$ClO$_4$ system. J. Phys. D: Appl. Phys. 23, 1307-1314.

63. Agrawal, R.C. and Pandey, G.P. 2008. Solid polymer electrolytes: Materials and designing and all-solid-state battery applications: An overview. Journal of Physics D: Applied. Physics. 41, 223001-223018.

64. Ibrahim, S., Yasin, S.M.M., Ahmad, R. and Johan, M.R. 2012. Conductivity, thermal and morphology studies of PEO based salted polymer electrolytes. Solid State Sci. 14, 1111-1116.

65. Chandra, A. and Thakur, K. 2013. Synthesis characterization and ion transport properties of hot-pressed solid polymer electrolytes (1-x) PEO:x KI. Chinese Journal of Polymer Science 31, 302-308.

66. Muthuraaman, B., Will, G., Wang, H., Moonie, P. and Bell, J. 2013. Increased charge transfer of poly (ethylene oxide) based electrolyte by addition of small molecule and its application in dye-sensitized solar cells. Electrochim. Acta 87, 526-531.

67. Bhide, A. and Hariharan, K. 2007. Ionic transport studies on $(PEO)_6:NaPO_3$ polymer electrolyte plasticized with PEG400. European Polymer Journal, 43, 4253-4270.

68. Lee, J.Y., Bhattacharya, B., Kim, D.W. and Park, J.K. 2008. Poly(ethylene oxide)/ poly(dimethylsiloxane) blend solid polymer electrolyte and its dye-sensitized solar cell applications. J. Phys. Chem. C 112, 12576-12582.

69. Ito, Y., Kanehori, K., Miyauchi, K. and Kudo, T. 1987. Ionic conductivity of electrolytes formed from $PEO-LiCF_3SO_3$ complex low molecular weight poly(ethyleneglycol). Journal of Materials Science 22, 1845-1849.

70. Nagasubramanian, G. and Stefano, S.D. 1990. 12-crown-4 ether-assisted enhancement of ionic conductivity and interfacial kinetics in polyethylene oxide electrolytes. Journal of Electrochemical Society 137, 3830-3835.

71. Bandara, L.R.A.K., Dissanayake, M.A.K.L. and Mellander, B.E. 1996. Solid State Ionics – New Developments. World Scientific, Singapore.

72. Rajendran, S., Ramesh Prabhu, M. and Usha Rani, M. 2008. Li ion conduction behavior of hybrid polymer electrolytes based on PEMA. Journal of Applied Polymer Science 110, 2802-2806.

73. Isa, K.B.M., Othman, L. and Osman, Z. 2011. Comparative studies on plasticized and unplasticized polyacrylonitrile (PAN) polymer electrolytes containing lithium and sodium salts. Sains Malaysiana 40, 695-700.

74. Prajapati, G.K., Roshan, R. and Gupta, P.N. 2010. Effect of plasticizer on ionic transport and dielectric properties of $PVA - H_3PO_4$ proton conducting polymeric electrolytes. J. Phys. Chem. Solids 71, 1717-1723.

75. Hema, M., Selvasekarapandian, S., Arunkumar, D., Sakunthala, A. and Nithya. H. 2009. FTIR, XRD and ac impedance spectroscopic study on PVA based polymer electrolyte doped with NH_4X (X = Cl, Br, I). J. Non-Cryst. Sol. 355, 84-90.

76. Hema, M., Selvasekarapandian, S., Hirankumar, G., Sakunthala, A., Arunkumar, D. and Nithya, H. 2010. Laser Raman and ac impedance spectroscopic studies of PVA: NH_4NO_3 polymer electrolyte. Spectrochim. Acta Part A 75, 474-478.

77. Kadir, M.F.Z., Majid, S.R. and Arof, A.K. 2010. Plasticized chitosan – PVA blend polymer electrolyte based proton battery. Electrochim. Acta 55, 1475-1482.

78. Awadhia, A. and Agrawal, S.L. 2007. Structural, thermal and electrical characterisation of $PVA:DMSO:NH_4SCN$ gel electrolytes. Solid State Ionics 178, 951-956.

79. Polak, A.J., Petty-weeks, S. and Beuhler, A.J. 1986. Applications of novel proton conducting polymers to hydrogen sensing. Sensors and Actuators 9, 1-7.

80. Daniel, M.F., Desbat, B. and Lassegues, J.C. 1988. Solid state protonic conductors: Complexation of poly(ethylene oxide) or poly(acrylic acid) with NH_4HSO_4. Solid State Ionics 28-30, 632-636.

81. Rekukawa, M., Kawahara, M., Sanui, K. and Ogata, N. 1996. 12th International Conference. Solid State Ionics. Greece.

82. Singh, R.P., Gupta, P.N., Agrawal, S.L. and Singh, U.P. 1989. Solid State Ionics. *In*: G. Nazari, R.A. Huggins and D.F. Shriver (eds.), Mater. Res. Soc. Pittsburgh, 361.

83. Angell, C.A., Liu, C. and Sanchez, E. 1993. Rubbery solid electrolytes with dominant cationic transport and high ambient conductivity. Nature 362, 137-139.

84. Angell, C.A., Xu, K., Zhang, S.S. and Videa, M. 1996. Variations on the salt-polymer electrolyte theme for flexible solid electrolytes. Solid State Ionics 86-88, 17-28.

85. Feng, L. and Cui, H. 1996. A new solid-state electrolyte: Rubbery 'polymer-in-salt' containing $LiN(CF_3SO_2)_2$. J. Power Sources 63, 145-148.

86. Ivory, D.M., Miller, G.G., Sowa, J.M., Shacklette, L.W., Chance, R.R. and Baughman, R.H. 1979. Highly conducting charge-transfer complexes of poly(p-phenylene). J. Chem. Phys. 71, 1506.

87. Peighambardoust, S.J., Rowshanzamir, S. and Amjadi, M. 2010. Review of the proton exchange membranes for fuel cell applications. Int. J. Hydrogen Energy 35, 9349-9384.

88. Smitha, B., Sridhar, S. and Khan, A.A. 2005. Solid polymer electrolyte membranes for fuel cell applications: A review. J. Membrane Science 259, 10-26.

89. Oszcipok, M., Zedda, M., Hesselmann, J., Huppmann, M., Wodrich, M., Junghardt, M. and Hebling, C. 2006. Portable proton exchange membrane fuel-cell systems for outdoor applications. J. Power Sources 157, 666-673.

90. Rhoo, H.J., Kim, H.T., Park, J.K. and Hwang, T.S. 1997. Ionic conduction in plasticized PVCPMMA blend polymer electrolytes. Electrochim. Acta 42, 1571-1579.

91. Stephan, A.M. 2006. Review on gel polymer electrolytes for lithium batteries. Eur. Polym. J. 42, 21-42.

92. Ross-Murphy, S.B. 1987. Polymer network—Principles of their formation, structure and properties. Stepto, R.F.T. (ed.). Blackie Academic and Professional, London, 288.

93. Scrosati, B. 1993. Applications of Electroactive Polymer. Chapman & Hall. London.

94. Vondriak, J., Sedlarikova, M., Velicka, J., Klapste, B., Novak, V. and Reiter, J. 2001. Gel polymer electrolytes based on PMMA. Electrochim. Acta 46, 2047-2048.

95. Sekhon, S.S., Arora, N. and Singh, H.P. 2003. Effect of donor number of solvent on the conductivity behaviour of nonaqueous proton-conducting polymer gel electrolytes. Solid State Ionics 160, 301-307.

96. Singh, B. and Sekhon, S.S. 2005. Polymer electrolytes based on room temperature ionic liquid: 2,3-dimethyl-1-octylimidazolium triflate. Journal of Physical Chemistry: B 109, 16539-16543.

97. Missan, H.P.S., Chu, P.P. and Sekhon, S.S. 2006. Ion conduction mechanism

in nonaqueous polymer electrolytes based on oxalic acid: Effect of plasticizer and polymer. J. Power Sources 158, 1472-1479.

98. Saikia, D. and Kumar, A. 2004. Ionic conduction in P(VdF-HFP)/PVdF-(PC+DEC)-LiClO$_4$ polymer gel electrolytes. Electrochim. Acta 49, 2581-2589.

99. Kamisan, A.S., Kudin, T.I.T., Ali, A.M.M. and Yahya, M.Z.A. 2011. Electrical and physical studies on 49% methyl-grafted natural rubber-based composite polymer gel electrolytes. 57, 207-211.

100. Tripathi, S.K., Gupta, A. and Kumari, M. 2012. Studies on electrical conductivity and dielectric behaviour of PVdF-HFP-PMMA-NaI polymer blend electrolyte. Bulletin of Materials Science 35, 969-975.

101. Zhou, D., Fan, L.Z., Fan, H. and Shi, Q. 2013. Electrochemical performance of trimethylolpropane trimethylacrylate-based gel polymer electrolyte prepared by in situ thermal polymerization. Electrochim. Acta 89, 334-338.

102. Weston, J.E. and Steele, B.C.H. 1982. Effects of inert fillers on the mechanical and electrochemical properties of lithium salt-poly (ethylene oxide) polymer electrolytes. Solid State Ionics 7, 75-79.

103. Appetecchi, G.B. and Passerini, S. 2000. PEO-carbon composite lithium polymer electrolyte. Electrochim. Acta 45, 2139-2145.

104. Itoh, T., Miyamura, Y., Ichikawa, Y., Uno, T., Kubo, M. and Yamamoto, O. 2003. Composite polymer electrolytes of poly(ethylene oxide)/BaTiO$_3$/Li salt with hyperbranched polymer. J. Power Sources 119-121, 403-408.

105. Bronstein, L.M., Karlinsey, R.L., Ritter, K., Joo, C.G., Stein, B. and Zwanziger, J.W. 2004. Design of organic–inorganic solid polymer electrolytes: Synthesis, structure, and properties. J. Mater. Chem. 14, 1812-1820.

106. Wieczorek, W., Florjanczyk, Z. and Stevens, J.R. 1995. Composite polyether based solid electrolytes. Electrochim. Acta 40, 2251-2258.

107. Przyluski, J., Siekierski, M. and Wieczorek, W. 1995. Effective medium theory in studies of conductivity of composite polymeric electrolytes. Electrochim. Acta. 40, 2101-2108.

108. Croce, F., Curini, R., Martinelli, A., Persi, L., Ronci, F., Scrosati, B. and Caminiti, R. 1999. Physical and chemical properties of nanocomposite polymer electrolytes. J. Phys. Chem. B 103, 10632-10638.

109. Croce, F. and Scrosati, B. 2003. Nanocomposite lithium ion conducting membranes. Ann. NY Acad. Sci. 984, 194-207.

110. Panero, S., Scrosati, B., Sumathipala, H.H. and Wieczorek, W. 2007. Dual-composite polymer electrolytes with enhanced transport properties. J. Power Sources 167, 510-514.

111. Stergiopoulos, T., Arabatzis, I.M., Katsaros, G. and Falaras, P. 2002. Binary polyethylene oxide/titania solid-state redox electrolyte for highly efficient nanocrystalline TiO$_2$ photoelectrochemical cells. Nano Letters 2, 1259-1261.

112. Skaarup, S., West, K. and Christiansen, B.Z. 1988. Mixed phase solid electrolytes. Solid State Ionics 28-30, 975-978.

113. Wieczorek, W. 1992. Temperature dependence of conductivity of mixed-phase composite polymer solid electrolytes. Mater. Sci. Eng. 15, 108-114.

114. Chandra, A., Srivastava, P.C. and Chandra, S. 1995. Ion transport studies in PEO:NH$_4$I polymer electrolytes with dispersed Al$_2$O$_3$. Journal of Materials Science 30, 3633-3638.

115. Wieckzorek, W., Zalewska, A., Raducha, D., Florjancyk, Z., Stevens, J.R., Ferry, A. and Jacobsson, P. 1996. Polyether, poly(N,N-dimethylacrylamide), and LiClO$_4$ composite polymeric electrolytes. Macromolecules 29, 143-155.

116. Choi, B.K. and Shin, K.H. 1996. Effects of SiC fillers on the electrical and mechanical properties of (PEO)$_{16}$ LiClO$_4$ electrolytes. Solid State Ionics 86-88, 303-306.

117. Sun, H.Y., Sohn, H.J., Yamamoto, O., Takeda, Y. and Imanishi, N. 1999. Enhanced lithium-ion transport in PEO-based composite polymer electrolytes with ferroelectric BaTiO$_3$. Journal of Electrochemical Society 146, 1672-1676.

118. Croce, F., Curini, R., Martinelli, A., Persi, L., Ronci, F., Scrosati, B. and Caminiti, R. 1999. Physical and chemical properties of nanocomposite polymer electrolytes. Journal of Physical Chemistry B 103, 10632-10638.

119. Kumar, B. and Scanlon, L.G. 1999. Polymer-ceramic composite electrolytes: Conductivity and thermal history effects. Solid State Ionics 124, 239-254.

120. Rajendran, S., Mahendran, O. and Krishnaveni, K. 2003. Effect of CeO$_2$ on conductivity of PMMA/PEO polymer blend electrolytes. Journal of New Materials for Electrochemical Systems 6, 25-28.

121. Li, Z., Su, G., Gao, D., Wang, X. and Li, X. 2004. Effect of Al$_2$O$_3$ nanoparticles on the electrochemical characteristics of P(VdF-HFP)-based polymer electrolyte. Electrochim. Acta 49, 4633-4639.

122. Saikia, D. and Kumar, A. 2005. Ionic transport in P(VdF-HFP)-PMMALiCF$_3$SO$_3$-(PC+DEC)-SiO$_2$ composite gel polymer electrolyte. European Polymer Journal 41, 563-568.

123. Li, Z., Su, G., Wang, X. and Gao, D. 2005. Micro-porous P(VdF-HFP)-based polymer electrolyte filled with Al$_2$O$_3$ nanoparticles. Solid State Ionics 176, 1903-1908.

124. Kalyana, S.N.T. and Subramania, A. 2007. Nano-size LiAlO$_2$ ceramic filler incorporated porous PVdF-co-HFP electrolyte for lithium-ion battery applications. Electrochim. Acta 52, 4987-4993.

125. Ramesh, S. and Wen, L.C. 2010. Investigation on the effects of addition of SiO$_2$ nanoparticles on ionic conductivity, FTIR, and thermal properties of nanocomposite PMMA–LiCF$_3$SO$_3$–SiO$_2$. Ionics 16, 255-262.

126. Low, S.P., Ahmad, A., Hamzah, H. and Rahman, M.Y.A. 2011. Nanocomposite solid polymeric electrolyte of 49% poly(methyl methacrylate)-grafted natural rubbertitanium dioxide-lithium tetrafluoroborate (MG$_{49}$-TiO$_2$-LiBF$_4$). Journal of Solid State Electrochemistry 15, 2611-2618.

127. Kumar, D. and Hashmi, S.A. 2010. Ion transport and ion-filler-polymer interaction in poly(methyl methacrylate)-based, sodium ion conducting, gel polymer electrolytes dispersed with silica nanoparticles. J. Power Sources 195, 5101-5108.

128. Deka, M. and Kumar, A. 2010. Enhanced electrical and electrochemical properties of PMMA–clay nanocomposite gel polymer electrolytes. Electrochim. Acta 55, 1836-1842.

129. Masoud, E.M., El-Bellihi, A.A., Bayoumy, W.A. and Mousa, M.A. 2013. Effect of LiAlO$_2$ nanoparticle filler concentration on the electrical properties of PEOLiClO$_4$ composite. Materials Research Bulletin 48, 1148-1154.

130. Yuan, L.X., Wang, Z.H., Zhang, W.X., Hu, X.L., Chen, J.T., Huang, Y.H. and

Goodenough, J.B. 2011. Development and challenges of LiFePO$_4$ cathode material for lithium-ion batteries. Energy Environ. Sci. 4, 269-284.

131. Alejandro, F. 2015. Rechargeable Lithium Batteries: From Fundamentals to Applications. Woodhead Publishing Series in Energy.

132. Zhang , H., Yu, X. and Braun, P.V. 2011. Three-dimensional bicontinuous ultrafast-charge and discharge bulk battery electrodes. Nat. Nanotechnol. 6, 277-281.

133. Chung, S.-Y., Bloking, J.T. and Chiang, Y.M. 2002. Electronically conductive phospho-olivines as lithium storage electrodes. Nat. Mater. 1, 123-128.

134. Liu, H. and Tang, D. 2009. The effect of nanolayer AlF$_3$ coating on LiMn$_2$O$_4$ cycle life in high temperature for lithium secondary batteries. Russ. J. Electrochem. 45, 762-764.

135. Kang, B. and Ceder, G. 2009. Battery materials for ultrafast charging and discharging. Nature 458, 190-193.

136. https://www.targray.com/articles/ev-battery-market-forecast-separator-materials

137. Williard, N., He, W., Hendricks, C. and Pecht, M. 2013. Lessons learned from the 787 dreamliner issue on lithium-ion battery reliability. Energies 6, 4682-4695.

138. Agubra, V.A. and Fergus, J.W. 2014. The formation and stability of the solid electrolyte interface on the graphite anode. J. Power Sources 268, 153-162.

139. Väyrynen, A. and Salminen, J. 2012. Lithium ion battery production. J. Chem. Thermodyn. 46, 80-85.

140. Wen, J., Yu, Y. and Chen, C. 2012. A review on lithium-ion batteries safety issues: Existing problems and possible solutions. Materials Express 2, 197-212.

141. Li, J., Huang, J. and Cao, M. 2018. Properties enhancement of phase-change materials via silica and Al honeycomb panels for the thermal management of LiFeO$_4$ batteries. Appl. Therm Eng. 131, 660-668.

142. https://www.livescience.com/62179-tesla-fire-cleanup-danger.html

143. https://money.cnn.com/2018/05/17/news/companies/electric-car-fire-risk/index.html

144. Goli, P., Legedza, S., Dhar, A., Renteria, J. and Balandin, A.A. 2013. Graphene-enhanced hybrid phase change materials for thermal management of Li-ion batteries. J Power Sources 7, 99-118.

145. Hussain, A., Abidi, I.H., Tso, C.Y., Chan, K.C., Luo, Z. and Chao, C.Y.H. 2018. Thermal management of lithium ion batteries using graphene coated nickel foam saturated with phase change materials. Int. J. Therm. Sci. 124, 23-35.

146. Qi, G.Q., Yang, J., Bao, R.Y., Liu, Z.Y., Yang, W., Xie, B.H. and Yang, M.B. 2015. Enhanced comprehensive performance of polyethylene glycol based phase change material with hybrid graphene nanomaterials for thermal energy storage. Carbon NY 88, 196-205.

147. Wu, W., Yang, X., Zhang, G., Ke, X., Wang, Z., Situ, W., Li, X. and Zhang, J. 2016. An experimental study of thermal management system using copper mesh-enhanced composite phase change materials for power battery pack. Energy 113, 909-916.

Processes and Technologies for the Recycling and Recovery of Spent Lithium Ion Batteries

Zhi Sun[1,2*], Weiguang Lv[1,2], Zhonghang Wang[1], Xiaohong Zheng[1], Hongbin Cao[1] and Yi Zhang[1]

[1] Division of Environment Technology and Engineering, National Engineering
   Laboratory for Hydrometallurgical Cleaner Production & Technology, Institute
   of Process Engineering, Chinese Academy of Sciences, Beijing 100190, China
[2] University of Chinese Academy of Sciences, Beijing 100190, China

1. Generation and Risks of Spent Lithium Ion Batteries

With the rapid increase in lithium ion batteries (LIBs) production and application in consumer electronics (CE) and electric vehicles (EV), spent LIBs have become the fastest growing solid waste by products since the 2000s. In 2017, electric vehicles sales surpassed one million worldwide for the first time [1]. It is predicted that 400 million spent LIBs are waiting to be disposed of by 2020 [2, 3]. Especially, driven by the booming development of EV industry in the EU, U.S., China, and Japan in recent years, the generation of spent LIBs will be further accelerated. The data in Fig. 1 shows that the sales of EV in China and USA reached 290 thousand in 2019 and 853 thousand in 2016. The huge amount of spent LIBs pose serious challenges to urban waste & safety management and have become a global concern in terms of improvement and resource recycling [4]. However, not only recycling technologies for industrial applications, but collection systems of spent LIBs also need to be further researched in the future [5].

In spent LIBs, the toxic and flammable fluorine-containing organic electrolytes and heavy metals, such as nickel, cobalt, and copper, exist a

*Corresponding author: sunzhi@ipe.ac.cn

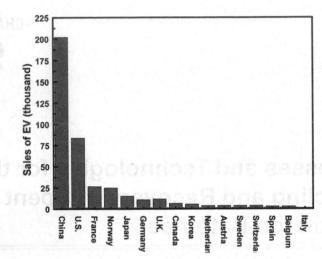

Fig. 1. Sales of the global electric car (EV) market by country in 2016 [6].

potential hazard to the environment and human health [7-9]. Some of the typical damages from spent LIBs have been shown in Table 1 [10]. The organic electrolytes might cause fires and contamination to the groundwater, if directly discarded or simply landfilled. In addition, the halogen-containing lithium salts, such as $LiBF_4$, $LiClO_4$, and $LiPF_6$, contained in the organic electrolytes are strong corrosive and unstable, and readily react with moisture in the air, releasing HF and dangerous phosphorus-containing gases, as described by reactions (Eqs 1, 2).

$$LiPF_6 \rightarrow LiF + PF_5 \qquad (1)$$

$$PF_5 + H_2O \rightarrow POF_3 + 2HF \qquad (2)$$

As for cathode materials, the main hazard comes from heavy metals. As known, the active material in the cathode is mainly a solid solution of metal oxides, which include lithium oxide (Li_2O), cobalt oxide (CoO_2), nickel oxide (NiO or NiO_2), manganese oxide (MnO_2) and aluminium oxide (Al_2O_3). When contact with water, acids, oxidants or reductants is made, the heavy metal elements easily ionise and contaminate groundwater and soils.

In addition to the above-mentioned hazards, the safety issues caused by the residual power and the lithium dendrites in spent LIBs also cannot be ignored. During over charging lithium ions tend to deposit on the anode to form lithium dendrites . These deposits can pierce the diaphragm and form an internal short circuit, and can also react violently with water and release hydrogen gas as well as huge amounts of heat, which can cause fires, explosions and other safety problems.

Table 1. Potential risks of spent LIBs [10]

Component	Material	Potential risk
Cathode	$LiCoO_2$ NCM NCA $LiMn_2O_4$ $LiNiO_2$	The cathodes include cobalt and nickel oxides which are regarded as carcinogens and can also cause severe intoxication. An acute reaction may occur when nickel content is high. The cathodes can degrade into toxic metal ions or lithium-containing oxides, as a consequence.
Anode	Carbon	A mixture of carbon dust and air can explode or burn, generating CO_2 or CO. A strong alkali solution can also be generated, or spontaneous combustion can occur, when inlaid lithium meets H_2O.
Electrolyte	$LiPF_6$	This is a strong corrosive reagent and can degrade into PF_5 when heated, generating HF in the presence of H_2O, and combusting in the presence of P_2O_5.
		Lithium hexa fluorophosphate is a strong corrosive agent and can release HF in the presence of H_2O or acid, or may generate Li_2O and B_2O_3 on combustion or heating.
		It is a strong oxidant, and can undergo combustion in the presence of LiCl, O_2, and Cl_2. It can dissolve in H_2O and releases HF in the presence of acid.
	$LiBF_4$	Lithium tetra fluoroborate can undergo combustion in the presence of CO, CO_2, SO_2, and HF. It can also generate HF in the presence of oxidants or acids.
	$LiClO_4$	Lithium perchlorate is a strong oxidant and can undergo combustion in the presence of LiCl, O_2, and Cl_2.
	$LiAsF_6$	Lithium hexafluoroarsenate can dissolve in H_2O and can release HF in the presence of acid.
	$LiCF_3SO_3$	Lithium trifluoromethanesulfonate can combust in the presence of CO, CO_2, SO_2, and HF. It can also generate HF in the presence of oxidants or acids.
Electrolyte solution	PC, EX, DMC, DEC, EC, DME and EMC	The carbonates are highly flammable and form an explosive atmosphere with aerial oxygen. The decomposition products possibly include methane, ethane, carbon monoxide and propene, which cause fire and explosion hazards.
Separator	PP and PE	Polypropylene and polyethylene can undergo combustion in the presence of carbon monoxide, aldehydes, and organic acids.
Binder	PVdF, PTFE	This can generate HF when heated.

Nevertheless, spent LIBs are also a potential "urban mine", considering that the contents of Li, Co, Ni, and other valuable metals are much higher in comparison with those in conventional ore sources. For example, of all the LIBs applied in CE and EV, the weight ratio of valuable metals range from 26 to 76%, of which 20% of total weight is copper and aluminium, and the contents of lithium and cobalt are 1-15% and 2-30%, respectively [10]. Furthermore, the increasing mining of natural ores for these metals is leading to severe shortages, creating a potential market for recycling spent LIBs.

2. State-of-the-art for Spent LIBs Recycling

Generally, collected spent LIBs are firstly disassembled to separate casting and foils, followed with mechanical/thermal treatment as well as dissolution in organic solvent/supercritical liquid to separate active materials from Al/Cu foils based on the differences on physicochemical properties [11-14]. Then, various pyro-/ hydro-/ bio-/electrometallurgical methods are used and combined to recover valuable metals, mainly rare and precious metals, from spent cathodes, as shown in Fig. 2. In this section, the development of spent LIBs recycling will be systematically reviewed and discussed, and the recycling methods are divided into two categories of pretreatment and recycling, of which the recycling of cathode materials, anode materials, and electrolyte are introduced in turn.

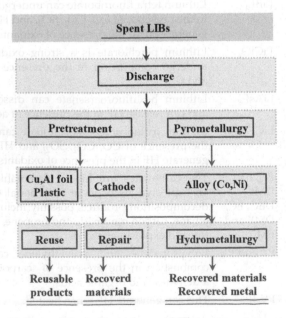

Fig. 2. General flow sheet of spent LIBs treatment processes.

2.1 Pre-treatment Method

Pre-treatment processes are generally required to remove the case, concentrate the valuable fraction, and reduce treatment burden of subsequent leaching procedures. General flow sheet of pre-treatment method for disposing spent LIBs is shown in Fig. 3. Normally, for safety reasons, it is necessary to release the residual power in the spent LIBs before pre-treatment, by either immersing in a salt solution, in water containing ion powder or by pressing batteries to evoke short-circuits inside the batteries while it is still closed (it has to be noted that pre-discharging is also considered as part of the pre-treatment sometimes) [15-18]. Afterwards, discharged batteries are disassembled manually or mechanically to obtain the cathode, anode and other components. Finally, solvent dissolution, mechanical method, ultrasonic-assisted separation, or thermal treatment are used to further separate cathode powders (black mass) from the other components.

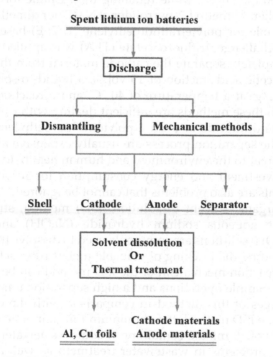

Fig. 3. General flow sheet of pretreatment process.

2.1.1 Solvent Dissolution Method

Solvent dissolution method uses organic or inorganic solvents to separate active materials from Al/Cu foils, based on different technical principles. The former uses an organic solvent to weaken the adhesion between the binder and cathode/anode powders, and the latter uses strong alkaline solutions, like NaOH solution to dissolve aluminium foil and thus recover the cathode materials.

In an organic solvent dissolution (OSD) method, the selection of an optimum organic solvent (OS) is the key [19]. The dissolution efficiency in this method is greatly affected by the type of binder used as well as the rolling method of electrodes. For PVdF-based electrodes, dimethylformamide (DMF), N,N-dimethylacetamide Dimethylacetamide (DMAC), dimethylsulfoxide (DMSO) are often adopted as OSs in addition to the most commonly used N-methyl pyrrolidone (NMP) [20, 21]. Zhou et al. [22] chose (dimethylformamide) DMF to dissolve polyvinylidene difluoride (PVDF), and they showed that the solubility of PVdF in dimethylformamide (DMF) was 176 g/L at 60 °C. Song et al. [23] effectively separated lithium cobalt oxide ($LiCoO_2$) powder from an aluminium foil in a N-methyl-2-pyrrolidone (NMP) and N,N-dimethylformamide (DMF) solution heated to 70 °C, while retaining the metallic form of the foil. However, neither (N-methyl-2-pyrrolidone) NMP nor dimethylformadide (DMF) is suitable for polytetrafluoroethylene (PTFE) based electrodes. To solve this challenge, trifluoroacetate (TFA) was applied by Zhang et al. [24] to completely separate the cathode material from the aluminium foil with an acetic acid fraction of 15 vol.%, a liquid-to-solid ratio (L/S ratio) of 8 mL/g, at a temperature of 40 °C, and a reaction time of 180 min. Although these methods are efficient dealing with cathodes with a polyvinylidene difluoride or even a polytetrafluoroethylene binder, the OSs used in the separation process are usually expensive and toxic, and thus pose a threat to the environment and human health. In addition, the equipment investment and energy consumption for recycling of used organic solvents are also problems that cannot be ignored.

In an inorganic solvent dissolution (ISD) method, strong alkaline solutions, like aqueous sodium hydroxide (NaOH) and potassium hydroxide (KOH) solutions are commonly used to dissolve the aluminium foil while avoiding the leaching of valuable metals from active materials [20, 25]. The reaction mechanism of the ISD method can be expressed as Eqs (3, 4) [26]. Simple operations and a high separation efficiency are the main advantages of this method in comparison with the OSD method. However, in the ISD method, the aluminium foil cannot be recycled, and a large amount of aluminate-containing alkaline wastewater is generated, bringing new problems in wastewater treatment as well as equipment investment [19].

$$Al_2O_3 + 2NaOH + 3H_2O \rightarrow 2Na[Al(OH)_4] \qquad (3)$$

$$2Al + 2NaOH + 6H_2O \rightarrow 2Na[Al(OH)_4] + 3H_2 \qquad (4)$$

2.1.2 Mechanical Methods

Mechanical pretreatment is applied to disintegrate batteries into components, and concentrating the black mass [27]. During mechanical pretreatment, a series of mechanical processes involving crushing, sieving, magnetic separation, fine crushing, and classification are performed to separate materials based on the difference of their physical properties such as density, conductivity and magnetic properties [28]. Zhang et al. [29] found that spent LIBs manifested excellent selective crushing properties, and the crushed batteries could be divided by sieving them into three fractions with different particle sizes: a Al-enriched fraction (>2 mm), a Cu- and Al-enriched fraction (0.25–2 mm), and a Co- and graphite-enriched fraction (<0.25 mm). Furthermore, it was found that the black mass obtained from the fraction with particle size <0.25 mm retained their original crystalline structure and chemical state. However, the surface of these powders was coated with a layer of hydrocarbons hindering subsequent flotation operations. Shin et al [30] proposed a combined process to recover metals from spent LIBs, in which mechanical separation is used to obtain cathode powders. First, particles enriched with Lithium cobalt oxide were obtained through crushing, sieving, and magnetic separation. Then, residual aluminium particles were separated out by finely grinding the Lithium cobalt oxide enriched particles.

In comparison with the solvent dissolution methods, the use of proper mechanical pre-treatment is necessary to achieve good separation of active materials from the other components at lower operating costs. While problems still exist during practical application, separation efficiency is not always achieved by using mechanical methods alone. In addition, the decomposition of lithium hexafluorophosphate, diethylene chloride (DEC), and propylene carbonate (PC) during mechanical processes is dangerous for the environment. Although many pre-treatment methods have been developed by researchers, challenges still exist regarding pre-treatment of spent LIBs.

2.1.3 Thermal Treatment Method

Thermal treatment is mainly used to remove organic components and graphite before the metal-leaching procedure, and it is usually carried out in the presence of oxygen (incineration) or in the absence of oxygen (pyrolysis) [31]. Research indicated that PVdF was the main cause hindering the separation of cathode or anode powders from foils [16]. During the thermal treatment, organic components, like the PVdF binder decomposed, and the black mass was easily separated from the foils by direct sieving [32, 33]. It was reported that PVdF binders begin to decompose at 350 °C, while other components (e.g., acetylene black, conductive carbon, etc.) would not until the temperature reaches 600 °C [13]. Sun and Qiu [34] proposed a novel method to separate cathode materials by means of vacuum pyrolysis. The electrolyte and binder evaporated or decomposed to weaken the adhesion of cathode materials and collectors through vacuum pyrolysis. The investigations indicated that the cathode materials were hard to strip away from the collectors, when the pyrolysis temperature was < 450 °C. The stripping efficiency increased with an increase in temperature from 500 to 600 °C. While by further increasing the pyrolysis temperature, the aluminium foil became fragile, making it difficult to separate the cathode material from the collectors. Yang et al. [35] proposed a reducing thermal treatment process to achieve the separation of cathode materials from Al foils. The results showed that cathode materials could be clearly separated from current collectors by controlling the reduction temperatures. In addition, the reducing treatment changed the molecular structures of active cathode materials facilitating the leaching of metals. The advantages of thermal treatment are simple operations and high production rate, accompanied with the release of poisonous gases which is a major disadvantage, and also an urgent problem needed to be solved in practical applications.

2.1.4 Ultrasonic-assisted Separation

Ultrasonic treatment is considered to be an effective method for stripping cathode materials from aluminium foils because of its cavitation effect [36-38]. Li et al. [38] found that the cathode materials could hardly be separated from collectors by using either mechanical agitation or ultrasonic cleaning alone. However, almost all of the cathode materials could be stripped from the collectors by combining these two methods. This might be because the cavitation effect of ultrasonic cleaning could generate greater pressure to destroy the combine between binders and foils. And the rinsing effect of mechanical agitation further promoted the separation of cathode materials from the collectors. He et al. [37] indicated that ultrasonic cleaning could promote the dissolution of binder, which enhanced the stripping efficiency. Cathode materials obtained from the ultrasonic cleaning process exhibited a low degree of agglomeration, which facilitates subsequent leaching processes.

2.2 Recycling of Cathode Materials

Cathode materials of spent LIBs contain most of the valuable metals, including Li, Ni, Co, Mn to mention a few.. Thus, the recycling activities mainly focus on dealing with cathode materials for their potential commercial value. For metal recovery, there are four main routes: pyrometallurgy, hydrometallurgy, bio metallurgy and electrometallurgy. There are also plenty of reports of the use of a combination of various techniques. Afterwards, final products are obtained using various separation and purification steps from the crude alloy or leachate by using chemical precipitation, solvent extraction or electrochemical techniques. The development of each route for the metal extraction process and its separation and purification will be briefly introduced and reviewed in the following sections.

2.2.1 Metal Extraction

2.2.1.1 Pyrometallurgical

Pyrometallurgical processes, notably smelting, is a traditional method to recover metals from spent LIBs. Some of the pyrometallurgical processes include incineration, smelting in a plasma arc furnace or blast furnace, sintering, melting and reactions in a gas phase at high temperatures [39]. These processes mainly consist of concentrating metals in metallic phase and rejecting other materials as slag or in the gaseous phase [39].

Normally, almost all kinds of spent LIBs can be directly treated through a pyrometallurgical process without any pre-treatment. From a technological point of view, the main concern is not the metallurgical process itself, which is very simple and mature, but rather the cleaning of the off-gases and dust, given that the load contains a significant amount of organic components and graphite. In addition, in a typical pyrometallurgical process, Li will end up in the slag, and will need to be further recycled by a combination with other methods, like hydrometallurgical leaching. To extract Li in the forthcoming steps, carbothermal reduction method was proposed by Xu et al. [40-42], in which the mixed spent LIBs could be converted to metal oxide, pure metal, or lithium carbonate. In the subsequent leaching step, Li in the carbonate could be directly leached out in water. However, the treatment capacity of this method was limited by the solubility of lithium carbonate (Li_2CO_3). To enhance the extraction efficiency of Li from lithium carbonate, Wang et al. used carbon dioxide (CO_2) to change the lithium carbonate into lithium bicarbonate ($LiHCO_3$) [43]. Other than these methods, in recent years, regeneration methods are being widely observed for simplicity and minimal environmental impact. For example, Nie et al. [44] regenerated lithium cobalt oxide, $LiCoO_2$ from spent $LiCoO_2$ with supplementary lithium carbonate, Li_2CO_3 at 800-900 °C for 12 h. The regenerated cathode materials can meet commercial requirements. Short technology avoids many risks of environmental pollution and reduces the loss of valuable metals compared with other technologies, though much more energy is required.

2.2.1.2 Hydrometallurgical Leaching

In comparison with pyrometallurgical methods, hydrometallurgical leaching is more commonly used to achieve simultaneous extraction of multiple metals at much lower temperatures [45, 46]. During a hydro-leaching process, valuable metals in the black mass were leached as ions in acidic or alkaline solutions. In early studies, traditional strong inorganic acids such as hydrochloric acid [47], sulfuric acid [48, 49], nitric acid [50], and phosphoric acid [51] are usually used to leach valuable metal elements from black mass powders, and reducing agents like hydrogen peroxide [52], sodium thiosulphate $Na_2S_2O_8$ [53] even current collectors [54] are often added to enhance the leaching processes. However, despite the high reaction efficiency, these leaching processes usually have no reaction selectivity, resulting in extremely complicated leachates when dealing with complex raw materials such as NCM or mixed spent LIBs, and as a result multiple separation and purification are often needed to achieve effective separation and extraction of different components. In addition, during leaching, the corrosion of the reaction equipment by the strong acidic solution and the formation of acid mist during the reaction are all problems that need to be overcome. To solve these problems, organic acids like citric acid [55], acidic acid [56], formic acid [57], oxalic acid [58] and acetic acid [59, 60] are widely adopted to recover specific metals from black mass selectively [61-63]. Meanwhile, organic reductants, like cellulose, glucose, and sucrose are also researched in many investigations [64]. However, in comparison with inorganic acids, organic acids are less reactive and much more expensive, resulting in a much lower treatment capacity and a much higher treatment cost, and this further limits its industrial applications [65].

Except for leaching in acidic solutions, the leaching of value metals from spent LIBs can also be achieved in alkaline solutions with ammonium ions i.e NH_4^+ [66]. For example, Zheng et al. [67] applied an ammonia-ammonium sulphate system for leaching of valuable metals from spent cathode materials. The results showed that more than 98.6% of Ni, Co and Li were leached with only 1.36% Mn removed in leachate. After that, the mechanism of ammonium leaching was analysed in many researches [67]. However, low disposing capacity limits the further development and application of ammonium leaching.

Although plenty of attention has been paid to recover valuable metals from the spent cathode materials, as listed in Table 2, the mechanism of the leaching processes has rarely been studied at molecular or atomic levels except for some primary comparison on leaching efficiencies, leaching selectivities as well as the energy consumption of some of them. [2, 65]. No quantitative evaluation of these methods has been made systematically to provide guidance for relevant researchers as well as the industrial technologists.

2.2.1.3 Biometallurgical Leaching

Biometallurgical leaching is a process utilizing bacterium to extract valuable metals from solid to liquid phase [68]. Mishra et al. [69] used chemo lithotrophic and acidophilic bacteria and acidithiobacillus ferrooxidans as leaching bacteria, while the leaching efficiencies of Co and Li were far from satisfying with a pH of 2.5 at 30 °C, even after adding Fe^{2+} as catalyst and increasing the leaching time. Zeng et al. and Chen et al. [70, 71] found that Ag^+ was an efficient catalyst to enhance the leaching efficiency of Co, which could reach 98.4% within seven days with 0.02 g/L Ag^+ added. Similarly, Cu^{2+} is also an effective catalyst for the acidithiobacillus ferrooxidans leaching process [72]. With 0.75 g/L Cu^{2+} added, the leaching efficiency of Co could reach 99.9% in six days. Mixing different bacteria, such as acidophilic sulfur-oxidizing bacteria (SOB) and iron-oxidizing bacteria (IOB), could also enhance the leaching of spent LIBs [73, 74]. Besides, fungal leaching was found to be a favorable technology compared with bacterial leaching [75]. Various organic acids are produced during the intensification of the leaching process by the fungi metabolite. In the meantime, growth over a pH range, tolerance for toxic materials and conducting with a high leaching rate are also important advantages of fungi [75, 76].

Compared to conventional methods, the operational conditions of the biometallurgical process are mild, consuming less energy. The main limitations of biometallurgy for industrial applications is the slow kinetics and low pulp density [65]. Niu [77] indicated that the bioleaching efficiency of Co and Li decreased from 52% to 10% and 80% to 37%, respectively, when pulp density increased from 1% to 4%. Of course, high leaching efficiencies of spent cathodes can be achieved in a high pulp density by controlling reaction conditions, like reaction temperature, mixed energy substrates, and pH [77]. However, the long reaction time is still difficult to overcome. Though, biometallurgical methods have a significant effect on energy-saving, there is still a long way to go for the recycling of LIBs.

2.2.1.4 Electrochemical Leaching

Electrochemical leaching can be separated into direct leaching and indirect leaching. Direct leaching utilizes an external source enhancing the potential of the system to intensify the leaching process. Meng et al. [94] used spent $LiCoO_2$ cathode materials as anode, platinum plate as cathode, and malic acid as electrolyte, achieving the leaching of Co and Li. The result indicated that the leaching efficiency of Co and Li could reach 90%, with an external electro source and no reductant. However, the spent materials contained in the anode were so little and the consumption of the electro source was so high that electroleaching had little value for industrial application. Prabaharan et al.[95]developed a complete leaching and extraction method to dispose of spent cathode materials with an electrochemical method. The recovery of Co, Mn and Cu can reach above

Table 2. The cases of leaching spent LIBs using different leaching reagents

Raw material	Reagent	T (°C)	T (min)	S/L (g/L)	Leaching efficiency (%)				Reference
					Co	Mn	Ni	Li	
Inorganic Acid Leaching									
Spent LIBs	1.75 mol/L HCl	50	90	20	99.0			100.0	[78]
LCO	4 mol/L HCl	80	30	50	90.6			93.1	[79]
LiFePO$_4$ and LiMn$_2$O$_4$	6.5 mol/L HCl + 5 vol.% H$_2$O$_2$	30	60	20		90.0		92.2	[80]
LiNi$_x$Mn$_y$Co$_z$O compounds	4 mol/L H$_2$SO$_4$ + 5 vol.% H$_2$O$_2$	65-70	120	50	96.0	96.0	99.0		[81]
Spent LIBs (mixture)	1 mol/L H$_2$SO$_4$ + 0.075 M NaHSO$_3$	95	240	20	91.6	96.0		96.7	[82]
LCO from laptop computers	2 mol/L H$_2$SO$_4$ + 5 vol.% H$_2$O$_2$	75	60	100	70.0			99.1	[83]
LCO	1.5 M H$_3$PO$_4$ + 0.02 M glucose	80	120	20	98.0			100.0	[84]
LCO from mobile phones	2% H$_3$PO$_4$ + 2 vol.% H$_2$O$_2$	90	60	8	99.0			88.0	[85]
LCO	1 mol/L HNO$_3$ + 1.7 vol.% H$_2$O$_2$	75	60	20	95.0			95.0	[86]
Alkaline Leaching									
Li(Ni$_{1/3}$Co$_{1/3}$Mn$_{1/3}$)O$_2$	4 mol/L NH$_3$-1.5 mol/L (NH$_4$)$_2$SO$_4$ +0.5 M Na$_2$SO$_4$	80	300	10	80.7			95.3	[67]

(Contd.)

Organic Acid Leaching									
LCO and CoO	1 mol/L Oxalate + 5 vol.% H_2O_2	80	120	50	98.0	98.0			[87]
Spent LIBs (LiCoO$_2$)	2 mol/L Citric acid + 0.6 g/g H_2O_2 (H_2O_2/spent LIBs)	70	80	50	96.0	98.0			
NCM	3 M trichloroacetic acid + 4 vol% H_2O_2	60	30	50	92.0	90.0	100.0	93.0	[89]
LCO	1 mol/L Oxalic acid	95	150	15	97.0	98.0			[58]
LCO	1 mol/L Iminodiacetic acid + 0.02 M ascorbic acid	80	360	2	99.0	90.0			[90]
LCO	1 mol/L Maleic acid + 0.02 M ascorbic acid	80	360	2	99.0	96.0			[90]
LCO	0.5 mol/L Glycine + 0.02 M ascorbic acid	80	360	2	91.0	—			[91]
LCO	1.5 mol/L Succinic acid + 4 vol.% H_2O_2	70	40	15	100.0	96.0			[92]
LCO and NCM (523)	2 mol/L L L-tartaric acid + 4 vol.% H_2O_2	70	30	17	98.6	99.3	99.3	99.1	[93]

96%. On the other hand, indirect leaching depends mainly on the products from the reaction, when potential is added onto the system externally. Boxall et al. [96] developed an indirect electrochemical leaching process to intensify the leaching process, whose main mechanism is the electrolysis of NaCl solution, illustrated in Eqs (5 to 8). Cl_2 is released at the anode, and hydrogen is formed at the cathode. Though, the method reduces the cost of the leaching reagent, the consumption of electro energy cannot be ignored, since the potential of Eq. (5) is very high. The economic and environmental effects need to be considered, as well, in the future.

$$2Cl^- \rightarrow Cl_2 + 2e^- \; E^0 = -1.360 \text{ V} \tag{5}$$

$$H_2O \rightarrow 0.5 \, O_2 + 2e^- + 2H^+ \; E^0 = -1.230 \text{ V} \tag{6}$$

$$Cl_2 \, (g) + H_2O \, (l) \rightarrow HOCl \, (aq) + HCl \, (aq) \tag{7}$$

$$2H_2O + 2e^- \rightarrow H_2 + 2OH^- \; E^0 = -0.828 \text{ V} \tag{8}$$

2.2.2 Separation and Purification

2.2.2.1 Solvent Extraction

Solvent extraction, also known as liquid-liquid extraction, is a process to separate different metal ions or compounds based on their uneven distribution in two phases (organic and aqueous). The process is commonly performed after the leaching process to separate partial metal ions (objective metal ions or impurities) from leachate, when used in the recycling of valuable metals from spent LIBs [97]. Solvent extraction is very effectively used in separating metals with similar characteristic, such as Co and Ni, W and Mo, Cr and V [98]. In the meantime, the extractants can also be repeatedly used in the extraction process. It is difficult for other methods, like chemical precipitation and electrochemical separation, to obtain a similar effect. Figure 4 shows the general process of solvent extraction, including extraction stage, scrubbing stage, stripping stage and regeneration stage. After that, the solution from stripping stage is transported to produce products through precipitation or electrowinning. The common extractants include acidic extractants (Cyanex 301, D2EHPA, Cyanes 272) and basic extractants (Alamine 336, TOA) [99]. The factors influencing solvent extraction mainly include concentrations of object metals, pH of solutions, temperature, extraction time, and organic/aqueous ratio (O/A). For example, the PC-88A has high extraction efficiency of Co and Ni at pH 4.5, but nearly no efficiency under pH 3 [100, 101]. The extraction efficiency of Co with Cyanex 272 increased from 10% to 90% with the pH increasing from 1 to 5 [102]. The main extraction mechanism of solvent extraction is based on the interfacial chemical reactions, in which protons are released from the reaction between acidic extractants and solution at the O/A interface. The mechanism can be described as in Eqs (9 to 11):

$$M_{Aq}^{2+} + A_{Org}^{-} + 2(HA)_{2Org} \rightarrow MA_2 \cdot 3HA_{Org} + H_{Aq}^{+} \qquad (9)$$

$$MOH_{Aq}^{+} + A_{Org}^{-} + 2(HA)_{2Org} \rightarrow M(OH)A_2 \cdot 3HA_{Org} + H_{Aq}^{+} \qquad (10)$$

where $A_{Org}^{-} + 2HA_{2Org}$ represents the saponification reaction as

$$Na_{Org}^{+} + \frac{1}{2}HA_{2Org} \rightarrow NaA_{Org} + H_{Aq}^{+} \qquad (11)$$

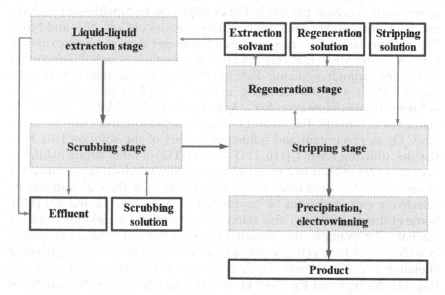

Fig. 4. General flow sheet of solvent extraction.

When extractants are used to separate single metal ions from leachates, multiple extraction stages result in the extraction process. To reduce the routine process, a co-extraction method was developed by Yang et al.[103]. In the co-extraction process, about 85% of the Ni, 99% of the Co and 100% of Mn are co-extracted from leachate with D2EHPA, leaving Li in the raffinate. After scrubbing and stripping stages, Ni, Co and Mn ions are co-precipitated with NaOH and NH₃·H₂O to fabricate the precursor of the cathode materials. As for Li, it is recovered as Li₂CO₃ from the raffinate by precipitation.

The advantages of solvent extraction are the short reaction time, high selectivity and gentle operational conditions. Meanwhile, the operation of solvent extraction has high automation in industrial applications. However, the loss of remaining metal ions in solution is one of the important disadvantages. As Yang et al. [103] reported, about 20% of Li is lost in the solvent extraction process. Meanwhile, the high cost of solvent is also a shortcoming limiting the development of solvent extraction. In the future, more investigations should be focused on researching new cheaper extractants and simplifying the routines.

2.2.2.2 Chemical Precipitation

Chemical precipitation is a method to separate metals from liquids as solids (hydroxides, oxides or salts) [104, 106]. The separation mechanism of chemical precipitation relies on the different solubilities of metal compounds at certain pH levels. For example, the hydroxides of Fe^{3+} have much lower solubility compared with hydroxides of Co^{2+}, Ni^{2+} and Mn^{2+}. The hydroxides of Fe^{3+} can exist in a lower pH than the hydroxides of Co^{2+}, Ni^{2+} and Mn^{2+}. It is very effective to remove Fe^{2+} or Fe^{3+} from other metal ions through oxidizing Fe^{2+} to Fe^{3+} at an appreciable pH. There is nearly no MnC_2O_4 solid occurring in solution until the pH of the solution is higher than 6. However, NiC_2O_4 and CoC_2O_4 have little solubility in acidic environments. The Ni and Co can be separated from Mn by using $Na_2C_2O_4$ as precipitant and adjusting the pH of the solution [102-109]. Besides, utilizing $KMnO_4$ [110, 111], NaClO [112] or other strong oxidizing agents can be utilised to convert Mn^{2+} into MnO_2, which can achieve the separation of Mn from other metal ions. Meanwhile, the Co^{2+} can also be selectively extracted from Ni^{2+} with NaClO as in Eqs (12 and 13) [112]. Some of the reagents can also selectively precipitate metal ions from the leachate. For example, the dimethylglyoxime reagent (DMG, $C_4H_8N_2O_2$) is widely used to selectively precipitate Ni^{2+} as nickle dimethylglyoxime chelating precipitate from Co^{2+}, Ni^{2+} and Mn^{2+} ions mixed solutions (Eq. 14). As reported by Chen et al. [107], nearly 96% of Ni^{2+} could be precipitated within 20 minutes at room temperature (Eq. 14). As for other precipitants including sodium carbonate (Na_2CO_3), sodium chromate $Na_2C_2O_4$, oxalic acid $H_2C_2O_4$ and sodium phosphate (Na_3PO_4), many of their transition metal salts have little solubility in solution. Therefore, it is hard to use them to separate single metals from solution.

$$2Co^{2+} + ClO^- + 2H_3O^+ \rightarrow 2Co^{3+} + Cl^- + 3H_2O \qquad (12)$$

$$2Co^{3+} + 6HO^- \rightarrow Co_2O_3 + 3H_2O \qquad (13)$$

$$6C_4H_8N_2O_2 + Ni_3(Cit)_2 \rightarrow 3Ni(C_4H_7N_2O_2) + 2H_3Cit \qquad (14)$$

Though, many precipitants have no selectivity for transition metals, a cute fabrication method is utilizing their characteristic to achieve the co-precipitation of many metals, like Ni, Co and Mn. The products from co-precipitation are the precursors of cathode materials [113, 114]. Therefore, chemical precipitation is also an effective method to obtain products from liquor, including not only single metal salts, but also multiple metal salts. The common precipitants include sodium hydroxide (NaOH), potassium hydroxide (KOH) [115, 116]. $Na_2C_2O_4$ and $H_2C_2O_4$ can also react with transition metals to produce metal salts. Sodium carbonate (Na_2CO_3) and sodium phosphate (Na_3PO_4) can react with Li^+ to produce Li_2CO_3 [117] or Li_3PO_4 [118, 119]. As for the co-precipitation method, it has been the main method to obtain precursors of cathode materials at the laboratory and industrial scales. For example, Li [120] and Yang [121] directly synthesized regenerated cathode materials from leachates, which had been purified and pH balanced, through co-precipitation, hydrothermal and calcination process. Perez et al. [122] fabricated organophosphonate coordination polymers with phosphonate organic linkers as precipitants. Therefore, the advantages of simple operation, low energy consumption and low cost of co-precipitation processes have resulted in their frequent utilisation for synthesizing new materials and will remain important for recycling industries in the future.

2.2.2.3 Electrochemical Extraction

Electrochemical methods are frequently used in metallurgical areas, but ignored for recycling spent LIBs in recent years. There has been insufficient research about using electrochemical methods in recovering cobalt and nickel from spent LIBs, and none at all for recycling of manganese has been reported. Lupi et al. investigated the change of charge efficiency and metal ions concentration in electrodeposition with a leachate, which has a low concentration of objective metal ions [123]. The research indicated that electrochemical methods could be used in the recycling of spent LIBs. To further illustrate the mechanism of Co and Ni during the electro-crystallization process, Freitas et al. investigated the influence of pH on the nucleation and growth mechanism during Co and Co-Cu electrodeposition processes [124, 125]. Besides, they also indicated that the reduction of Co^{2+} was carried out with the formation of adsorbed hydrogen. Meanwhile, the electro-dissolution of cobalt is different at a different pH. It is a one-step oxidation process at a pH of 2.7, but a two-step oxidation process with intermediary Co^{2+} at a pH of 5.4 [126]. Besides, synthesizing new materials, like Co_3O_4 [127] and $Co(OH)_2$ [128] from leachates through electrochemical methods are also potential methods for the recycling of valuable metals [129].

However, all the researches conducted have revealed the same disadvantage limiting the practical application. They choose a leachate with low concentration of valuable metals which doesn't match the leachate from leaching spent LIBs in industrial production. At the same time they do not focus on the recycling efficiency of valuable metals, which is one of the significant parameters for the recycling process. Though the electrochemical process has a potential practical application in the future, it needs more research to be further investigated.

2.3 Recycling of Anode Materials

The recycling of anode and electrolyte are also essential parts for establishing a comprehensive recovery technology for spent LIBs. However, the recycling of anodes has been ignored due to the low value of graphite in recent years. Most of the researches focused on synthesizing new functional materials from spent graphite, utilizing the characteristic of graphite. Others are focussed on extracting the residual Li in spent anodes through traditional metallurgical technologies. After the acid or alkali is removed, the graphite is 100% pure with no impurity. Then, the graphite can be used as a raw material to obtain new materials. For example, it was directly used as a catalyst to achieve the reduction of oxygen and degrade pollutants in an electro-Fenton process by Cao et al. [130]. Wang et al. [131] synthesized Graphene oxide (GO) through Hummer's method from spent LIB-Graphite. These products demonstrate excellent activity compared to materials fabricated from commercial graphite in catalytic ozonation for the removal of organic pollutants [131]. Besides, decorating spent graphite can also be used to enhance the activity. Zhang et al. [132] fabricated mesocarbon microbead from spent graphite and then coated magnesium hydroxide nanoparticles on it. The composite exhibits excellent adsorbent activity in wastewater treatment [132]. Obviously, the structure of spent anodes is easily destroyed in the recycling process. It is hard to make it arrive to its initial level of activity. There are many defects existing in the spent anodes and disposed spent anodes. However, if the defect can be utilized to enhance the reaction activity, it will have better use than raw materials. As for extracting valuable metals from spent anodes, Guo et al. [133] used a traditional acid leaching method, achieving the recycling of Li from spent anodes.

2.4 Recycling of Electrolyte

Though, the electrolyte is one of the main constituents of spent LIBs, its recycling has been ignored in recent years, for the following reasons. To start with, due to the volatile nature, inflammability, and toxic properties of the electrolyte components, their recycling needs to be conducted in a safe and closed environment. Secondly, the components of electrolyte are so complex that it is hard to be separate and purify them. Also the value of the recycling process can't match the utility of the extraction process. In the end, some of electrolyte is strongly absorbed on the separator and it is hard to separate active materials completely [3, 65], .

The supercritical carbon dioxide extraction process is the only method, which has been widely researched for the recycling of electrolyte in recent years. The advantages of this method are high selectivity, efficient separation of many electrolyte components, no pre-clean up steps, and no pre- or post-concentration steps [134]. Dai et al. systematically investigated the recycling of electrolyte and reclaimed electrolyte from spent LIBs through four steps including supercritical CO_2 extraction, weakly basic anion exchange resin deacidification, molecular sieve dehydration, and components supplement [135]. The reclaimed electrolyte has a good ionic conductivity of 0.19 mS·cm^{-1} at 20 °C. However, the stability of reclaimed electrolyte is expected to be enhanced in the future. The capacity of batteries with reclaimed electrolyte faded fast in the initial cycles. Besides, the products from electrolyte decomposition were also detected and analysed, which is a significant analysis for the recycling of electrolyte. In the meantime, it also enhances the importance of preventing possibly toxic and contaminative chemicals from the spent electrolyte.

Besides, other researches also reported more investigations about the application of SC CO_2 in the recycling of electrolyte [14, 136]. However, the strict specifications of the equipment required and the stability of reclaimed electrolyte still limit the practical application of SC CO_2. Methods with better efficiency need to be investigated in the future. Simultaneously, good pretreatment methods are also needed to separate the electrolyte from the separator and active materials. The recycling of the electrolyte is a significant step to achieve the close-loop of LIBs.

3. State-of-the-art for Industrial Application

The recycling of spent LIBs has been investigated widely all over the world, and some have been successfully applied at an industrial scale, as summarized in Table 3. Interestingly, in comparison with other methods that are intensively studied and reported, pyrometallurgical methods are more suitable for industrial application, due to their superior advantages of ease of operation and high adaptability with raw materials. Also for many recycling companies, pyrometallurgical technology is more easily combined with the existing production lines and equipment. While, as stated in section 2.2.1.1, only crude alloy or metal could be obtained with low purity by barely using pyrometallurgical technologies. Thus, to enhance the extraction efficiency and obtain products with higher purity, hydrometallurgical/electrochemical separation and purification techniques are usually employed after the pyrometallurgical processing. This strategy has been successfully developed and applied by Umicore, Sumitomo-Sony (Japan), Accurec GmbH (Germany), and others, as demonstrated in Fig. 5.

Unlike pyrometallurgical processes, in hydrometallurgical technologies, spent LIBs have to be first pretreated before extraction of valuable metals. Figure 6 shows a typical hydrometallurgical process applied by Recupyl. After mechanical treatment, the waste is dissolved in H_2SO_4, followed with extraction of cobalt as cobalt hydroxide, $Co(OH)_2$, from the leachate using sodium hypochlorite, $NaClO$. After removing other

Table 3. Summary of current LIB recycling industrial technologies throughout the world [65, 137-139]

Company	Country	Capacity (tonnes/a)	Technology description	Product
Umicore (VAL'EAS™)	U.S.	7000	It combines pyrometallurgical and hydrometallurgical technologies. Spent LIBs are fed to a single shaft furnace. The alloy containing Co, Ni from furnace is further processed through leaching process. Li and Al mainly exist in the slag and need to be further treated through hydrometallurgical technologies.	$CoCl_2$
S.N.A.M	France	300	The first stage involves sorting techniques. Then the roughly separated products are disposed of by pyrolysis pretreatment, crushing, and sieving.	
Batrec AG	Switzerland	200	The spent LIBs are stored and shredded under CO_2 atmosphere. Then, the scrap is neutralized with moist air and further treated with a hydrometallurgical process.	
Inmetco	U.S.	6000	The scrap is processed in a rotary hearth furnace and further refined in an electric arc furnace.	Alloy (Co/Ni/Fe)
Sumitomo-Sony	Japan	150	Electrolytes and plastics are removed through calcination. A pyrometallurgical process is used to recover an alloy containing Co-Ni-Fe. A hydrometallurgical process is conducted to recover Co.	CoO
AkkuSer Ltd	Finland	4000	A two-phase crushing line is designed. Magnetic separation, and other separation methods follow it. Then, the scrap is delivered to smelting plants followed by leaching.	Metal powder

Toxco	U.S.	4500	It combines mechanical methods, like shredding, milling and screening and hydrometallurgical methods, like leaching, and precipitation.	CoO, Li_2CO_3
Recupyl Valibat	France	110	Mechanical methods are used in the first stage. Then, hydrometallurgical methods are used to obtain cobalt hydroxide, $Co(OH)_2$ and lithium phosphate, Li_3PO_4.	$Co(OH)_2$, Li_3PO_4.
Accurec GmbH	Germany	6000	A vacuum furnace is used first. Then, mechanical methods are used to separate different materials. After that, scrap is fed to an electric furnace. Slag is disposed through hydrometallurgical methods.	Co alloy, Li_2CO_3
AEA	U.K.	-	The first stage uses organic solvents to remove electrolyte, solvent, and binder. Leaching of cathode materials is carried out by electrolyzing.	$LiOH$, CoO
Glencore plc. (former Xstrata)	Canada/Norway	7000	It combines pyrometallurgical and hydrometallurgical methods.	Alloy (Co/Ni/Cu)
Onto process	U.S.	-	The preliminary step involves discharge, electrolyte recovery, refurbishing, and ball milling. A supercritical fluid is used to separate different materials.	Cathode powder
LithoRec process	Germany	-	It combines similar mechanical and hydrometallurgical methods.	CoO, Li salt
Green Eco-manufacture Hi-Tech Co	China	20000	Hydrometallurgical methods including leaching, purifying and leaching-resynthesizing are used.	Co powder
Bangpu Ni/Co High-Tech Co	China	3600	Hydrometallurgical methods including leaching, purifying and leaching-resynthesizing are used.	Cathode material, Co_3O_4

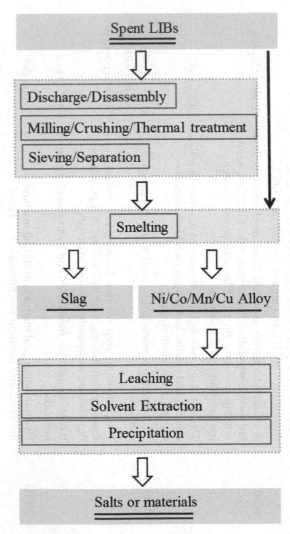

Fig. 5. The general flow sheet of industrial technology with the combine of pyrometallurgical and hydrometallurgical methods.

metal ions, lithium remaining in the solution is precipitated as lithium carbonate, Li_2CO_3 by using carbon dioxide, CO_2 gas. In comparison with pyrometallurgical ones, hydrometallurgical processes usually have much longer routes, and this results in more loss of valuable metals and more secondary pollution. To solve this issue, Brump established a short-term technology to recycle valuable materials from spent LIBs in a closed-loop, but, unfortunately, the details of this process have not been reported, and the drawbacks of hydrometallurgical methods described in previous sections are still awaited to be overcome in the future.

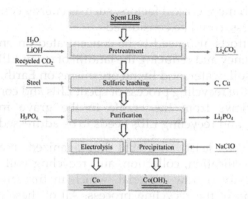

Fig. 6. The general flow sheet of industrial technology
only through hydrometallurgical methods

4. Conclusions and Prospects

Globally, several companies, including AEA Technology (U.K.), SNAM
(France), Toxco (American), and Umicore (Belgium), have established
complete production lines to recycle metals from spent LIBs. For the
complex process involving traditional methods, their economic benefit
has little competitiveness in comparison to new shorter processes, such
as selective extraction, regeneration, and repairing processes. However,
most of these methods are still investigated at laboratory scale. It is hard
to dispose the mixed waste with different spent LIBs materials, which are
the main waste collected by companies. Besides, many mechanisms of
physical and chemical changes in the recycling process need to be deeply
explored. A complete system of recycling LIBs needs to be discussed and
established. For future research in the process of recycling spent LIBs,
focus should be on the following aspects:

- The mechanisms of the leaching process are still inconclusive.
  Significant work still needs to be done to guide the selection of leaching
  reagent and operating conditions. For example, changes in the crystal
  structure of the leaching process need to be deeply researched. It will
  be very helpful for providing insight into the reaction mechanism
  during leaching.

- A closed-loop process with little steps and high efficiency, especially
  selective extraction methods, need to be established in the future.
  They are important to establish a green and economic technology for
  the recycling of waste.

- Most studies of disposing spent LIBs only focus on the kinetics and
  influence of operational parameters on leaching process, whereas
  more efforts need to be paid for establishing a complete evaluation

system in which more critical factors, such as energy consumption of the whole process, could be considered.

- Furthermore, efforts also need to be concentrated on enhancing the collecting efficiency and reducing the landfill of spent LIBs, which are hazardous for soil, water, and living organisms on Earth.
- Not restricting the recycling process of spent LIBs and comprehensive and holistic views from the cradle to the grave in designing, manufacturing, and recycling LIBs needs to be addressed.

Last but not the least, a global homogenized mechanism in manufacturing, classification, collection, and recycling will significantly reduce the complexity of raw material, which in turn minimizes the energy consumption of the recycling process. All of these aspects have a lasting influence on achieving economic and eco-friendly recycling of spent LIBs.

References

1. Harper, G. et al. 2019. Recycling lithium-ion batteries from electric vehicles. Nature 575(7781), 75-86.
2. Gao, W. et al. 2018. Comprehensive evaluation on effective leaching of critical metals from spent lithium-ion batteries. Waste Manag. 75, 477-485.
3. Zhang, X. et al. 2018. Toward sustainable and systematic recycling of spent rechargeable batteries. Chem. Soc. Rev. 47(19), 7239-7302.
4. Sun, Z. et al. 2017. Toward sustainability for recovery of critical metals from electronic waste: The hydrochemistry processes. ACS Sustainable Chemistry & Engineering 5(1), 21-40.
5. Gu, F. et al. 2017. An investigation of the current status of recycling spent lithium-ion batteries from consumer electronics in China. Journal of Cleaner Production 161, 765-780.
6. Network, C.I.I. 2017. Global new energy passenger vehicle industry market status and production and sales forecast. 2017; Available from: http://www.chyxx.com/industry/201705/526702.html.
7. Horeh, N.B., Mousavi, S.M. and Shojaosadati, S.A. 2016. Bioleaching of valuable metals from spent lithium-ion mobile phone batteries using Aspergillus niger. J. Power Sources 320, 257-266.
8. Winslow, K.M., Laux, S.J. and Townsend, T.G. 2018. A review on the growing concern and potential management strategies of waste lithium-ion batteries. Resources, Conservation and Recycling 129, 263-277.
9. Oliveira, L. et al. 2015. Key issues of lithium-ion batteries – From resource depletion to environmental performance indicators. Journal of Cleaner Production 108, 354-362.
10. Zeng, X., Li, J. and Liu, L. 2015. Solving spent lithium-ion battery problems in China: Opportunities and challenges. Renewable and Sustainable Energy Reviews 52, 1759-1767.

11. da Costa, A.J. et al. 2015. Beneficiation of cobalt, copper and aluminum from wasted lithium-ion batteries by mechanical processing. International Journal of Mineral Processing 145, 77-82.

12. Guan, J. et al. 2017. Mechanochemical process enhanced cobalt and lithium recycling from wasted lithium-ion batteries. ACS Sustainable Chemistry & Engineering 5(1): 1026-1032.

13. Hanisch, C. et al. 2015. Recycling of lithium-ion batteries: A novel method to separate coating and foil of electrodes. Journal of Cleaner Production 108, 301-311.

14. Mönnighoff, X. et al. 2017. Supercritical carbon dioxide extraction of electrolyte from spent lithium ion batteries and its characterization by gas chromatography with chemical ionization. J. Power Sources 352, 56-63.

15. Natarajan, S. and Aravindan, V. 2018. Recycling strategies for spent Li-ion battery mixed cathodes. ACS Energy Letters 3(9), 2101-2103.

16. Zhang, G. et al. 2018. Enhancement in liberation of electrode materials derived from spent lithium-ion battery by pyrolysis. Journal of Cleaner Production 199, 62-68.

17. Ojanen, S. et al. 2018. Challenging the concept of electrochemical discharge using salt solutions for lithium-ion batteries recycling. Waste Manag. 76, 242-249.

18. Li, J., Wang, G. and Xu, Z. 2016. Generation and detection of metal ions and volatile organic compounds (VOCs) emissions from the pretreatment processes for recycling spent lithium-ion batteries. Waste Manag. 52, 221-227.

19. Zheng, X. et al. 2018. A mini-review on metal recycling from spent lithium ion batteries. Engineering 4(3), 361-370.

20. Zhang, X. et al. 2016. Sustainable recycling and regeneration of cathode scraps from industrial production of lithium-ion batteries. ACS Sustainable Chemistry & Engineering 4(12), 7041-7049.

21. Xu, Y. et al. 2015. Application for simply recovered $LiCoO_2$ material as a high-performance candidate for supercapacitor in aqueous system. ACS Sustainable Chemistry & Engineering 3(10), 2435-2442.

22. Zhou, X. et al. 2010. Recycling of electrode materials from spent lithium-ion batteries. 2010 4th International Conference on Bioinformatics and Biomedical Engineering. IEEE: Chengdu, China.

23. Song, D. et al. 2014. Heat treatment of $LiCoO_2$ recovered from cathode scraps with solvent method. J. Power Sources 249, 137-141.

24. Zhang, X. et al. 2014. A novel process for recycling and resynthesizing $LiNi_{1/3}Co_{1/3}Mn_{1/3}O_2$ from the cathode scraps intended for lithium-ion batteries. Waste Manag. 34(9), 1715-1724.

25. Chen, J. et al. 2016. Environmentally friendly recycling and effective repairing of cathode powders from spent $LiFePO_4$ batteries. Green Chem. 18(8), 2500-2506.

26. Ferreira, D.A. et al. 2009. Hydrometallurgical separation of aluminium, cobalt, copper and lithium from spent Li-ion batteries. J. Power Sources 187(1), 238-246.

27. Ekberg, C. and Petranikova, M. 2015. Lithium batteries recycling (Chapter 7). pp. 233-267. *In*: Lithium Process Chemistry. Chagnes, A. and Światowska, J. (Eds). Elsevier: Amsterdam.

28. Dutta, D. et al. 2018. Close loop separation process for the recovery of Co, Cu, Mn, Fe and Li from spent lithium-ion batteries. Separation and Purification Technology 200, 327-334.

29. Zhang, T. et al. 2014. Chemical and process mineralogical characterizations of spent lithium-ion batteries: An approach by multi-analytical techniques. Waste Manag. 34(6), 1051-1058.

30. Shin, S.M. et al. 2005. Development of a metal recovery process from Li-ion battery wastes. Hydrometallurgy 79(3-4), 172-181.

31. Barik, S.P., Prabaharan, G. and Kumar, B. 2016. An innovative approach to recover the metal values from spent lithium-ion batteries. Waste Manag. 51, 222-226.

32. Chen, L. et al. 2011. Process for the recovery of cobalt oxalate from spent lithium-ion batteries. Hydrometallurgy 108(1-2), 80-86.

33. Zeng, X. and Li, J. 2014. Innovative application of ionic liquid to separate Al and cathode materials from spent high-power lithium-ion batteries. J. Hazard Mater. 271, 50-56.

34. Sun, L. and Qiu, K. 2011. Vacuum pyrolysis and hydrometallurgical process for the recovery of valuable metals from spent lithium-ion batteries. J. Hazard Mater. 194, 378-384.

35. Yang, Y. et al. 2016. Thermal treatment process for the recovery of valuable metals from spent lithium-ion batteries. Hydrometallurgy 165, 390-396.

36. Li, L. et al. 2012. Ascorbic-acid-assisted recovery of cobalt and lithium from spent Li-ion batteries. J. Power Sources 218, 21-27.

37. He, L.P. et al. 2015. Recovery of cathode materials and Al from spent lithium-ion batteries by ultrasonic cleaning. Waste Manag. 46, 523-528.

38. Li, J. et al. 2009. A combined recovery process of metals in spent lithium-ion batteries. Chemosphere 77(8), 1132-1136.

39. Veit, H.M. and Bernardes, A.M. 2015. Electronic waste recycling techniques. *In*: Electronic Waste Recycling. Kasper, A.C. et al. (Eds). Switzerland: Springer, Cham.

40. Xiao, J., Li, J. and Xu, Z. 2017. Recycling metals from lithium ion battery by mechanical separation and vacuum metallurgy. J. Hazard Mater. 338, 124-131.

41. Li, J., Wang, G. and Xu, Z. 2016. Environmentally-friendly oxygen-free roasting/wet magnetic separation technology for in situ recycling cobalt, lithium carbonate and graphite from spent $LiCoO_2$/graphite lithium batteries. J. Hazard Mater. 302, 97-104.

42. Xiao, J., Li, J. and Xu, Z. 2017. Novel approach for in situ recovery of lithium carbonate from spent lithium ion batteries using vacuum metallurgy. Environ. Sci. Technol. 51(20), 11960-11966.

43. Hu, J. et al. 2017. A promising approach for the recovery of high value-added metals from spent lithium-ion batteries. J. Power Sources 351, 192-199.

44. Nie, H. et al. 2015. $LiCoO_2$: Recycling from spent batteries and regeneration with solid state synthesis. Green Chem. 17(2), 1276-1280.

45. Vieceli, N. et al. 2018. Hydrometallurgical recycling of lithium-ion batteries by reductive leaching with sodium metabisulphite. Waste Manag. 71, 350-361.

46. Lv, W. et al. 2018. A sustainable process for metal recycling from spent lithium-ion batteries using ammonium chloride. Waste Manag. 79, 545-553.
47. Takacova, Z. et al. 2016. Cobalt and lithium recovery from active mass of spent Li-ion batteries: Theoretical and experimental approach. Hydrometallurgy 163, 9-17.
48. Peng, C. et al. 2018. Selective reductive leaching of cobalt and lithium from industrially crushed waste Li-ion batteries in sulfuric acid system. Waste Manag. 76, 582-590.
49. Meshram, P., Pandey, B.D. and Mankhand, T.R. 2015. Recovery of valuable metals from cathodic active material of spent lithium ion batteries: Leaching and kinetic aspects. Waste Manag. 45, 306-313.
50. Xu, J. et al. 2008. A review of processes and technologies for the recycling of lithium-ion secondary batteries. J. Power Sources 177(2), 512-527.
51. Yang, Y. et al. 2017. A closed-loop process for selective metal recovery from spent lithium iron phosphate batteries through mechanochemical activation. ACS Sustainable Chemistry & Engineering 5(11), 9972-9980.
52. Zeng, X., Li, J. and Singh, N. 2014. Recycling of spent lithium-ion battery: A critical review. Critical Reviews in Environmental Science and Technology 44(10), 1129-1165.
53. Higuchi, A. et al. 2016. Selective recovery of lithium from cathode materials of spent lithium ion battery. Jom 68(10), 2624-2631.
54. Joulié, M. et al. 2017. Current collectors as reducing agent to dissolve active materials of positive electrodes from Li-ion battery wastes. Hydrometallurgy 169, 426-432.
55. Li, L. et al. 2018. Process for recycling mixed-cathode materials from spent lithium-ion batteries and kinetics of leaching. Waste Manag. 71, 362-371.
56. Gao, W. et al. 2018. Selective recovery of valuable metals from spent lithium-ion batteries – Process development and kinetics evaluation. Journal of Cleaner Production 178, 833-845.
57. Gao, W. et al. 2017. Lithium carbonate recovery from cathode scrap of spent lithium-ion battery: Closed-loop Process. Environ. Sci. Technol. 51(3), 1662-1669.
58. Zeng, X., Li, J. and Shen, B. 2015. Novel approach to recover cobalt and lithium from spent lithium-ion battery using oxalic acid. J. Hazard Mater. 295, 112-118.
59. Natarajan, S., Boricha, A.B. and Bajaj, H.C. 2018. Recovery of value-added products from cathode and anode material of spent lithium-ion batteries. Waste Manag. 77, 455-465.
60. Li, L. et al. 2018. Economical recycling process for spent lithium-ion batteries and macro- and micro-scale mechanistic study. J. Power Sources 377, 70-79.
61. Nayaka, G.P. et al. 2018. Effective and environmentally friendly recycling process designed for $LiCoO_2$ cathode powders of spent Li-ion batteries using mixture of mild organic acids. Waste Manag. 78, 51-57.
62. Li, L. et al. 2017. Sustainable recovery of cathode materials from spent lithium-ion batteries using lactic acid leaching system. ACS Sustainable Chemistry & Engineering 5(6), 5224-5233.
63. Golmohammadzadeh, R., Rashchi, F. and Vahidi, E. 2017. Recovery of lithium and cobalt from spent lithium-ion batteries using organic acids: Process optimization and kinetic aspects. Waste Manag. 64, 244-254.

64. Chen, X. et al. 2018. Organic reductants based leaching: A sustainable process for the recovery of valuable metals from spent lithium ion batteries. Waste Manag. 75, 459-468.

65. Lv, W. et al. 2018. A critical review and analysis on the recycling of spent lithium-ion batteries. ACS Sustainable Chemistry & Engineering 6(2), 1504-1521.

66. Ku, H. et al. 2016. Recycling of spent lithium-ion battery cathode materials by ammoniacal leaching. J. Hazard Mater. 313, 138-146.

67. Zheng, X. et al. 2017. Spent lithium-ion battery recycling – Reductive ammonia leaching of metals from cathode scrap by sodium sulphite. Waste Manag. 60, 680-688.

68. Ijadi Bajestani, M., Mousavi, S.M. and Shojaosadati, S.A. 2014. Bioleaching of heavy metals from spent household batteries using Acidithiobacillus ferrooxidans: Statistical evaluation and optimization. Separation and Purification Technology 132, 309-316.

69. Mishra, D. et al. 2008. Bioleaching of metals from spent lithium ion secondary batteries using Acidithiobacillus ferrooxidans. Waste Manag. 28(2), 333-338.

70. Zeng, G. et al. 2013. Influence of silver ions on bioleaching of cobalt from spent lithium batteries. Minerals Engineering 49, 40-44.

71. Chen, S.Y. and Lin, J.G. 2009. Enhancement of metal bioleaching from contaminated sediment using silver ion. J. Hazard Mater. 161(2-3), 893-899.

72. Zeng, G. et al. 2012. A copper-catalyzed bioleaching process for enhancement of cobalt dissolution from spent lithium-ion batteries. J. Hazard Mater. 199-200, 164-169.

73. Xin, B. et al. 2009. Bioleaching mechanism of Co and Li from spent lithium-ion battery by the mixed culture of acidophilic sulfur-oxidizing and iron-oxidizing bacteria. Bioresour. Technol. 100(24), 6163-6169.

74. Xin, Y. et al. 2016. Bioleaching of valuable metals Li, Co, Ni and Mn from spent electric vehicle Li-ion batteries for the purpose of recovery. Journal of Cleaner Production 116, 249-258.

75. Ren, W.X. et al. 2009. Biological leaching of heavy metals from a contaminated soil by Aspergillus niger. J. Hazard Mater. 167(1-3), 164-169.

76. Bahaloo-Horeh, N. and Mousavi, S.M. 2017. Enhanced recovery of valuable metals from spent lithium-ion batteries through optimization of organic acids produced by Aspergillus niger. Waste Manag. 60, 666-679.

77. Niu, Z. et al. 2014. Process controls for improving bioleaching performance of both Li and Co from spent lithium ion batteries at high pulp density and its thermodynamics and kinetics exploration. Chemosphere 109, 92-98.

78. Barik, S.P., Prabaharan, G. and Kumar, L. 2016. Leaching and separation of Co and Mn from electrode materials of spent lithium-ion batteries using hydrochloric acid: Laboratory and pilot scale study. Journal of Cleaner Production 147, 37-43.

79. Zhang, P. et al. 1998. Hydrometallurgical process for recovery of metal values from spent lithium-ion secondary batteries. Hydrometallurgy 47(2-3), 259-271.

80. Huang, Y. et al. 2016. A stepwise recovery of metals from hybrid cathodes of spent Li-ion batteries with leaching-flotation-precipitation process. J. Power Sources 325, 555-564.

81. Gratz, E. et al. 2014. A closed loop process for recycling spent lithium ion batteries. J. Power Sources 262, 255-262.

82. Meshram, P., Pandey, B.D. and Mankhand, T.R. 2015. Hydrometallurgical processing of spent lithium ion batteries (LIBs) in the presence of a reducing agent with emphasis on kinetics of leaching. Chemical Engineering Journal 281, 418-427.

83. Jha, M.K. et al. 2013. Recovery of lithium and cobalt from waste lithium ion batteries of mobile phone. Waste Manag. 33(9), 1890-1897.

84. Meng, Q., Zhang, Y. and Dong, P. 2017. Use of glucose as reductant to recover Co from spent lithium ions batteries. Waste Manag. 64, 214-218.

85. Pinna, E.G. et al. 2017. Cathodes of spent Li-ion batteries: Dissolution with phosphoric acid and recovery of lithium and cobalt from leach liquors. Hydrometallurgy 167, 66-71.

86. Lee, C.K. and Rhee, K.-I. 2002. Preparation of $LiCoO_2$ from spent lithium-ion batteries. J. Power Sources 109(1), 17-21.

87. Sun, L. and Qiu, K. 2012. Organic oxalate as leachant and precipitant for the recovery of valuable metals from spent lithium-ion batteries. Waste Manag. 32(8), 1575-1582.

88. Chen, X. et al. 2015. Sustainable recovery of metals from spent lithium-ion batteries: A green process. ACS Sustainable Chemistry & Engineering 3(12), 3104-3113.

89. Zhang, X. et al. 2015. A closed-loop process for recycling $LiNi_{1/3}Co_{1/3}Mn_{1/3}O_2$ from the cathode scraps of lithium-ion batteries: Process optimization and kinetics analysis. Separation and Purification Technology 150, 186-195.

90. Nayaka, G.P. et al. 2016. Use of mild organic acid reagents to recover the Co and Li from spent Li-ion batteries. Waste Manag. 51, 234-238.

91. Nayaka, G.P. et al. 2016. Recovery of cobalt as cobalt oxalate from spent lithium ion batteries by using glycine as leaching agent. Journal of Environmental Chemical Engineering 4(2), 2378-2383.

92. Li, L. et al. 2015. Succinic acid-based leaching system: A sustainable process for recovery of valuable metals from spent Li-ion batteries. J. Power Sources 282, 544-551.

93. He, L. et al. 2017. Recovery of lithium, nickel, cobalt, and manganese from spent lithium-ion batteries using L-tartaric acid as a leachant. ACS Sustainable Chemistry & Engineering 5(1), 714-721.

94. Meng, Q., Zhang, Y. and Dong, P. 2018. Use of electrochemical cathode-reduction method for leaching of cobalt from spent lithium-ion batteries. Journal of Cleaner Production 180, 64-70.

95. Prabaharan, G. et al. 2017. Electrochemical process for electrode material of spent lithium ion batteries. Waste Manag. 68, 527-533.

96. Boxall, N.J. et al. 2018. Multistage leaching of metals from spent lithium ion battery waste using electrochemically generated acidic lixiviant. Waste Manag. 74, 435-445.

97. Joo, S.H. et al. 2016. Extraction of manganese by alkyl monocarboxylic acid in a mixed extractant from a leaching solution of spent lithium-ion battery ternary cathodic material. J. Power Sources 305, 175-181.

98. Joo, S.-H. et al. 2016. Selective extraction and separation of nickel from cobalt, manganese and lithium in pre-treated leach liquors of ternary

cathode material of spent lithium-ion batteries using synergism caused by Versatic 10 acid and LIX 84-I. Hydrometallurgy 159, 65-74.

99. Torkaman, R. et al. 2017. Recovery of cobalt from spent lithium ion batteries by using acidic and basic extractants in solvent extraction process. Separation and Purification Technology 186, 318-325.

100. Wang, F. et al. 2016. Recovery of cobalt from spent lithium ion batteries using sulphuric acid leaching followed by solid–liquid separation and solvent extraction. RSC Advances 6(88), 85303-85311.

101. Virolainen, S. et al. 2017. Solvent extraction fractionation of Li-ion battery leachate containing Li, Ni, and Co. Separation and Purification Technology 179, 274-282.

102. Jha, A.K. et al. 2013. Selective separation and recovery of cobalt from leach liquor of discarded Li-ion batteries using thiophosphinic extractant. Separation and Purification Technology 104, 160-166.

103. Yang, Y., Xu, S. and He, Y. 2017. Lithium recycling and cathode material regeneration from acid leach liquor of spent lithium-ion battery via facile co-extraction and co-precipitation processes. Waste Manag. 64, 219-227.

104. Pant, D. and Dolker, T. 2016. Green and facile method for the recovery of spent lithium Nickel Manganese Cobalt Oxide (NMC) based lithium ion batteries. Waste Manag. 60, 689-695.

105. Pagnanelli, F. et al. 2016. Cobalt products from real waste fractions of end of life lithium ion batteries. Waste Manag. 51, 214-221.

106. Ordoñez, J., Gago, E.J. and Girard, A. 2016. Processes and technologies for the recycling and recovery of spent lithium-ion batteries. Renewable and Sustainable Energy Reviews 60, 195-205.

107. Chen, X. et al. 2015. Separation and recovery of metal values from leach liquor of waste lithium nickel cobalt manganese oxide based cathodes. Separation and Purification Technology 141, 76-83.

108. Chen, X. et al. 2016. An atom-economic process for the recovery of high value-added metals from spent lithium-ion batteries. Journal of Cleaner Production 112, 3562-3570.

109. Chen, X. et al. 2015. Hydrometallurgical recovery of metal values from sulfuric acid leaching liquor of spent lithium-ion batteries. Waste Manag. 38, 349-356.

110. Wang, R., Lin, Y. and Wu, S. 2009. A novel recovery process of metal values from the cathode active materials of the lithium-ion secondary batteries. Hydrometallurgy 99(3-4), 194-201.

111. Chen, X. et al. 2015. Separation and recovery of metal values from leaching liquor of mixed-type of spent lithium-ion batteries. Separation and Purification Technology 144, 197-205.

112. Joulié, M., Laucournet, R. and Billy, E. 2014. Hydrometallurgical process for the recovery of high value metals from spent lithium nickel cobalt aluminum oxide based lithium-ion batteries. J. Power Sources 247, 551-555.

113. Senćanski, J. et al. 2017. The synthesis of Li(Co Mn Ni)O_2 cathode material from spent-Li ion batteries and the proof of its functionality in aqueous lithium and sodium electrolytic solutions. J. Power Sources 342, 690-703.

114. Sa, Q. et al. 2015. Synthesis of high performance LiNi$_{1/3}$Mn$_{1/3}$Co$_{1/3}$O$_2$ from lithium ion battery recovery stream. J. Power Sources 282, 140-145.

115. Pegoretti, V.C.B. et al. 2017. Thermal synthesis, characterization and electrochemical study of high-temperature (HT) $LiCoO_2$ obtained from $Co(OH)_2$ recycled of spent lithium ion batteries. Materials Research Bulletin 86, 5-9.

116. Yun, J.-Y. et al. 2016. Fabrication of nanosized cobalt powder from Cobalt(II) hydroxide of spent lithium ion battery. Applied Surface Science 415, 80-84.

117. Meng, Q., Zhang, Y. and Dong, P. 2018. A combined process for cobalt recovering and cathode material regeneration from spent $LiCoO_2$ batteries: Process optimization and kinetics aspects. Waste Manag. 71, 372-380.

118. Cai, G. et al. 2014. Process development for the recycle of spent lithium ion batteries by chemical precipitation. Industrial & Engineering Chemistry Research 53(47), 18245-18259.

119. Wang, X. et al. 2018. Hydrothermal preparation and performance of $LiFePO_4$ by using Li_3PO_4 recovered from spent cathode scraps as Li source. Waste Manag. 78, 208-216.

120. Li, L. et al. 2014. Synthesis and electrochemical performance of cathode material $Li_{1.2}Co_{0.13}Ni_{0.13}Mn_{0.54}O_2$ from spent lithium-ion batteries. J. Power Sources 249, 28-34.

121. Yang, Y. et al. 2016. Synthesis and performance of spherical $LiNi_xCo_yMn_{1-x-y}O_2$ regenerated from nickel and cobalt scraps. Hydrometallurgy 165, 358-369.

122. Perez, E. et al. 2016. Recovery of metals from simulant spent lithium-ion battery as organophosphonate coordination polymers in aqueous media. J. Hazard Mater. 317, 617-621.

123. Lupi, C., Pasquali, M. and Dell'era, A. 2005. Nickel and cobalt recycling from lithium-ion batteries by electrochemical processes. Waste Manag. 25(2), 215-220.

124. Freitas, M.B.J.G., Celante, V.G. and Pietre, M.K. 2010. Electrochemical recovery of cobalt and copper from spent Li-ion batteries as multilayer deposits. J. Power Sources 195(10), 3309-3315.

125. Freitas, M.B.J.G. and Garcia, E.M. 2007. Electrochemical recycling of cobalt from cathodes of spent lithium-ion batteries. J. Power Sources 171(2), 953-959.

126. Garcia, E.M. et al. 2008. Electrodeposition of cobalt from spent Li-ion battery cathodes by the electrochemistry quartz crystal microbalance technique. J. Power Sources 185(1), 549-553.

127. Falqueto, J.B. et al. 2018. Photocatalytic properties of Co/Co_3O_4 films recycled from spent Li-ion batteries. Ionics 24(7), 2167-2173.

128. Barbieri, E.M.S. et al. 2014. Recycling of spent ion-lithium batteries as cobalt hydroxide, and cobalt oxide films formed under a conductive glass substrate, and their electrochemical properties. J. Power Sources 269, 158-163.

129. Barbieri, E.M.S. et al. 2014. Recycling of cobalt from spent Li-ion batteries as β-$Co(OH)_2$ and the application of Co_3O_4 as a pseudocapacitor. J. Power Sources 270, 158-165.

130. Cao, Z. et al. 2018. Efficient reuse of anode scrap from lithium-ion batteries as cathode for pollutant degradation in electro-Fenton process: Role of different recovery processes. Chemical Engineering Journal 337, 256-264.

131. Wang, Y. et al. 2018. Tailored synthesis of active reduced graphene oxides from waste graphite: Structural defects and pollutant-dependent reactive radicals in aqueous organics decontamination. Applied Catalysis B: Environmental 229, 71-80.
132. Zhang, Y. et al. 2016. Mesocarbon microbead carbon-supported magnesium hydroxide nanoparticles: Turning spent Li-ion battery anode into a highly efficient phosphate adsorbent for wastewater treatment. ACS Appl. Mater. Interfaces 8(33), 21315-21325.
133. Guo, Y. et al. 2016. Leaching lithium from the anode electrode materials of spent lithium-ion batteries by hydrochloric acid (HCl). Waste Manag. 51, 227-233.
134. Grützke, M. et al. 2014. Supercritical carbon dioxide extraction of lithium-ion battery electrolytes. The Journal of Supercritical Fluids 94, 216-222.
135. Liu, Y. et al. 2017. Purification and characterization of reclaimed electrolytes from spent lithium-ion batteries. The Journal of Physical Chemistry C 121(8), 4181-4187.
136. Grützke, M. et al. 2015. Aging investigations of a lithium-ion battery electrolyte from a field-tested hybrid electric vehicle. J. Power Sources 273, 83-88.
137. Valio, J. 2017. Critical Review on Lithium Ion Battery Recycling Technologies. Aalto University, Helsinki, Finland.
138. Knights, B.D.H. and Saloojee, F. 2015. Lithium Batteries Recycling. Metallurgical Consultancy & Laboratories, South Africa.
139. Marinos, D. 2014. An Approach to Beneficiation of Spent Lithium-ion Batteries for Recovery of Materials. University of Colorado, Colorado.

Batteries Charge Controller and Its Technological Challenges for Plug-in Electric and Hybrid Electric Vehicles

C. Bharatiraja

Department of Electrical and Electronics Engineering, Kattankulathur Campus, SRM Institute of Science and Technology, Kattankulathur, Tamil Nadu - 603203, India

Email: bharatic@srmist.edu.in

1. Introduction

Li-ion batteries have been commercialized in the last two decades for Electrical Vehicles (EV) and other applications. The technology is considered relatively mature on the basis of the current battery chemistry. Li-ion batteries have been dominantly used in electronic devices, including cell phones and laptop computers, and are starting to play an increasing role in electric vehicles. Li-ion batteries in the future will also be considered for application in sustainable energy grids to store sustainable energy generated from renewable sources.

2. The Existing State of Electric-Car Battery Technology

The value supply chain of EV batteries consists of seven important steps: component production (including raw materials); cell manufacturing; module production; integration of modules into the battery pack, assembly of the battery pack into the vehicle; use during the life of the vehicle; and reuse and recycling. The focus would be on the first four steps, which make up the manufacture of battery packs for use by OEMs.

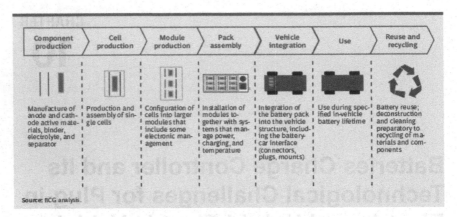

Fig. 1. Value chain for electric-car batteries comprises seven steps.

Table 1. Companies using different battery technologies

S. No.	Company	Country	Vehicle model	Battery technology
1.	BMW	Germany	X6 Mini E (2012)	Li-ion
2.	BYD	China	E6	Li-ion
3.	Chrysler	USA	Chrysler 200C EV	Li-ion
4.	Daimler Benz	Germany	ML450, S400 Smart EV (2010)	NiMH Li-ion
5.	Ford	USA	Escape, Fusion, MKZ HEV Escape HEV	NiMH Li-ion
6.	GM	USA	Chevy-Volt Saturn Vue Hybrid	NiMH NiMH
7.	Honda	Japan	Civic, Insight	NiMH
8.	Hyundai	South Korea	Sonata	Lithium polymer
9.	Mitsubishi	Japan	iMiEV (2010)	Li-ion
10.	Nissan	Japan	Altima Leaf EV (2010)	NiMH Li-ion
11.	Tesla	USA	Roadster (2009)	Li-ion
12.	Think	Norway	Think EV	Li-ion, Sodium/Metal chloride
13.	Toyota	Japan	Prius, Lexus	NiMH

2.1 Key Concern Parameters for Good Performance

Safety: The most important criterion for EV batteries is the Safety. Even a single battery hazard could turn public opinion against electric mobility and deteriorate industry development for months or decades. The major concern in this area is avoiding thermal runaway, a major feedback loop whereby chemical reactions triggered in the cell exacerbate heat release, potentially resulting in a fire.

Life Span: Battery's lifespan can be estimated by two factors: cycle stability and overall age. Cycle stability is the number of times a battery can be fully charged and discharged before deteriorating to 80 percent of its original capacity at full charge. Overall age is the number of years a battery can be expected to be made into use. Today's batteries do meet the cycle stability requirements of EVs under various test conditions.

Performance: The expectation that the owner of an electric vehicle should be able to drive it both in extreme hot summer temperatures and at subzero winter temperatures constitutes substantial engineering challenges. Batteries can be optimized for either high or low temperatures, but it is difficult to engineer them to function over a wide range of temperatures without incurring performance degradation. One elucidation might be for OEMs to rate batteries for particular climates.

Specific Energy and Specific Power: The specific energy of batteries— that is, their capacity for depositing energy per kilogram of weight is still only one percent of the specific energy of gasoline. Unless there is a major breakthrough, batteries will continue to restrict the driving range of electric vehicles to some 160 to 190 miles between charges. Battery cells today can achieve a nominal energy density of 140 to 170 watt-hours per kilo-gram (Wh/kg), compared with 13,000 Wh/kg for gasoline.

Charging Time: Long charging times constitute another technical challenge and a commercial barrier that must be disentangled. It takes almost ten hours to charge a 15-kWh battery by plugging it into a standard 120V DC outlet. Fast charging methods that engage more sophisticated charging terminals can reduce this period significantly. These charging systems do come with an additional price and weight, as they require enhanced cooling systems on board the vehicle.

The Cost Challenge: The United States Advanced Battery Consortium has set a price target of $250 per kWh. But even if battery makers are able to achieve the technical challenges outlined above, battery prices may remain above that target.

3. EV Standard by SAE

The Society of Automotive Engineers (SAE) Information Report publishes some specific standards for Electric Vehicles that identify and define the preferred technical guidelines relating to safety for vehicles that contain High Voltage (HV), such as Electric Vehicles (EV), Hybrid Electric Vehicles (HEV), Plug-In Hybrid Electric Vehicles (PHEV), Fuel Cell Vehicles (FCV) and Plug-In Fuel Cell Vehicles (PFCV) during normal operation and charging, as applicable. Guidelines in this document do not necessarily address maintenance, repair, or assembly safety issues.

The Standardization by SAE includes various criteria and benefits:

1. Voluntary & Collaborative Effort
2. Leverages Industry Expertise
3. Provide Foundational Elements
4. Speeds Technology Advancement
5. Addresses Common Pinch Points
6. Enhance safety
7. Create common language
8. Facilitate trade through reduced regulations
9. Harmonize global markets
10. Permit common interfaces

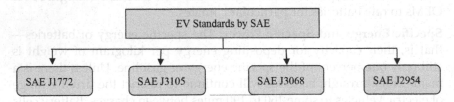

Fig. 2. Standards defined by the Society of Automotive Engineering.

The recommend connector, Portal Charging methods and AC Depot Conductive Charging are represented below.

3.1 SAE J1772 Revision 8

Revision edition of SAE J1772 has raised the voltage and current limit of the SAE Combo.

Connector as shown in Fig. 3. It is in practice since October 2017. The features of the connector are listed below:

- Current limit from 200 A to 350 A
- Voltage limit from 500 Vdc to 1000 Vdc
- 350 kW Max Power

3.2 SAE J3105 Overhead and Portal Charging

The SAE J3105 overhead and portal charging station is in practice since 2016. The SAE J3105 document will standardize the interface between the infrastructure and the bus targeted towards in-route DC charging, for

Fig. 3. SAE J1772 voltage and current limit SAE combo connector.

example to recharge at transit bus station during a short stop. The DC Power Levels (Voltage Range: 250-1,000 DC Volts) permissible are up to 1 MW. Figure 10.4 shows the SAE J3105 overhead and portal charging for the bus.

Fig. 4. The SAE J3105 overhead and portal charging for the bus.

3.3 SAE J3068 AC Depot Conductive Charging

The AC Depot Conductive is a promising charging method for EV systems. The charging system utilises an on board charging method. Figure 5 shows the SAE J3068 AC Depot Conductive Charging.

Depot Charging Three-Phase AC (J-3068) is targeted towards charging at commercial and industrial locations or other places where three-phase AC power is available and is preferred at such locations (160 A, 480 V, AC 3-phase = 133 kW). It defines a conductive power transfer method

Fig. 5. SAE J3068 AC Depot Conductive charging.

including a digital communication system. It also covers the functional and dimensional requirements for the vehicle inlet, supply equipment outlet, and mating housings and contacts.

3.4 SAE J2954 Electric Vehicle Wireless Power Transfer

Wireless Power Transfer (WPT) is the transmission of electrical power from the power source to an electrical load without the use of physical connectors. Wireless power transfer is the advanced technology adopted for the upcoming electrical vehicles. WPT technology can be used as a solution in eliminating many charging hazards and drawbacks related to cables. Wireless power transfer has developed rapidly in recent years. The transfer distance increases from several millimeters to several hundred millimeters. This advancement makes WPT very attractive to the electric vehicle (EV) charging applications in both stationary and dynamic charging scenarios. Wireless power transfer provides an inherent electrical isolation and reduces on board charging cost, weight and volume. It also completely eliminates the existing high tension power transmission lines and cables, towers and substations between the generating station and consumers to facilitate the interconnection of electrical generation plants on a global scale. Charging becomes the easiest task due to wireless energy transfer to the electric vehicle. The drivers just need to park their car and leave for a stationary WPT system. The battery capacity of EVs with wireless charging could be increased to 20% or less compared to EVs with conductive charging.

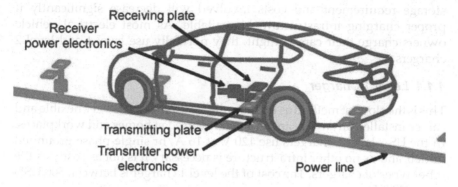

Fig. 6. Electric vehicle wireless power transfer.

SAE J2954 establishes minimum performance, interoperability and safety criteria for wireless charging of EVs/PHEVs. The considerations for WPT charging system mainly focused on the following criteria.

- Inductive Charging Interoperability
- Automated Charging
- Power Transfer Communications
- Smart Grid Interoperability
- Automatic Shutdown Capability
- Autonomous Parking/Charging

4. Levels of Charging

There exist three standard charging levels used to charge EVs. All electric cars can be charged with level 1 and level 2 stations. These types of power chargers offer the same charging power as the ones installed at home. Level 3 chargers – also called DCFC or fast charging stations – are much more powerful chargers than level 1 and 2 stations, meaning you can charge an EV much faster with them and that being said, some vehicles cannot charge at level 3 charging stations. Knowing the vehicle's capabilities is therefore very important.

Chargers are required to convert the AC power from grid to DC power which is supplied to the battery in the electric vehicle, and this is done with the help of a rectifier and for fast charging a AC to DC converter is also used. Chargers play a very important role in an electric vehicle as they impact the battery life and charging time.

4.1 Chargers Classification

Based on the power level of the charger many parameters are affected such as charging time, cost and impact on the grid. The on-board energy storage requirement and costs involved will decrease significantly if proper charging infrastructure is available. As most electrical vehicle owners charge their cars at night, they normally use level 1 and level 2 chargers.

4.1.1 Level 1 Charger

This is the slowest method as this supplies low power. It is very flexible and can be installed anywhere and is generally used at homes and workplaces. In the USA level 1 chargers use 120 V at 16 A/hr single-phase grounded outlets and as no other infrastructure is necessary it is not as costly as the other types of chargers. The cost of the level 1 charger is between 500 USD and 900 USD.

4.1.2 Level 2 Charger

This type of charger is specifically installed for charging of electric vehicles in both private and public locations. It consists of specific equipment which has to be installed at the location for charging, the equipment primarily consisting of power electronic converters. Some car manufacturers such as Tesla give these power electronics on-board, thus only requiring an outlet for charging. Level 2 chargers work at a voltage of 240 at 80A/hr. These type of chargers are preferred as they can charge a car overnight. The installation cost is between $1000 and $3000.

4.1.3 Level 3 Charger

Level 3 chargers are able to charge an electric vehicle in less than an hour and are only commercially viable as they are very costly to set up , their cost ranging from $30000 to $160000. The operational voltage is about 480 V and higher. These type of charging stations are similar to gas stations for normal cars. These staions have a large impact on the local grid and cause heavy harmonic distortions and losses in local transformers. Regular maintainance of transformers is required to avoid any malfunction and to ensure safe operation. Integrating a level 3 charger with a smart grid would help negate some of the distortions caused.

Table 2 shows the various charging levels, corresponding charging time and applications of different levels of charging.

The Level 1 and Level 2 chargers are On Board chargers with power level varying from 1.4 kW to 19.2 kW whereas the Level 3 charger is an Off-Board charger which is used commercially i.e., charging station. The charging time of Level 3 charger (0.2 to 0.5 hour) is very less when compared to the Level 1 (4 to 11 hours) and Level 2 chargers (1 to 4 hours). Level 1 and Level 2 chargers are mostly used for office/home/private/public outlet charging. Table 3 shows the different characteristics of AC Level and DC Level Charging based on the SAE 1772 standard.

Table 2. Charging power levels

Power level	Level 1		Level 2		Level 3	
Vehicle Terminology	PHEV (5 kWh-15 kWh)	EV (16 kWh-50 kWh)	PHEV (5 kWh-15 kWh)	EV (16 kWh-30 kWh)	EV (3 kWh-50 kWh)	EV (20 kWh – 50 kWh)
Charger placement	On Board Single Phase		On Board Single Phase or Three Phase			Off Board Three Phase
Application	Office or Home Charging		Private or Public outlet charging			For Commercial use (Charging Station)
Supply interface	At convenience outlets		Dedicated EV supply equipment			Dedicated EV supply equipment
Power level	1.4 kW (12 A)	1.9 kW (20 A)	4 kW (17 A)	8 kW (32 A)	19.2 kW (80 A)	50 kW 100 kW
Charge time	4 to 11 hours	11 to 36 hours	1 to 4 hours	2 to 6 hours	2 to 3 hours	0.4 to 1 hour 0.2 to 0.5 hour

Table 3. Characteristics of AC/DC level charging (SAE 1772)

Power type	Level	Voltage	Current	Power	Charging time		Phase
					PHEV	BEV	
AC	Level 1	120 V	12 A	1.4 kW	7 hrs	17 hrs	1 phase
			16 A	1.9 kW			
	Level 2	240 V	80 A	19.2 kW	3 hrs	7 hrs	1 phase and 3 phase
	Level 3	-	-	>20 kW	Under development	Under development	Under development
DC	Level 1	200-500 V	<80 A	40 kW	22 min	1.2 hrs	3 phase
	Level 2	200-500 V	<200 A	100 kW	10 min	20 min	3 phase
	Level 3	200-600 V	<400 A	240 kW	Under development	Under development	Under development

Table 4. Public charging stations

Level 1	Level 2	Level 3
Level 1 charging is the technical jargon for plugging your car into an ordinary household outlet.	Level 2 supplies 240 V, like what an electric dryer or oven uses along with current range of 80 A.	Level 3 supplies 3 phase ac voltage along with 120 A current.

Table 5. Parameters of electric vehicles

EV manufacturers		Toyota Prius	Chevrolet Volt	Tesla Roadster	Nissan Leaf	Mitsubishi Motors
Battery		4.4 kWh (Li-Ion)	16 kWh (Li-Ion)	53 kWh (Li-Ion)	24 kWh (Li-Ion)	16 kWh (Li-Ion)
Range		14 miles	40 miles	245 miles	100 miles	96 miles
Connecter type		SAE J1772	SAE J1772	SAE J1772	SAE J1772 JARI/TEPCO	SAE J1772 JARI/TEPCO
Level 1 Charging	Demand	1.4 kW	0.96-1.4 kW	1.8 kW	1.8 kW	1.5 kW
	Charge time	3 hrs	5-8 hrs	30+ hrs	12-16 hrs	7 hrs
Level 2 Charging	Demand	3.8 kW	3.8 kW	9.6-16.8 kW	3.3 kW	3 kW
	Charge time	2.5 hrs	2-3 hrs	4-12 hrs	6-8 hrs	14 hrs
DC Fast Charging	Demand	-	-	-	>50 kW	50 kW
	Charge time	-	-	-	15-30 min	30 min

The standard SAE 1172 defines particular standards for AC/DC level charging. AC level charging can be used for both single phase and three phase supplies whereas DC level charging is for three phase supply only. The minimum time for charging in AC level is 3 hours while the minimum time for charging in DC level is 10 minutes. In DC level charging the maximum power is 240 kW while in AC level charging the maximum power is more than 20 kW and less than 40 kW. Table 5 describes the parameters of different electric vehicles around the globe.

From the charging characteristics and infrastructures of some manufactured PHEVs and EVs off all the car manufacturing companies, Tesla Roadster has the highest range of 245 miles and it has the highest battery capacity of 53 kWh. The type of battery used in all these cars is lithium ion (Li-ion) which is considered the most efficient till date. The connector type in all the above cars is also the same i.e. SAE J1772. The charging time among the above models is the least for Toyota Prius which is 2-3 hours with a power demand of 1.4 kW.

Chargers can be classified into two types depending on the power flow- unidirectional and bidirectional chargers.

• **Unidirectional Chargers**

When using unidirectional chargers, power can only be transferred from the grid to the electric vehicle, while the vehicle can't transfer power to the electric grid. A general unidirectional charger has a filter, AC/DC and DC/DC converter. Nowadays these components are joined into a single stage unit to make them compact and cheap. Controlling impact of EV charging on the grid is easier while using unidirectional chargers due to ease of control of these chargers. Unidirectional chargers overcome most of the drawbacks of bidirectional chargers such as cost, performance and safety. Research in this area is focused on exploring the impact of using this charger on the grid and to explore optimal charging strategies.

• **Bidirectional Chargers**

Bidirectional chargers can take supply from the grid and send power to the electric vehicle or it can also transmit power from the electric vehicle to the grid. For these two to take place two parts are used in the charger. A bidirectional AC to DC converter is used to control power factor and current is controlled through a DC to DC bidirectional converter. When the charging of electric vehicle takes place, the charger draws sinusoidal current with a defined phase angle, to control both the real power and reactive power. Similarly during discharge the charger must return current in the same sinusoidal form. Isolated or nonisolated circuit chargers configurations are used to design these chargers. In single-phase nonisolated configuration, buck converter is used to reduce output voltage in comparison to the input.

One major drawback of bidirectional charging is possible degradation of battery due to frequent cycling, and other drawbacks are higher cost and metering issues. Interconnection issues and anti-islanding issues need to be looked into while using this configuration. A lot of safety measures are required for a successful bidirectional power flow implementation. The charger infrastructure used for bidirectional/unidirectional configuration is compared below.

The comparisons in Table 6 include challenges faced, merits, price, ease of control and power flow. Use of bidirectional chargers helps boost spinning reserves and enable energy balance. Bidirectional chargers are only feasible for level 2 infrastructure as in level 1 charging power level is low and in level 3 charging the connection time is too low to allow substantial reverse power flow.

Table 6. Unidirectional and bidirectional charger infrastructure comparison

Parameters	Bidirectional chargers & infrastructure	Unidirectional chargers & infrastructure
Switches and power flow	Power flow is two-way from grid to vehicle (G2V) and vehicle to grid (V2G MOSFET), IGBT, GTO	Power flow is only one-way from grid to vehicle (G2V), Diode Bridge and unidirectional converter.
Power level	Level 2	Levels 1, 2 and 3
Situation	Not available	Available
Cost	High price	Low price, and no additional cost required
Requirements and challenges	• Effective communication a must to minimize impact on power quality • Suitable sensors and smart metering • Proper coordination is essential • Additional investment and cost • Significant loss of energy • Considerable stress on device	Connection to power grid
Control	Complex, extra drive control circuits	Simple, active control of charging current
Distributed system	Necessary updates and further investment is required	No further updating of system or investment of money is required

(Contd.)

Isolation and safety	Either isolated or non-isolated. Non-isolated has the advantage of having a simpler structure with higher efficiency and reduced cost along with a high degree of reliability. • High degree of safety • Protection from islanding • Issues related to interconnection	Non-isolated or isolated.
Effect on battery	Degradation from frequent cycling	No degradation due to discharge
Benefits	• Creates a flexible and efficient energy system • Adequate spinning reserves • Reactive power support • Load following • Energy balance	• Services are provided based on variable charge rates and reactive power • Absorbs or supplies reactive power using current phase angle control, which can be done without discharging the battery • Frequency and voltage control

Chargers can be classified into two types depending on the placement of the power electronic converters.

(1) *On-board:* In this type of charger, power electronic converters such as the rectifier and the chopper are situated inside the electric vehicle but due to this configuration there is a limitation in the amount of power supplied to the electrical vehicle. This power level is similar to the power level of a level 1 charger's power supply. At the heart of any electric (EV) or plug-in hybrid (HEV) vehicle lays the high-voltage (200 to 450 VDC) battery and its associated charging system. The on-board charger (OBC) provides the means to recharge the battery from the AC mains either at home or from outlets found in private or public charging stations. From a 3.6 kW single-phase to a 22 kW three-phase high-power converter, today's OBCs must have the highest possible efficiency and reliability to ensure rapid charging times as well as meeting the limited space and weight requirements.

(2) *Off-board:* These types of chargers have power electronic converters outside the car, due to which fast charging can take place. This type of charger is costly. There are "off-board" DC chargers around the world that do not require the use of the onboard vehicle charge, because they rectify the AC power into DC power, which goes directly to the battery.

These chargers are known as:

- CHAdeMO - worldwide
- CCS-Combo2 - Europe
- SAE-CCS-Combo1 - North America
- GB/T - China

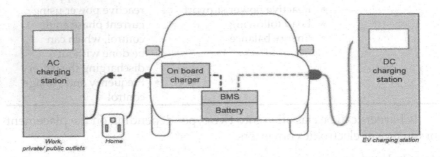

Fig. 7. On-board and off-board charging.

- On-board charging notable points
  - The on-board charger takes AC input from your charging station and converts it to direct current to charge a 350V or 650V battery
  - Low transfer of energy (kWh)
  - Problem of heating of battery pack is not a concern
  - Slow charging
  - On-board rectification is used to manage battery status via battery management system
  - Recharge at any place without any additional equipment
- Off-board charging notable points
  - The off-board charging station takes DC input from the charging station and provides DC output to the vehicle
  - High transfer of energy (kWh)
  - Need to address battery heating issue
  - Fast charging

> o Require charging stations of proper rating
> o Converter circuits are present outside the vehicle in order to provide the rated power values

4.2 Infrastructure for V2G

Four charging modes are possible for an EV. Depending on the charging mode, the infrastructure required for V2G operation will change. In charging modes 1, 2 and 3, the battery charger of the EV is directly connected to the AC mains through a cable as shown in Fig. 8. Battery chargers are situated in the onboard and/or in the electric vehicle supply equipment (EVSE).

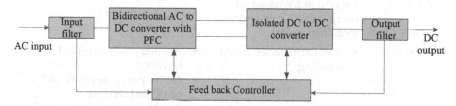

Fig. 8. Generic configuration of a battery charger.

Level 1 charging connects the onboard charger to the household circuit using a standard three-prong household connection and rated up to 16 A. Level 2 charging is a permanently wired EVSE and up to 60 Amps. A variety of circuit topologies and control methods have been developed for battery chargers. These have an input power factor correction (PFC) circuit and filters on both the sides. As there are many regulations governing the harmonics drawn by the battery charger, filters are essential. The BPC is used to exchange power from or to the grid. Depending on the maximum current rating of the MOSFETs used in the converter, four operating modes can be identified. If the battery charger is situated in the EVSE, then Q absorption and Q generation are possible even when the EV is not connected to the charger.

In mode 3, charging an EVSE is used. The BPC can be placed at a central location or a distributed BPC can be used. Especially, if fast charging is provided, the distributed architecture is preferred as current in the LV side of the MV/LV transformer is of the order of hundreds of amperes. Therefore, it is important to use the shortest possible connection cable for EVs.

In bidirectional V2G type power losses and emissions are minimized and voltage levels are also regulated, but in this type battery degrades quickly. In unidirectional V2G type operation cost is low and a serious drawback is the limited service available.

Table 7 summarizes the benefits, drawbacks and services of each V2G type. It is worth mentioning that a successful implementation of Vehicle to Grid technology in a smart grid will lead to greater integration of renewable energies into the power grid.

Table 7. Comparison of unidirectional V2G and bi-directional V2G

V2G Types	Uni-directional	Bi-directional
Services	Uni-directional V2G technology controls the EV charging rate in a single power flow direction from the grid to EV based on energy scheduling and incentive system.	Bi-directional V2G refers to the dual direction power flow between EV and the power grid to achieve numerous benefits.
Descriptions	• Ancillary service – load levelling	• Ancillary service – load levelling • Active power support and peak load shaving • Reactive power support • Harmonic filtering • Augments integrating of renewable energy resources
Merits	• Reduced power losses • Increased profit • Reduced operational expenses • Minimizes emission	• Reduced power losses • Increased profit • Reduced operational expenses • Minimizes emission • Avoids power grid overloading • Improves load profile • Regulates voltage level • Increases renewable energy generation
Demerits	• Service availability is limited	• Battery degradation • Infrastructure hardware is complex. • Investment cost is high • Social barriers

5. Battery Movement System for EV

With the advent of battery technology, their efficiency, energy density, and specific power density have all increased manifolds. However, what has also increased is the vulnerability of the cell chemistries towards accidents and faults due to factors which hardly mattered during operation of older, more robust chemistries – like lead acid among others. So, to be able to utilize the new, vast horizons that newer cell chemistries have opened up for the world of energy storage, we need to use the batteries with safety and precaution.

This precautionary layer is what comprises the battery management system (BMS). Its primary function is to safeguard the battery during regular, or in some fleeting scenarios, overboard operation and protect the battery and the end-user from any potential hazard posed by the easily enraged chemistry. The BMSs have gained large utility in the field of energy management because they allow practical application of volatile, though, relatively efficient chemistries like the lithium-based batteries, or super capacitors – in gadgets, appliances, RC vehicles, robotics, storage of excess power for the grid or even renewable resources, however, more predominantly in hybrid and electric vehicles.

5.1 Overview BMS

BMS stands for Battery Management System, and is an electronic device utilized wherever Li-ion chemistry cells are in operation – appliances, electric vehicles or general energy storage.

The purpose of a BMS is to:

- Provide battery safety and longevity, a prerequisite for Li-ion.
- Determine state-of-function in the form of state-of-charge and state-of-health.
- Caution against any imminent errors. This could be high temperature, cell imbalance or calibration.
- Indicate end-of-life when the capacity falls below the user-set target threshold.

Traditionally BMS has not been used in Lead-Acid batteries. Most other methods of chemical energy storage have some form of electronics associated with them. Ni-Cd, Na-Air, Zn-Air batteries are also other examples. Super capacitors and Fuel cells also use Battery Management Systems [9].

The primary functions are battery protection and showing state-of-charge (SoC). Capacity is the prime indicator of battery *state-of-health (SoH)* and should be part of the battery management system (BMS). Knowing SoC and SoH provides the *state-of-function (SoF)* – the ultimate measure of battery-readiness; however, technologies to measure it are still being worked upon. The BMS also provides protection during charging and discharging. It puts the battery off-circuit, if set limits are exceeded or if a failure occurs.

Some of the established BMS standards are the SM Bus (System Management Bus) used for mostly portable applications, as well as the

CAN Bus (Controller Area Network) and the simpler LIN Bus (Local Interconnect Network) for automotive applications [6].

Essentially, a system comprising various devices is responsible for managing a rechargeable battery at the cell level, at the cell module level, and at the pack level. Typically, a BMS has two control layers:

- Battery Monitoring Unit (BMU)
  - Monitoring at the cell level – Voltage, current and temperature
  - Performs cell balancing – active or passive
  - Acts as a slave to the master BMS
- Master/Supervisory BMS
  - Top level controller
  - Centre point of the complete battery monitoring architecture
  - Performs battery control via a pre-established algorithm
  - Communicates with other automotive systems

5.2 Salient Features of a Generic Battery Management System

- Data logging – On-board memory (Flash/SD Card), Off-board memory (Cloud), Streaming (through GPRS)
- Geo-tagging – Geo-fencing
- Communication – Over CAN with vehicle
- Isolation Measurement
- Battery cooling control
- Cell balancing

Considering an electric vehicle as a particular application for BMS, we observe that an EV does not consist of a one unit battery pack but rather as one which has been divided and segregated into stacks or modules of smaller, more manageable sizes. Each of these stacks are then shorted with each other using busbars or high voltage connectors in series or in parallel. This improves the serviceability of the battery pack while keeping in check fault propagation to the entire pack. Each of these stacks has one BMS slave module, each monitoring and balancing each cell of each stack.

As an example, we can study battery stack arrangement in a Tesla Model-S. Model-S modules comprise 3400 mAh cells arranged in a 6s74p configuration, rated at 500 amps, and a peak of 750 amps. They have integrated an ingenious liquid cooling and heating system to keep the cells from crossing their threshold during regular operation; however, they can also be air cooled in regular or light applications. They have included connector with cell level connectivity for BMS systems and two integrated thermistors. Each pack consists of 444 cells, and each cell is independently protected by fuses on each terminal.

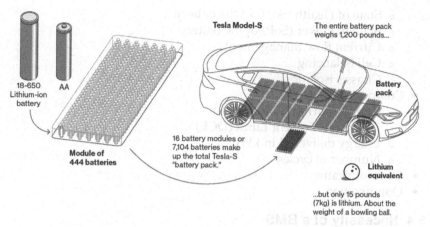

Fig. 9. Representation of cell stacking in a Tesla Model-S

Table 8. Various parameters of cells used by Tesla (21700 NCA)

Pack capacity	232 Ah, 5.3 kWh
Module height	7.874 cms
Module width	30.224 cms
Module length	66.548 cms
Module weight	24.94 kgs
Voltage nominal	3.8 V/Cell, 22.8 V/Module
Charge voltage cut-off	4.2 V/Cell, 25.2 V/Module
Discharge cut-off	3.3 V/Cell, 19.8 V/Module
Max. discharge current (10 sec.)	750 ps

5.3 Functions of a BMS

Its primary function is to safeguard the battery during regular, or even during fault case operation and protects the battery and the end-user from any potential hazard posed by the irritable chemistry.

Its secondary function is to communicate with the peripheral and auxiliary systems around the vehicle through CAN bus or LIN bus and relay or receive the precise parameter values without significant translational lag.

A BMS has several different functions, such as [1]:
- Monitoring of
  - o Total and individual cell voltage
  - o Temperature
  - o State of Charge (SoC) of the battery

- State of Health (SoH) of the battery
- State of Power (SoP) of the battery
- Current flow management
- Cell balancing
- Chassis isolation
- Computing
  - Charge Current Limit (CCL)
  - Discharge Current Limit (DCL)
  - Energy delivered in kWh
  - Number of cycles
- Communication
- Optimization

5.4 Necessity of a BMS

For safe and proper working of modern-day batteries which are comparatively volatile in nature a BMS is needed to keep the pack from over utilizing itself and monitoring the chemical composition.

A BMS is the quintessential utility in the power and driving circuit because:

- Batteries should be prevented from operating beyond recommended operating conditions because
  - Voltage has upper and lower limits
  - Current has upper limit
  - Temperature has an upper and sometimes a lower limit
- There needs to be a method to disconnect the battery in case of a severe problem
  - MOSFET/IGBT based disconnect circuit
  - Contactor based disconnect circuit
- Battery needs to be used in a controlled and safe manner [2]
  - EV batteries are governed by functional safety (ISO-26262)

5.5 Procedure to Design a BMS

Design and fabrication of a BMS from scratch for each application is a cumbersome task and requires proper research and development about the technical aspects that need to be included in the system because every stack design is very specific and for efficient use of time and resources, only what is essential should be added – for instance not every use will need a real time monitoring of SoC, even though it is of utmost importance in an EV.

However, to keep the commercial viability of the BMS as a product, it needs to be mass produced and the features need to be generic and should cover all possible test scenarios.

Table 9. Classification of SoC measurement and different mathematical methods [4]

Type of SoC measurement methods	Mathematical method	Description
Direct Measurement Method	Open circuit voltage	The open circuit voltage (OCV) and its State of Charge (SoC) are linear in nature. $V_{oc}(t) = a_1 \times SoC(t) + a_0$ This relation can't be same as the OCV-SoC characteristics differs among batteries.
	Terminal voltage	The terminal voltage of the battery pack is proportional (linearly) to the state of charge (SoC). This is because the EMF of the battery pack is also proportional to the terminal voltage which in turn is linearly proportional to the state of charge (SoC).
	Impedance	Impedance method provides knowledge of several parameters, the magnitude depends on the State of Charge (SoC) of the battery pack. Even if the parameters and variations of impedance with State of Charge (SoC) are not the same for all the batteries, it is crucial to perform impedance experiments for identification and utilization of parameters for estimation of the State of Charge (SoC) of the battery pack.
	Impedance-Spectroscopy	Impedance spectroscopy is generally used to measure impedances of ac frequencies at various discharging and charging currents. By least-squares regression of measured impedance values the model impedance can be obtained. State of Charge may be indirectly concluded by measuring the present battery pack impedances and comparing them with available impedances at various levels of State of Charge.

(Contd.)

Table 9. (*Contd.*)

Type of SoC measurement methods	Mathematical method	Description
Book-keeping estimation	Coulomb-Counting	In order to estimate the State of Charge, Coulomb-Counting integrates the discharge current with respect to time and measures the discharging current of a battery pack $SoC(t) = SoC(t-1) + \dfrac{I(t)}{Q_n}\Delta t$
	Modified Coulomb-Counting	Modified Coulomb-counting method is the improvement over Coulomb-counting. The corrected current is used to improve the accuracy. A function of discharging current gives you the corrected current. The relation between the corrected current and discharging current of battery pack is given by: $I_c(t) = k_2 I(t)^2 + k_1 I(t) + k_0$ where k_2, k_1 and k_0 are experimental constant values. and $I(t)$ is the discharging current as a function of time. State of Charge (*SoC*) is calculated by: $SoC(t) = SoC(t-1) + \dfrac{I_c(t)}{Q_n}\Delta t$ Compared to the conventional Coulomb-counting method, the accuracy of this method is superior.

| Adaptive systems | BP-Neural Network | It is the popular type in artificial neural networks. Its ability to self-learn as well as organize and nonlinear map makes BP-Neural Networks the first choice in the estimation of the State of Charge (SoC). It predicts the current SoC with the help of recent history of voltage, current, and the ambient temperature of a battery. This architecture has an input, an output, and a hidden layer. Input layer has three neurons for terminal voltage, discharge current, and temperature for State of Charge (SoC), the hidden layer has g neurons, and output layer has only one.

The neuron in the hidden layer is given by

$$neti_j = \sum_{i=1}^{3} x_i v_{ij} + b_j$$

The hyperbolic tangent function is used as an activation function applied to the neuron in hidden layer and is given by.

$$h_j = f(neti_j) = \frac{1 - e^{-2neti_j}}{1 + e^{-2neti_j}}$$

The neuron O in the output layer is

$$net\ o = \sum_{i=1}^{g} h_i w_i + k$$ |
| Radial Basis Function neural networks | | The Radial Basis Function neural networks are used to estimate systems with incomplete information. They are used to analyze the relationships between reference sequence and the comparative in a given set. The Radial Basis Function neural networks have been used in SoC estimation. |

(Contd.)

Table 9. (*Contd.*)

Type of SoC measurement methods	Mathematical method	Description
	Support vector machine	The support vector machine helps in pattern recognition. It is a nonlinear estimation system as well as more robust than a least-squares estimation system because it is insensitive to minor changes. The SVM based estimator not only removes the drawbacks of the Coulomb counting SoC estimator but also produces accurate SoC estimates.
	Fuzzy neutral network	Fuzzy neural network (FNN) has been used in many applications, especially in identification of unknown systems. In nonlinear system identification, FNN can effectively fit the nonlinear system by calculating the optimized coefficients of the learning mechanism.
	Kalman filter	Using real-time measurement road data to estimate the SoC of battery would normally be difficult or expensive to measure. The application of the Kalman filter method is shown to provide verifiable estimations of SoC for the battery via the real-time state estimation.
Hybrid method	Coulomb counting and EMF combination	A new SoC estimation method that combines direct measurement method with the battery EMF measurement during the equilibrium state and book-keeping estimation with Coulomb counting method during the discharge state has been developed and implemented in a real-time estimation system

| Coulomb counting and Kalman filter combination | Wang et al. proposed a new SoC estimation method, denoted as "Kalman-Ah method," which uses the Kalman filter method to correct for the initial value used in the Coulomb counting method. In Kalman-Ah method, the Kalman filter method is used to make the approximate initial value converge to its real value. Then the Coulomb counting method is applied to estimate the SoC for a long working time. The SoC estimation error is 2.5% when compared with the real SoC obtained from a discharge test. This compares favorably with an estimation error of 11.4% when using Coulomb counting method. |
| Per-unit system and EKF combination | Kim and Cho described the application of an EKF combined with a per-unit (PU) system to the identification of suitable battery model parameters for the high accuracy SoC estimation of a lithium-ion degraded battery. To apply the battery model parameters varied by the aging effect, based on the PU system, the absolute values of the parameters in the equivalent circuit model in addition to the terminal voltage and current are converted into dimensionless values relative to a set of base values. The converted values are applied to dynamic and measurement models in the EKF algorithm. |

Ideally, the design and fabrication process should proceed in the given chronological order:

- Get the customer requirement
  Each BMS needs to be customized to the specific demands and configurations of each battery pack arrangement for the most efficient design and fabrication.
- Create a system architecture
  The system architecture decides the base design of any BMS and hence is an important step in deciding processing and handling of parameter data.
- Select major components
  There are various ICs out in the market that are capable of monitoring a cell. We have to decide which is the most suitable for our specific purpose. Suppose, to monitor six cells we do not want to use an IC that has the capacity to monitor 12 or more than six cells because it would add to the already complex circuit and then result in wastage of time – for designing the circuit; money – higher cell monitoring capacity implies an increased production cost; and effort – that goes into soldering the components onto the PCB for the BMS.
- Design Schematic
  After each component is decided, it's time to connect them all together into a coherent unit. Schematic is designed to represent how each component interacts with every other component.
- Design PCB
  Post schematic, to obtain the PCB, we must design the board file of the schematic.
- Do the DFMEA
  Design failure mode and effect analysis (DFMEA) is a set of pre-defined boundary rules used to evaluate the design and simulated performance of the PCB design.
- Get the embedded code ready
  The algorithm for the master unit of the BMS should be ready to be boot loaded to the IC.
- Fabricate and test the PCB
  The BMS PCB once fabricated, needs to be tested for continuity errors or mounting mismatching.
- Test in real battery pack
  Once the PCB is cleared of all possible errors and faults, it is tested with the battery pack assembly it was originally designed for.
- Extensive lab testing
  The PCB is rigorously tested repeatedly in an isolated environment for surety of design and reliability.
- Safety tests

All the errors are now purposefully simulated physically on a customized electrical jig designed for the BMS, and tests like overcharge, over-discharge, short circuit, and overheating are executed to determine if the design is fail-proof.

- Performance tests
  Here we see how the BMS performs in real life situation, that is, its performance characteristics during its operation inside the vehicle – during starting when current demand is high, or during regeneration when power flows back to the battery stack. However, all the situations are simulated outside a vehicle, on a jig.
- Accuracy tests
  The values of all the parameters read by the BMS are cross-verified to check the credibility of the values displayed and stored by the BMS.
- Documentation is the key
  Each step of the way during the entire process of designing and fabricating the BMS PCB, should be well documented – each error, problem, solution, roadblock, technical difficulty and the eventual way around. This makes it way easier for the second iteration to be designed and implemented.
- Extensive field testing
  The particular design of the BMS should now be tested after being fit in the vehicle, like planned – connected to the battery pack, receiving feedback from all the sensors and well calibrated.
- Functional safety documentation
  All the safety and precautionary measures should also be well documented. The design should be according to the ISO standards for seamless clearance for use in public and commercial service.

5.6 Topologies for BMS Architecture

In which configuration will the BMS be planned and designed is a very crucial question because it will determine the entire embedded hardware system design of the BMS. Various topologies have already been declared for BMS design and each comes with its own set of pros and cons. On the basis of hardware design, few topologies are listed below [8].

- Master slave star topology:
  This topology consists of entire battery pack being reorganised into modules with one slave assigned to each module. High voltage batteries can be built by putting more modules together and since the main battery current bypasses the slave boards, it can also be used for high current batteries. Disadvantages are that the communications between the sensors and the slaves are analogue, thus prone to noise, and the large number of sensor wires, four per cell precisely, are required.

Slave boards:

- o Each board houses a temperature sensor, senses the voltage, and implements cell balancing for each individual cell.

Master board:

- o Multiple slaves are connected to the master which further monitors the time-integrated current to calculate the approximate coulomb flow, and then further uses voltage and temperature data and calculates the battery SoC.
- o The master board controls the main contractor for battery isolation, or the triggering of MOSFETs/IGBTs to initiate battery protection in response to data from the in-line current sensor, voltage and temperature data from the slave boards.
- o Processing of signals is shared among the master and the slaves by means of I2C buses.

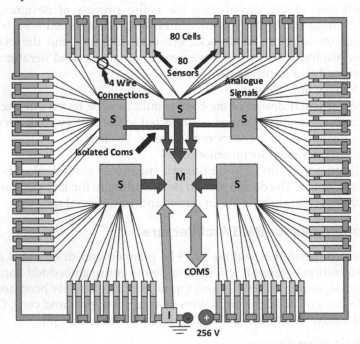

Fig. 10. Master-slave topology.

- • Daisy chain ring topology:

  Advantages of this topology are that the design is simpler and also the construction gives higher reliability in an automotive environment. Large number of mini-slave boards are needed, though and the difficulty of mounting them on some cell types, and that the master board has a higher processing load are some of the disadvantages.

o Slave boards are in connection with each cell via voltage and temperature sensors, as well as a switch to bypass current to mitigate cell balancing by shunting of charge.

o A transceiver for communication with capacitive isolation is there to enable digital reception/transmission of data.

o The slave is powered from the cell it monitors and the nodes of all the slaves are connected by a single data bus to the master which in turn does polling for latest parameter values.

Also, what is important for the architecture of a BMS is the chemistry for which it is intended to be used. It can be designed for any particular chemistry as the SoC and SoH calculating algorithms are dependent on the chemistry of the cell. Although, to keep its commercial value intact, the BMS must be designed and fabricated keeping in mind multi-chemistry support, so a BMS could be integrated with any of the more popular cell chemistries seamlessly. Some of the battery chemistries and their properties are listed as follows in Table 10.

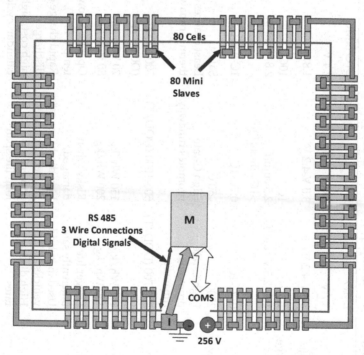

Fig. 11. Daisy chain ring topology.

• Passive balancers establish balancing across cells by diverting input current when the cell voltage reaches a pre-defined level. The cell voltage is not a good indicator of the cell's SoC and for certain

Table 10. Summary of various Li-ion chemistries and their properties

Li technology	LiMn$_2$O$_4$	LiCoO$_2$	Li$_2$TiO$_3$	LiNiMnCoO$_2$	LiFePO$_4$
Parameters	LMO	NCA	LTO NMC	NMC	LFP
Nominal volt	3.7 V	3.6 V	2.4 V	3.6 V	3.2 V–3.3 V
Operating range	3.0 V–4.2 V	3.0 V–4.2 V	1.8 V–2.85 V	3.0 V–4.2 V	2.5 V–3.65 V
Rate of discharge	1C 10C 2.5 V cutoff	1C 2.5 cutoff >1C affects life	10C 2.8 V cutoff	1C, 2C	1C 25C < 2 V causes damage
Rate of charge	0.7C–1C	0.7C–1C	1C–2.85C	0.7C–1C	1C
Thermal runaway	250C High C-rate promotes runaway	150 C High C-rate promotes runaway	Safest Least chances of thermal runaway	210 C High C-rate promotes runaway	270 C
Lifecycle	300–700 (DOD/T)	500–1000 (DOD/T)	3000–7000 (DOD/T)	1000–2000 (DOD/T)	1000–2000 (DOD/T)
Energy density	110 Wh/kg–130 Wh/kg	150 Wh/kg–200 Wh/kg	70 Wh/kg–80 Wh/kg	150 Wh/kg–200 Wh/kg	95 Wh/kg–130 Wh/kg
Characteristics	Longer life & inherently safe, Used in hybrid vehicles, High discharge rates	Risky when damaged, Low discharge rates, Hi specific energy, power, long life span	Operate at low temp (–40°C), Rapid charge & discharge, high cost, high calendar, cycle life	Longer life & inherent safety, Less prone to heating, Lowest cost, High specific energy	Reduced risks of heating/fire. Much less volumetric capacity, higher self-discharge, moderate specific energy long lifespan

Application	US and Japanese EVs Nissan Leaf, Volt, Renault, Mitsubishi I-MIEV	Tesla Model S	Grid storage, EV, buses and ferries	German EVs VW, BMW i3, Fiat, Daimler Smart	Chevrolet spark, Coda, e-Bus, other Chinese EVs
Supplier	AESC, LG Chem, LiEnergy Japan	Panasonic	Toshiba, Altairnano, Tiankang	Samsung, Litech	A123, ATL, Calb, Lishen Tianjin, Saft, BYD

lithium based chemical compositions such as $LiFePO_4$ it is not even an indicator; thus, cell voltages equalization by passive regulators fails to balance SoC.

- Active balancers discharge the excess charge through a resistor, to achieve charge balance. Only the cell voltage is needed to enable the active regulators.

Physical construction wise BMS topologies fall in three categories:

- Centralized construction: just one controller is connected to the cells, rather the battery pack, via a dense connection of wires.
- Distributed construction: a BMS slave board is mounted at each cell, with just a single harness between the battery/module and a controller.
- Modular construction: one controller handles multiple cells, with inter-controller communication.

Centralized BMSs are best for cost efficiency and are least flexible as another module of cells cannot be added to the monitored network, and are also plagued by complicated wiring.

Distributed BMSs are quite expensive, though, they are easy to install, and offer a very neat assembly.

Modular BMSs lie somewhere in between the two aforementioned types and offer a midway between the pros and cons of the other two topologies.

BMS technology varies in complexity and performance, and cell balancing techniques also comprise the overall topology.

The requirements for a BMS in stationary applications (UPSs for servers) and mobile applications (such as electric vehicles) are pretty divergent, especially from the space and weight constraints point of view, so the hardware and software design must be planned accordingly. For EVs or PHEVs, the BMS is the part of a subsystem and does not work as a standalone device. It has to communicate with a load, charger (while on charge), emergency shutdown, and thermal management subsystems. Therefore, in an efficient vehicle design the BMS is very well integrated with each subsystem. Some small mobile applications (like electric wheelchairs, electric scooters, and fork lifts) have off-board charging hardware; however the on-board BMS must have real-time communication with the charger.

5.7 Major Challenges Associated with BMS Design

Even though embedded system designing has grown by leaps and bounds in overcoming several challenges of a diverse nature, newer challenges keep popping up as the designing gets more complex and intricate. With a system as vast as that of an electric vehicle, the software and hardware designing are bound to hit some roadblocks.

Some major challenges that baffle every embedded systems designer are [7]:

- Accurate measurement algorithm of the following parameters
  - Primary voltage drift and reference voltage drift
  - State of charge: The amount of energy left in a battery
  - State of health: Degradation of the battery caused due to aging
  - State of function: Battery operation readiness in terms of usable energy by observing SoC
  - Remaining capacity estimation
  - Battery malfunction
  - Ageing monitoring
  - Mechanical and chemical defect and fault detection
- Long term drift
  - Humidity
  - Temperature
- Isolation
  - HV stack isolation from chassis of the vehicle for safety
  - Isolated communication between battery/control module
- Satisfying the ISO26262 safety standard
  - Measurement unit not having ASIL rating

5.8 Features of a Hybrid EV BMS

A BMS designed for application in an EV or hybrid EV is not similar to BMSs designed for use in gadgets or energy storage banks, because the electrical, electronic, and the auxiliary systems on an EV are very extensive and should hence be meticulously interfaced with each other in terms of communication, interoperability, and flexibility of operation. Also, to an extent the software design determines the user experience of the vehicle through the infotainment system – necessary parameter readings regarding the vehicle must be on constant display with real time variations accounted for and to also establish efficient exchange of reliable information with other automotive systems.

So, the given features enhance the capabilities of an automotive BMS.

- **Battery Model**

The Battery Model takes into account the characteristics of the battery in response to various external and internal stimulus conditions and translates them into a software algorithm. This acts as a model for estimation and calculations of various crucial cell parameters. An essential function of the battery model is to calculate the SoC of the battery. The SoC is determined by integrating the current flow over a period of time, modified to take into consideration the many factors which bind the cell performance, then removing the result from the prior capacity of the fully charged battery. This battery model can be used for logging past history for troubleshooting and maintenance purposes or to predict how much distance the vehicle may run before the battery drains to nil. The remaining range of the battery pack, derived from usual driving or usage patterns, is calculated from the current SoC and the energy consumed and the miles covered since the previous charge (or alternatively from a previous long-term average). The distance travelled is derived from data provided by other sensors on the CAN bus. The accuracy of the range calculation is more important for EVs whose only source of power is the battery. HEVs and bicycles have an alternative "back up" source of power should the battery become completely discharged. The problem of losing all power when a single cell fails can be mitigated at the cost of adding four more expensive contactors which effectively split the battery into two separate units. If a cell should fail, the contactors can isolate and bypass the half of the battery containing the failed cell allowing the vehicle to limp home at half power using the other (good) half of the battery. Outputs from the model are sent to the vehicle displays also using the CAN bus [1].

- **Multiplexed Signals**

To reduce costs, instead of monitoring each cell in parallel, the Battery Monitoring Unit incorporates a multiplexing architecture which switches the voltage from each cell (input pairs) in turn to a single analogue or digital output line (see below). Cost savings can be realized by reducing the number of analogue control and/or digital sampling circuits and hence the components count to a minimum. The drawbacks are that only one cell voltage can be monitored at a time. A high speed switching mechanism is required to switch the output line to each cell so that all cells can be monitored sequentially.

The BMU also provides the inputs for estimating the SoH of the battery; however since the SOH changes only gradually over the lifetime of the battery, less frequent samples are needed. Depending on the method used to determine the SOH, sampling intervals may be as low as once per day. Impedance measurements for example could even be taken only in periods when the vehicle is not in use. Cycle counting of course can only occur when the vehicle is operational.

- **Personality Module or Demand Module**

The Demand Module is similar in some respects to the Battery Model in that it contains a reference model with all the tolerances and limits relevant to the various parameters monitored by the Battery Model. The Demand Module also takes instructions from the communications bus such as commands from the BMS to accept a regenerative braking charge or from other vehicle sensors such as safety devices or directly from the vehicle operator. This unit is also used to set and to monitor the vehicle operating mode parameters.

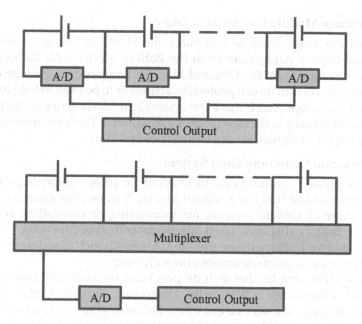

Fig. 12. Multiplexing to reduce components.

This module is sometimes called the Personality Module since it includes provision for programming into the system, all the custom requirements which may be specific to the customer's application. For example, the cell maker will recommend a temperature limit at which the battery must be automatically disconnected, for safety reasons. However, the car manufacturer may set two lower limits, one at which forced cooling may be switched on and another which lights up a warning light on the driver's instrument panel.

For HEV applications, the Personality Module interfaces with the engine Electronic Control Unit (ECU) via the CAN bus. Provision is made in this module for setting the desired system SoC operating range and the parameters for controlling the power sharing between the electric drive and the internal combustion engine.

The Demand Module also contains a memory block for holding all the reference data and for accumulating the historical data used for monitoring the battery SOH. Data to display the SOH or to switch on warning lights can be provided to the vehicle instrumentation module via the CAN bus. The outputs from the Demand Module provide the reference points for setting the operating conditions of the battery or triggering the action of protection circuits. Test access to the BMS for monitoring or setting system parameters and for downloading the battery history is provided through a standard RS 232 or RS 485 serial bus.

- **Standalone Module for Decision Logic**

The Decision Logic module compares the status of the measured or calculated battery parameters from the Battery Model with the desired or reference result from the Demand Module. Logic circuits then provide error messages to initiate cell protection actions or to be used in the various BMS feedback loops which drive the system to its desired operating point or isolate the battery in the case of unsafe conditions. These error messages provide the input signals for the Battery Control Unit.

- **Inter-system Communication System**

The BMS needs a communications channel for passing signals between its various internal functional circuit blocks. It must also interface with several external vehicle systems for monitoring or controlling remote sensors, actuators, displays, safety interlocks and other functions.

Automotive BMS therefore use the CAN bus which was designed for this purpose as its main communications channel.

The system should also include provision for standard automotive On-Board Diagnostics (OBD) with Diagnostic Trouble Codes (DTC) made available to the service engineer. This connection is important for identifying any external causes of battery failure.

- **HV Module – Battery Control Unit**

The Battery Control Unit contains all the BMS power electronics circuitry. It takes control signals from the Battery Monitoring Unit to control the battery charging process and to switch the power connections to individual cells.

Some of the possible functions of this unit are:

  o To control input current and voltage profiles during charging.
  o To provide charge to each cell for charge balancing on all the cells in the battery pack.
  o To isolate the battery modules from each other and from the control electronics during lapse or fault conditions.
  o To divert the charge generated during regenerative braking towards the battery according to the requirement.
  o To ground the excess charge generated during regenerative braking when the battery is full.
  o To adapt to changes in the vehicle requirements according to the driving conditions.

To provide such functional ability, each battery module in the battery stack will require high current switches capable enough to switch 200 Amps or above to provide the necessary power.

- **Progressive Control and Binary Control**

In its simplest form, the BMS provides a "binary" ON/OFF response to a fault or an out of tolerance condition such as an overload, merely isolating the battery completely by opening the main contactors. Progressive or variable control can however be provided in the case of an overload by utilizing the CAN Bus to call for a reduced demand on the battery.

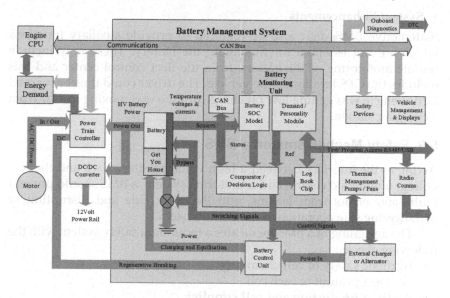

Fig. 13. Block diagram of signal flow and power flow in an EV.

- **Charge Balancing**

This is another essential function of the automotive BMS. As noted above it is required to compensate for weaknesses in individual cells which could eventually cause the failure of the complete battery. The reasons for cell balancing and how this is implemented are explained on the Cell Balancing page [3].

- **Reserve Storage – Limp Home Mode**

Although batteries are designed to be trouble free for multiple years, there is always the possibility that the battery could be rendered useless by the failure of a single cell. If a cell becomes open circuit, the battery is essentially dead. However, the BMS is designed to monitor the status of every cell and so the location of the faulty cell will automatically be identified. The battery is then split into two sections in series, each of which can be independently bypassed by disconnecting the section of the battery containing the faulty cell and switching a conducting bus bar in its place. This will allow the vehicle to reach the nearest source of power on remaining power using the good section of the battery as well as the links. The system will need two more expensive high-power contactors to implement this function. Although this investment may be well justified when the alternative could be a costly and dangerous breakdown on the motorway.

- **System Enhancements**

Automotive BMSs may also be required to provide auxiliary functions which are not essential for managing the battery. These may include remote monitoring of the battery from the fleet control center and this includes the GPS location of the vehicle. The driver could thus be warned if the vehicle was getting low on charge or if he was moving too far from a charging station.

5.9 Safety Mechanisms on a BMS

Acting as the interface between a high voltage battery pack and the low voltage embedded control and monitoring system, a BMS should prevent faults and errors from leaking into the other side and disrupting or endangering safe operation of the entire system.

The designing of a BMS inculcates a redundant safety system with the following checks:
- **Intrinsic safety of cell chemistry**
    o Design audit of each cell.
- **Audit of production and cell supplier**
    o Technical competence of technicians.
    o Process controls.
- **Internal safety devices (Cell level)**
    o Circuit Interrupt Device (CID) disconnects the circuit when inside pressure exceeds the limit.
    o Shut down propagating separators.
    o Pressure releasing vents.
- **External circuit devices**
    o Low power applications facilitated by PTC resistors [5].
    o Excess current withdrawal protection by fuses.
    o Individual cell and module isolation – using mechanical and electrical contactors and physical separation to prevent a fault from traveling further from the point of origination.
- **Software for BMS**
    o Monitors all important indicators related to control actions – load management, power disconnect, cooling among others.
    o Control process and execution.
    o Switch off for off-limit situation.
- **Fail safe back-up – BMS hardware**
    o Hardware actuated switch off when software failure occurs – set slightly above threshold.

Table 11. Summary of cell faults and remedial steps

Temperature code (°C)	Problem	Consequence	Action	Result	Time dependent impairment / Irreversible Damage
180 to 190	Oxidation of cell components	Thermal runaway	Containment (too late)	Cell destruction	
120 to 130	Cathode breakdown	Oxygen released	Pressure vent open	Pressure release	
	Pressure build up	Possible rupture	Pressure vent open	Pressure release	
	Electrolyte breakdown	Flammable gas released	Internal CID open	Current shut off	
	Separator melts	Short circuit / overheating	BMS	Current shut off	
60 to 80	Over temperature	SEI layer breakdown	BMS/separator shutdown	Current shut off	
	Over current	Overheating	Fuse or BMS	Current shut off	
	Over voltage	Overcharge/ overheating	PTC or BMS	Current off Cooling on	
	Low voltage	Anode dissolution	BMS	Current shut off	
−60 to 10	Low temperature	Lithium plating	BMS	Turn on heater	

Source: https://www.mpoweruk.com/bms.htm

o Battery switch off when failure of BMS power supply happens.
- **Battery casing**
  o Robust container including vents for passage of heat and gases in case of any mishap.
  o Physical isolation amongst two cells.

Table 11 lists some common cell faults and how the BMS tackles each to troubleshoot and to rid the system of errors – preventing any damage to the systems in case of a mishap.

5.10 Safety Disconnects Employed in a BMS

- **MOSFET triggered disconnect circuit**

BMS IC measures the voltage of each cell using a potential divider circuit. And the temperature sensors provide it with the temperature figures of each cell. The algorithms for calculation of cell SoC and SoH are bootloaded in the IC. For monitoring of circuit current, BMS IC utilizes a current sensor. It also outputs error signals to break the circuit in case of an anomaly. The calculated and measured values are relayed over to the microcontroller via SPI or I2C protocol. The BMS IC monitors the cell parameters and sends them to the microcontroller for external communication via CAN bus. Also, in case of any fault detection the microcontroller provides the

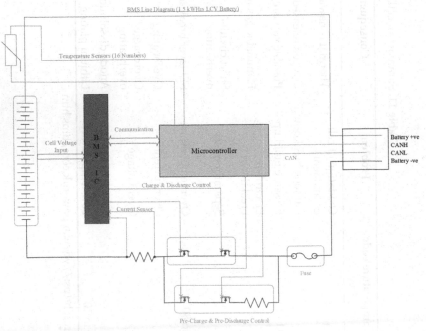

Fig. 14. BMS with a MOSFET triggered disconnect circuit.

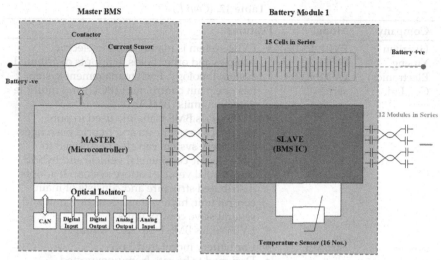

Fig. 15. BMS with a physical contactor triggered disconnect circuit.

MOSFETs with trigger pulses to break the circuit and prevent excessive damage to the cells or the circuitry.

- **Physical contact or triggered disconnect circuit**

In this architecture, the master unit senses current using a current sensor and opens the circuit by opening the contactor, caters to all the inputs from external circuits and provides outputs for access to the master system and communication with the peripherals via CAN. While the slave unit measures and monitors the cell parameters like voltage, current, and temperature and relays them to the master unit after processing them for calculation of SoC and SoH according to the pre-loaded algorithms, for communication with external components.

Table 12. Popular suppliers of plug-and-play BMS

Company	Model	Features
BYD Co., Ltd.	BYD BMS	• The battery management system also has a battery cell automatic equalization function. • Active monitoring • Battery balancing function used to achieve timely and automatic maintenance • Reduced time • Battery maintenance, prolongs the service life of the battery, and enhances various stages performance.

(Contd.)

Table 12. (*Contd.*)

Company	Model	Features
Huizhou Yineng Electronics Co., Ltd.	EV01, EV02, EV03, EV04, EV05 series	• EV02 system is mainly used in electric vehicles and other fields. It adopts distributed system topology. Each management system has one main control unit (BCU) and multiple detection units (BMU). • EV03 series BMS is mainly used in pure electric passenger cars and hybrid passenger car battery system can also be applied to pure electric commercial vehicle and hybrid commercial vehicle battery system. It adopts distributed structure and consists of main control unit, high voltage detection unit and several slave control units. • EV05 series BMS is up and coming.
Sosaley	carBMS	• Car battery monitored 24×7. • Designed to be safe from unexpected breakdowns. Data sent to the cloud using onboard GPRS. • A GPS on system makes it possible to track the vehicle.
Automotive Power	AMP BMS	• High flexibility. • Scalability for extension of battery pack configuration (multi-string, single string). • 400V/800V configurable. • Centralized or distributed design. • Optimized for cost, reliability, performance and safety. • Software abstraction (SIL capable).
Ewert Systems	ORION BMS	• Capability to measure up to 180 cells per unit in series connection. • Remote Cell Tap Expansion Modules can be connected for measurement of higher voltage battery packs (up to 800 V DC max). • Several configurations available in multiples of 12. • High EMI and noise immunity due to centralized design. • Performs passive cell balancing. • Calculates state of charge (SoC). • Uses automotive-grade locking mechanism connectors. • Calculates charge current limit (CCL) and discharge current limit (DCL). • Measure cell voltages from 0.5 V to 5.0 V. • Fully programmable dual CAN bus 2.0B interface. • Onboard diagnostic protocol support – OBD2.

6. Summary

In this chapter, we covered in detail, the intricacies of a Battery Management System and how it enables the safe application of volatile cell chemistries. We studied the various popular, conventional utilities of the BMS as well as the other non-conventional uses, especially its specific application in the automotive industry, with the growing significance of electric vehicles. The prime function of the BMS is to monitor the voltage, temperature, SoC, SoH and simultaneously balance the charge on cells while charging-which we know from the extensive article on SoC measurement techniques and SoH calculation algorithms. We also dug into the depths of the types of architectures which would be the most efficient according to specific demands of the application depending on the pros and cons of each topology. While, as noted, there are various plug-and-play BMSs available which could be bought off the shelf and used, but designing and fabricating an in-house BMS comprises its own set of challenges and difficulties. However meticulous the design might be, some sort of error is bound to creep in. We have also noted design of such a protection circuit to keep the BMS and the power circuit of cells, and busbars isolated from each other to prevent leakage of faults into the other side and snowballing into something major and serious, because as it is the new-age chemical compositions are not robust and will lead to short circuits or even fire, in case of grave accidents. A Battery Management System is bound to get more complex in design and implementation as newer, more efficient chemistries pop out into existence providing the end users greater degree of control in terms of which parameters to monitor and changing the nature of the BMS on-the-go to better deal with the transient conditions encountered during the operation of an electric vehicle.

References

1. Cheng, Review of battery management systems for electric vehicles. Energy Systems for Electric and Hybrid Vehicles 349-371.
2. Salehen, P.M.W., Su'ait, M.S., Razali, H. and Sopian, K. 2017. Battery management systems (BMS) optimization for electric vehicles (EVs) in Malaysia.
3. Battery Management Systems in Electric Vehicles." Advanced Battery Management Technologies for Electric Vehicles (December 21, 2018): 231 -248. doi:10.1002/9781119481652.ch8.
4. Pelegov, D. and Pontes, J. 2018. Main Drivers of Battery Industry Changes: Electric Vehicles – A Market Overview. Batteries 4(4), 65.
5. Bharatiraja, C. et al. 2018. Energy Management Strategy for Rural Communities' DC Micro Grid Power System Structure with Maximum Penetration of Renewable Energy Sources. Applied Sciences 8(4), 585.

6. Mahmoudzadeh Andwari, A., Pesiridis, A., Rajoo, S., Martinez-Botas, R. and Esfahanian, V. 2017. A review of Battery Electric Vehicle technology and readiness levels. Renewable and Sustainable Energy Reviews 78, 414-430.

7. Hannan, M.A., Hoque, M.M., Hussain, A., Yusof, Y. and Ker, P.J. State-of-the-Art and Energy Management System of Lithium-ion Batteries in Electric Vehicle Applications: Issues and Recommendations. IEEE

8. Xiong, R., Cao, J., Yu, Q., He, H. and Sun, F. 2017. Critical review on the battery state of charge estimation methods for electric vehicles. IEEE Access, vol. 6, 1832-1843, doi: 10.1109/ACCESS.2017.2780258. Access, vol. 6, pp. 19362-19378, 2018.

9. Manzetti, S. and Mariasiu, F. 2015. Electric vehicle battery technologies: From present state to future systems. Renew. Sustain. Energy Rev. 51, 1004-1012.

10. Suresh, K., Chellammal, N., Bharatiraja, C., Sanjeevikumar, P., Blaabjerg, F. and Nielsen, J.B.H. 2019. "Cost-efficient nonisolated three-port DC-DC converter for EV/HEV applications with energy storage," International Transactions on Electrical Energy Systems, vol. 29, no. 10, Jun. 2019.

11. Bharatiraja, C., Sanjeevikumar, P., PierluigiSiano, Ramesh, K. and Raghu., S. 2017. "Real Time Forecasting of EV Charging Station Scheduling for Smart Energy System", Energies, vol. 10, no. 377, pp. 1 to14, March.

12. Bharatiraja, C. et.al., 2020. A Hybridization of Cuk and Boost Converter Using Single Switch with Higher Voltage Gain Compatibility, Energies, vol. 13, no. 9, 2020.

Index

Printed in the United States
by Baker & Taylor Publisher Services

Printed in the United States
by Baker & Taylor Publisher Services